CW01221137

AROIDS

Aroids

PLANTS OF THE ARUM FAMILY

Deni Bown

Photography by Deni Bown

CENTURY
London Melbourne Auckland Johannesburg

Text copyright © Deni Bown 1988
Photographs copyright © Deni Bown 1988
Line drawings copyright © David Leigh 1988

All rights reserved

First published in 1988 by
Century Hutchinson Ltd,
Brookmount House, 62–65 Chandos Place,
Covent Garden,
London WC2N 4NW

Century Hutchinson Australia Pty Ltd
PO Box 496
16–22 Church Street
Hawthorn
Victoria 3122
Australia

Century Hutchinson New Zealand Limited
PO Box 40-086
Glenfield
Auckland 10
New Zealand

Century Hutchinson South Africa (Pty) Ltd
PO Box 337
Bergvlei
2012 South Africa

Typeset in Monophoto Bembo by
Vision Typesetting, Manchester
Printed and bound in Great Britain by
Mackays of Chatham Ltd,
Chatham, Kent

British Library Cataloguing in Publication Data
Bown, Deni
  Aroids: plants of the arum family.
  1. Araceae
  I. Title
  584'.64   QK495.A685
ISBN 0-7126-1822-8

Contents

	Foreword by Dr Simon Mayo (Royal Botanic Gardens, Kew)	7
	Acknowledgements	9
	Introduction	13
1	**Variations on a Theme** What are aroids and where do they grow?	18
2	**Of Tails and Traps and the Underworld** Mechanisms of reproduction.	37
3	**Woodlanders** Species of temperate woodland and higher altitudes of the tropics and subtropics.	55
4	**Aquatics and Amphibians** Species of wetlands and water.	73
5	**A Place in the Sun** Species of arid and seasonally dry regions.	100
6	**In the Shadows** Forest floor species of tropical rain forests.	119
7	**Towards the Light** Tropical climbers and epiphytes.	144
8	**The Titans** Giant tuberous species of the tropics.	173
9	**An Acquired Taste** Aroids as food plants.	188
10	**Acids and Crystals** The chemistry and toxicity of aroids; medicinal and folk uses.	208
	Check List of Aroid Genera	228
	Selected Bibliography	233
	Glossary	235
	Typical Aroid Leaf and Inflorescence Shapes	239
	References	240
	Index	250

DEDICATION
For my children: Anna, William and Robin
to whom even *Stenospermation* is a household word!

Foreword

Aroids, or *Araceae*, are plants which everybody knows but relatively few people recognize. As Deni Bown argues in this book, there is hardly an office or home in the land that lacks an aroid of some kind or other, but there is no clear general perception of this plant group comparable with, say, bromeliads or orchids. This situation exists for one reason above all – the lack of an illustrated general text which offers in sufficient detail an overview of the aroids as a whole.

Such a yawning gap in the horticultural and botanical literature has hampered many an inquisitive mind from acquiring a better understanding of the family. As in some remote and dangerous territory possibly harbouring unfriendly denizens, only the hardy and determined explorer can expect to discover anything worthwhile. The trail is beset by tiresome obstacles in the shape of reported but unobtainable books, old and outdated literature in diverse languages, impenetrable thickets of nomenclatural confusion, and indeed the plants themselves, often huge and untameable, or apparently infinitely variable, not lending themselves to the neat and tidy habits of the stamp collector. Aroids, indeed, are plants that demand passion and commitment from their lovers.

Deni Bown is just such an explorer. She is one of those rare individuals who, having encountered an apparently insurmountable obstacle, does not turn away but attempts to do something about it. Despite extremely unpromising circumstances, which would have defeated most people at the outset, she has, by a magnificent effort, produced a book of real and lasting worth. As far as I am aware, this is the first modern book devoted exclusively to aroids that deals in a general way with the family, setting out to say something about the majority of genera.

I would go so far as to say that this is probably the first truly accessible book in any language that treats the aroids in sufficient detail and yet with sufficiently broad scope to be really useful to a wide range of interested readers. Aroids are valued in temperate lands primarily for their use as foliage ornamentals in horticulture, and to a lesser extent as flowering plants. The book's appeal to horticulturists must therefore be a prime consideration and has been instrumental in determining its organization into sections defined on ecological lines; that is, a 'classification' of the subject material that well matches the interests of the grower. This, combined with Deni Bown's own experience in cultivating aroids, will ensure that gardeners find the book of enormous value.

Scientifically, this book is also important. It is a common misconception that scientists enjoy difficult technical treatises rather than an 'easy read'. The truth is rather the reverse, particularly when delving into a field that is related to but not one's own speciality. This book is not a taxonomic text, nor is it so intended, but all taxonomists of the family will want it on their shelves, and so will any chemist, zoologist, anatomist, cytologist, floral biologist, economic botanist, ethnologist, etc., who wants a general picture of the family, some members of which might be important in a current research project. Although not one of the really big families of flowering plants, the *Araceae* is nevertheless a large one, among the top thirty. At this order of magnitude few people can have a firm grasp of the group in all its many manifestations. A general text of the type Deni Bown has given us is an invaluable tool to enable researchers to focus their attention on a group of outstanding biological interest, giving an impression of its ecology, biology, taxonomy, and access to the technical literature.

The appearance of *Aroids* is timely, because these plants are now emerging from obscurity and being more widely recognized for their economic, scientific and horticultural worth. Today there is a very large number of people around the world for whom better knowledge of the aroids is an important professional concern, not to mention the greater number for whom these plants are a keen leisure interest. Aroids have outstanding potential as subsistence food crops. The recent publication of two books on edible tuberous aroids underlines the importance of aroids in helping to alleviate food shortages in humid tropical countries. There has also been a tremendous resurgence of interest in the taxonomy of the family during the last ten years, and progress has been such that radical new classifications are being proposed and actively worked out.

As a tropical forest family, *par excellence*, the aroids are greatly affected by the current massive destruction of natural tropical forest vegetation, and this lends an urgency to their collection and classification. Their horticultural appeal reinforces this need. Multimillion-dollar industries can emerge from the successful commercial exploitation of species with horticultural potential, as demonstrated by the Hawaiian anthurium industry. Where the species are native, commercial developments can give a positive incentive to the maintenance of natural populations and diversity. And of course spectacular new species are being discovered all the time. Deni Bown describes the recent discovery of the outstanding *Anthurium amnicola*; others that might be mentioned are *Anthurium superbum* and *Philodendron rugosum*.

There doubtless will be other books on aroids written in the future as more and more fascinating aspects of these extraordinary plants come under close scrutiny. But this is the one that I and others like me have been awaiting for many years, a truly pioneering contribution from the pen of a genuine enthusiast. Deni Bown has emerged from the forest with a prize and we have cause to be grateful to her.

Dr Simon Mayo
HERBARIUM, ROYAL BOTANIC GARDENS, KEW

Acknowledgements

As the aim of this book is to bring together strands of knowledge about the *Arum* family that until now have been widely separated, I have had to draw extensively on the learning of others. Being a novice (and non-scientist) in a specialized subject, I must first acknowledge all the authors, whether or not I have referred to their works, who have written on aspects of *Araceae* and who have thereby contributed to my understanding.

My deepest debt of gratitude is to Dr Simon Mayo of the Royal Botanic Gardens, Kew. I can well imagine the qualms he must have had as he witnessed my early attempts to grapple with the leviathan, but in spite of this, he was never once discouraging. Indeed, he has constantly pointed me in the right direction and been unstinting with his advice and constructive

ACKNOWLEDGEMENTS

criticism over the three years of my research and writing. Without his comments on the first and final drafts there would undoubtedly have been more errors and misinterpretations: as it is, those remaining are entirely my responsibility.

Most of the research has been undertaken at the Royal Botanic Gardens, Kew. I am most grateful to the Director and staff of the Herbarium and Library for placing such facilities at my disposal. Especial thanks to the librarians who helped me locate particularly elusive texts and to John Hale in the Tropical Department Nursery Collection, Dr Brinsley Burbidge (who tried to find the whereabouts of the missing painting of *Amorphophallus titanum*) and Peter Boyce whose knowledge of the Mediterranean and Middle Eastern aroids prompted additions to chapter 5.

The pursuit of araceous subjects and expertise has however led me to a number of other institutions. In the preliminary stages, to Cambridge University Botanic Gardens where I received invaluable help from Dr Peter Yeo, who made generous comments on the synopsis; and to the Department of Plant Sciences, Oxford University, at which Dr Caroline Pannell and the librarian, Mrs A. Townsend, were most helpful. Further afield, I would like to thank the staff of the following (in the order visited) for their help and friendliness: the Marie Selby Botanical Gardens, Florida – and in particular, Dr Bruce McAlpin, who provided much practical assistance and a home-from-home; Fairchild Tropical Gardens, Miami – to which I made a fleeting visit in awful weather under the guidance of Nancy Hammer; Missouri Botanical Gardens, St Louis – the domain of the redoubtable aroid worker, Dr Thomas Croat, who has given enthusiastic support to my project from the start, identified many 'mystery' species I photographed, made pertinent comments on the first draft of chapters 6 and 7, and whose energy and good humour have been as inspiring as his achievements; the Harold L. Lyon Arboretum of the University of Hawaii, to which I was welcomed with traditional Hawaiian warmth and given every assistance by Raymond Baker; the National Biological Institute and Kebun Raya in Bogor, Indonesia, at which I was guided by Dr Elizabeth Widjaya, who also accommodated me and showed me every kindness and consideration; the Biology Department of Andalas University and the PHPA (Nature Conservancy) in Sumatra, which provided guides in the Ulu Gadut, Lembah Harau and Gunung Sago areas; and last but by no means least, Munich Botanic Gardens, which has become a mecca for aroid lovers through the dedicated exploration, research and horticultural

ACKNOWLEDGEMENTS

prowess of Josef Bogner, who read through the first draft, gave me the benefit of his extensive knowledge and liberal hospitality, and went out of his way to assist me in photographing his remarkable collection, as well as answering numerous queries by post.

In addition, I would also like to thank the many others who have helped, in particular: John Banta of River Haven Nursery, Alva, Florida (now editor of *Aroideana*), who invited me into both his families – human and aroid – with equal spontaneity, and all the other members of the International Aroid Society in Florida who proffered help – especially Dr Monroe Birdsey, Luis Bueno, Joe Wright, Marilyn Johnson, Elaine Spear, Denis Rotalante, Jerry Horne and Richard Button who allowed me to visit and photograph their collections, and most of all, Bob See, who put me up and devoted every moment to making my stay fruitful and enjoyable; Barrie and Janet Findon of Anmore Exotics, Hampshire, who treated me to the finest plants and afternoon tea imaginable; Mr D. Richardson of Rochford Landscape Limited, Hertfordshire, who allowed me to take photographs at the nursery; Dr Douglas Knepp and Dianne Knepp who made me so much at home and entertained me so well in St Louis; Yves Laumonnier of Biotrop, Indonesia, who provided useful contacts and helped in a variety of ways – not least in meeting me at Djakarta airport; Tony Whitton of Environmental Manpower Development in Indonesia, who gave me a wonderful pep talk in a moment of culture shock; Mr and Mrs Jules Gervais of Kuaola Farms, Hawaii, who let me photograph their splendid anthuriums; Charlie and Paul Reppun, who patiently explained the differences between their many taro cultivars (and spelled the names!); John Akina, a traditional Hawaiian taro farmer and Sako Kanishiro of the *poi* factory in Waipio Valley, who gave me a glimpse into their worlds; and Dr Michael Grayum, Dr Alistair Hay and Dr Simon Mayo who allowed me to read and refer to their unpublished doctoral theses.

I should also like to thank the many others who have responded to pleas for information, but whom I have not had the pleasure of meeting: Linda Theus, membership secretary of the International Aroid Society; Marilynne Burbage of the Tropical Research and Development Institute, who furnished me with valuable information on the edible aroids; Janet Crossley, Staff Pharmacist of the Winchester Area Health Authority, who carried out a literature search on *Dieffenbachia* toxicity and obtained reprints of medical papers on the subject; A. Whitehead of the National

Poisons Information Service, Guy's Hospital, London, who obtained statistics on *Dieffenbachia* intoxication; Dr P. Hylands, Chelsea Department of Pharmacy, King's College, University of London; Dr Robert Mill of the Royal Botanic Garden, Edinburgh; Dr Paul Simons of *New Scientist*, who has introduced *Guardian* readers to aroids on more than one occasion; Dr Kerim Alpinar of the University of Istanbul; Dr Jacob Koach of the Hebrew University of Jerusalem; Professor H. Kamemoto of the Department of Horticulture, University of Hawaii; Larry Yamamoto of the Department of Agriculture in Hawaii; Ramon de la Peña of the College of Tropical Agriculture and Human Resources at the University of Hawaii; and Dr Dan Nicolson of the Department of Botany, Smithsonian Institution.

Warmest thanks are due to David Leigh, who has put so much effort into the line drawings and remained cheerful in the face of what must have seemed impossible demands; to Miles Jackson of Allen and Unwin, who first saw the potential in this book; and to my editors, Valerie Buckingham and Isabel Sutherland, who have allayed the worst of my anxieties as a first-time author.

In addition to professional advice and assistance, I am also aware that this book has been made possible by the support and encouragement I have received from relatives and friends during difficult phases – especially from Esme and John Bown, Tony Bown, and Judy Hammond who held the fort while I was travelling; Sara Hicks, my acupuncturist, who often had to pick up the pieces; and my companions in the search for *Amorphophallus titanum* – Jon (Erizon Samsuar), interpreter and friend through thick and thin in Sumatra and Herman Rutuwene, driver extraordinaire, whose attunement to our decrepit vehicle and skill in foreseeing and gently manoeuvring past potholes, landslides and less careful motorists almost inspired me instead to write a book on Zen and the art of Indonesian minibus driving.

Lastly, immense gratitude to my children, whose loyalty and devotion have never wavered, though at times I was not to be seen for aroids – and a special thank-you to William for taking the inside jacket photograph of me and for his help in organizing the colour plates and check list of genera.

Finally, in spite of appealing to many bodies for funding during the initial period of research, only the Manpower Services Commission responded, and for the first year I was the grateful recipient of an Enterprise Allowance.

Deni Bown MAY 1987

Introduction

Aroids are not the sort of plants you can be neutral about. Amongst the declarations of handsome and magnificent come a multitude of comments that they are curious, weird, rather obscene, revolting or even terrifying. But of course, before you can make these kinds of judgements, you have first to know what an aroid is and it has been very obvious to me, over many years of lunatic enthusiasm for this family of plants, that whenever I mention the word aroid in conversation, a look of blank incomprehension passes across the face of the listener.

Fortunately it is quite easy to change the puzzled expression to one of enlightenment. Aroid may not be a household word, but the images of certain aroids are familiar to almost everyone.

Perhaps best known is the Swiss cheese plant (*Monstera deliciosa*), with its unmistakable foliage. The huge heart-shaped leaves cut with slashes and holes are some of the most extraordinary in the plant world, but now so commonplace that we take them for granted. Then there are the virginal white arum or calla lilies (*Zantedeschia aethiopica*), produced in abundance by the floral trade for bouquets, wreaths and church decoration at Easter. Another florist's favourite is the flamingo flower (*Anthurium andraeanum*), symbol of many an exotic holiday location. The waxy scarlet blooms look exactly like plastic and it is hard to believe they are real. Less flamboyant but equally well-known are the wild aroids of temperate woodlands and hedgerows: the European arum or lords-and-ladies (*Arum maculatum* and *A. italicum*), with pale green, cowl-shaped inflorescences at bluebell time followed in late summer by clusters of bright red berries on bare stalks. Its equivalent in North America is Jack-in-the-pulpit (*Arisaema triphyllum*), which is similar in appearance, the conical 'petal' (the spathe) as the pulpit and the upright projection (the spadix) in the centre of it being Jack.

Once these examples are given, it becomes obvious that aroids are indeed very familiar plants in our homes and gardens and in the surrounding countryside, though they go under so many different guises – from modest woodland and marsh plants to mighty tropical climbers – that it is by no means straightforward to tell that they belong to the same family. In fact, most of us see aroids every day. If you have a few houseplants, there is almost certainly an aroid amongst them and a glance around many of the places we visit regularly – shops, offices, banks, waiting rooms and so forth – will reveal many more. It is estimated that over 50 million houseplants are sold annually and topping them all in popularity for years has been the Swiss cheese plant. Other aroids, such as philodendrons, dieffenbachias and aglaonemas, are close runners-up. Indeed, aroids are exploited more than any other family as houseplants for their outstanding foliage and their tolerance of the conditions found in most homes and workplaces. And in warm climates, aroids also form a large proportion of decorative plants used outdoors in landscape, gardens and containers. They are of importance too in the cut-flower trade: *Anthurium* blooms are a major product of the Hawaiian Islands.

The value of aroids is not limited to the ornamental. In the tropics various aroids are cultivated for food: 400 million people include taro (*Colocasia esculenta*) in their diets and several other species have significance or potential. Their importance is often overlooked because they are generally consumed where they are grown and rarely exported. As a result, they have so far largely been ignored by plant breeders and agronomists. Nevertheless, innumerable varieties have been selected by growers over the centuries, together with different methods of cultivating them. They are often an integral part of local culture, especially as some of them are also important in medicine and ritual. The use of aroids as food and medicine is particularly interesting in view of the fact that they are invariably toxic and require careful preparation for safe consumption.

But being able to spot an aroid when you see one and realizing their popularity and importance is only a start. Finding out more about them is no easy task. There are many excellent books on other families of plants but almost nothing on aroids. There is fleeting mention of the family in encyclopedic and reference works on plants (usually in connection with its ingenious pollination mechanisms), and manuals on gardening and houseplants will cover the best known species – and, of course, there are scientific papers in journals – but to the general plant lover and hobbyist aroids must so far have remained rather a mystery. It seems to be the most neglected of all plant families that are widely grown as ornamentals and it is hard (for the enthusiast at least) to comprehend such an omission on behalf of writers and publishers.

One reason for this could be that by conventional standards of beauty in flowers, aroids are unlikely to compare favourably with, say, roses or orchids – though the white arum lily, for its purity and simplicity, has many admirers. Aroid blooms are indeed very odd. They consist of a single 'petal' called the spathe which provides a backing or cover for a central flower-bearing protuberance, the spadix. This floral peculiarity is technically an inflorescence, not a flower (the real flowers being very tiny and often hidden by the spathe), and in its numerous and bizarre manifestations sets aroids apart from all other families of plants. It has also given rise to the special mystique enjoyed by many aroids, for the shape of this inflorescence has inescapable sexual connotations beyond the fact that it is the plant's reproductive apparatus. The characteristically upright spadix projecting from the shelter of the spathe has both fascinated and repelled human observers of all cultures throughout history. In the opinion of many it is indeed a more appropriate bloom for a fertility rite than for a buttonhole or a vase.

Unfortunately the shape is not the only obstacle to acclaim. Although a number of aroids are attractive and sweetly scented when in flower, many produce inflorescences which have dingy colours, smell simply terrible, have hairs on their surface, and actually raise the temperature of their tissues – not, generally speaking, activities which are acceptable in plants grown to bring pleasure to the senses! The size of aroid inflorescences is eccentric too: from a totally insignificant centimetre or two in a floating waterweed to the world's largest which, in the steaming jungles of Sumatra, exceeds the height of a man.

A further criticism from a cultural point of view is that many species produce what in horticultural circles are termed 'inconspicuous' blooms: that is, hard to distinguish amongst the leaves. They are usually described as curiosities. To the fanciful observer these may appear as little pixies or mice and offer a certain charm, but to many they suggest something far more sinister.

Nevertheless, if there are black marks against some aroids when they flower, this is more than redeemed by the beauty and astonishing variety of their foliage. Ironically, many plants which are adored for their fabulous flowers have unexceptional, even drab foliage, but this has been no impediment to their near-worship throughout the centuries. It would appear then that most people tend to prize flowers above leaves, however magnificently shaped or coloured. Foliage plants have so far been the poor relations.

INTRODUCTION

There is change in the air though and recent years have seen increasing interest in the use of beautiful foliage in ground cover, garden design, floral arrangements and in home and workspace decor. There is plenty of well-illustrated literature on the subject and many nurseries now specialize in foliage plants. And with such information and availability comes greater discrimination in the plant buyer.

In addition to greater sophistication on the part of growers, there is also an increasing fascination with the more novel aspects of plant life and cultivation; subjects as diverse as medicinal herbs, carnivorous plants, pip growing, bottle gardens, and botanical sex command a ready audience. This points to a demand for the unusual on both practical and imaginative levels. It would seem that aroids can at last come into their own with their unique blend of usefulness, strangeness and beauty.

If the rose – or the orchid – is queen of flowers, I would propose that the aroid is the king of foliage plants. The display is dazzling in form, size, hue and sheer variety, from perfect dark green hearts to marbled ovals, sleek arrowheads and huge shining shields. Some have such deeply cut and complex leaves that they resemble palms and others are so simple that they look more like reeds and bird's nest ferns. A few even have perfect holes through them – no one is sure why – but it is a feature almost unknown outside the family. As for colour, there is every shade from pure white to richest red, often with contrasting veining or a quite different colour in new leaves and undersurfaces. And variegation is nowhere as varied as in aroids: freckles, streaks, stripes, spots, patterned or randomly piebald, even to translucent refractive markings.

Textures are likewise full of surprises, anything from high gloss or velvet to the nearest thing in plants to the metallic and crystalline. And the interest doesn't stop at the leaves: many have the most beautiful stalks, mottled like reptilian skin and eggshells. Finally, if you can rid yourself of ideas of prettiness, the inflorescences can be appreciated for their sculptural quality, an almost space-age attention to line and form which is at the same time primitive in its symbolism.

From this brief survey of aroidean characteristics you will gather that they are plants of extremes. They have considerable curiosity value and visual impact. The invariably handsome foliage and the peculiar flowers give them a 'beauty and the beast' appeal. Anyone designing a setting for the Garden of Eden or a tropical paradise would be certain to include the elegant and luxuriant shapes of aroid leaves. On the other hand, it is no coincidence that the horrific (I almost wrote striking) botanical structures devised for film versions of *The Day of the Triffids* bear an uncanny resemblance to aroid inflorescences. This combination of characteristics — the alluring and the repellent — is, as in the fairy tale, inevitably disturbing yet a large part of the fascination of these plants.

If your imagination has been captured by the features outlined in this introduction, I hope the text and illustrations will enable you to explore further dimensions and serve as a general introduction to this impressive family of plants. I am aware that I have constantly had to sacrifice comprehensive detail to the anecdotal and inspirational in order that the book as a whole is readable rather than only referable. Nevertheless, I also hope that it will prove sufficiently informative to attract enthusiasts who already grow these remarkable subjects and who so far have been unable to obtain much in the way of books on their interests. Hopefully too, it may appeal to botanists and horticulturists who may have noticed the scarcity of publications on *Araceae*: perhaps it will inspire a more authoritative work on the various aspects which deserve a far more detailed approach.

Lastly, this book is intended for all those who find structure and design in nature an endless fascination. Once we have fallen under this spell, any natural history topic is a delight and our initial judgements and preferences are extended to a much wider reverence for the complexity and ingenuity of living things.

I
Variations on a Theme

What are aroids and where do they grow?

Aroids are first and foremost flowering plants, albeit strange ones.

Flowers are potent symbols in the human imagination. The very word 'flower' conjures up images of exuberant yet fragile creations of the natural world that inspire us or serve to express our feelings on special occasions. Many have universal appeal and their shapes, colours and scents pervade our lives on all levels, from horticulture, art and design to toiletries. Alas! few aroids produce blooms which fulfil such conventions of beauty. On first acquaintance at least, most people find them unusual, if not completely beyond the pale.

From a botanical point of view these considerations are irrelevant. Whatever a particular flower means to us personally or socially, there is no escaping the fact that in physical terms flowers are the highly practical and vital parts of a plant's reproductive anatomy. Far from putting us off, looking at flowers from this angle can open up whole new areas of discovery and appreciation. In this light, aroids are on a par with – and even excel – many other flowering plants in terms of interest. Being unconventional in structure and lifestyle, they are full of surprises – pleasant and unpleasant.

——— The spathe-and-spadix inflorescence ———

The hallmark of *Araceae* is the spathe-and-spadix inflorescence – a floral structure consisting of a petal-like leaf (the spathe) and a flower-bearing protuberance (the spadix). Though technically an inflorescence and not a flower – because it consists of a spike of several or many individual flowers – it functions very much like a single flower. The spadix is basically club-shaped and fleshy and the actual flowers are bractless and, in most cases, minute. An individual flower may be bisexual or unisexual, depending on the genus. If unisexual, the sexes are usually segregated into distinct zones, females at the base and males higher up, sometimes with sterile florets interspersed or at the boundaries. In a number of species the sterile florets are modified into extensions of the spadix. Lower down the spadix, out of

VARIATIONS ON A THEME

sight, these tend to be lateral protrusions resembling prongs or bristles which play a role in directing insects entering the inflorescence. At the apex they more often form an elongated and clearly visible appendage whose main purpose is to disperse odours.

The spathe might look (and behave) like a rather odd petal, but is in fact a modified leaf or bract. Many species betray this origin by having spathes which are green and leaf-like or which turn green and photosynthesize after flowering. The spathe protects the developing flower spike and in most species plays an important part in the pollination biology. More often than not, it decays after flowering, but in some it persists throughout fruit formation.

Aroid flowers are mostly too small to see well with the naked eye, and if we could see them clearly we would probably conclude that they bear little resemblance to what we are accustomed to call flowers. Whereas many plants go to town on their flowers, in aroids all the energy for adornment seems to have gone into spathes and spadices – not to mention foliage – leaving the flowers themselves insignificant and stripped down to the most fundamental components. They sit flat and often tightly packed on the spadix: the female parts consisting of a single pistil and an ovary and the male of highly condensed stamens with squat filaments. Bisexual flowers usually have very modest greenish or brownish tepals (a type of perianth

Unisexual and bisexual inflorescences
In both types female receptivity occurs first and only when it is over do the stamens extrude pollen. This avoids self-pollination.
a] ($\times 1$) In **Anthurium gracile** the spadix is covered in bisexual flowers. Female and male parts have separate phases but not all those of the same sex mature together and flowering is often prolonged.
b] ($\times 4$) The bisexual flower of *A. gracile* has a central pistil (female part) surrounded by the anthers (male parts) and four tepals.
c] ($\times \frac{2}{3}$) **Alocasia lowii** has unisexual flowers: the females at the base of the spadix and the males above. The terminal portion of the spadix (the appendix) and the zone between male and female flowers are covered by sterile flowers.
d] ($\times 2$) A male flower of *A. lowii*.
e] ($\times 2$) Female flowers of *A. lowii*.

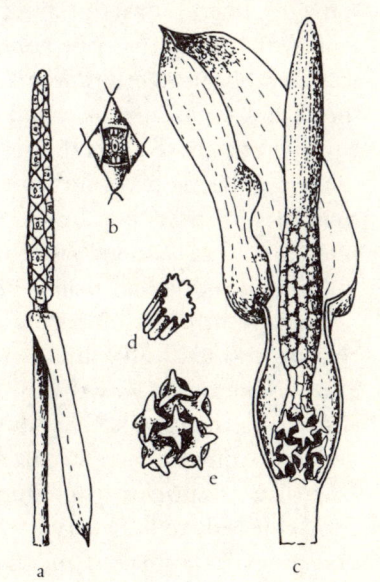

found in flowers without clearly differentiated sepals and petals) but commonly the flowers are naked – that is, devoid of a perianth to protect the developing reproductive parts.

The female flowers ripen first, a condition known as protogyny which is unusual among flowering plants but may be associated with beetle pollination.[1] Then, when female receptivity has ended, the male flowers release pollen. This system prevents self-pollination. The stages are easily observed in anthuriums (which have bisexual flowers), the flowers secreting a sticky solution during the female phase and later extruding pollen when the male parts mature. In species with bisexual spadices the flowering period is usually lengthy: first the female flowers opening in sequence up the spike and then the male, which may take weeks or even a month or more. In contrast, flowering in a unisexual spadix is often very brief, with both female and male phases over in a few days.

For all these generalizations, flowering in aroids is absurdly variable. All aroid inflorescences have a spathe and spadix (well, nearly all – the exception being the uppermost inflorescence of *Pothoidium* which is enclosed by a foliage leaf) but apart from this characteristic there is little consistency. Inflorescences may be single or in clusters, on short or long stalks (peduncles) and are produced annually or continuously or rarely, according to the species. There is a wide range of colours and textures; some bright and attractive, but predominantly shades of maroon and green which are frequently striped or mottled and waxy or fleshy. They also vary considerably in size: *Pistia* inflorescences measure a few millimetres and those of several *Amorphophallus* species are man-sized. Floral parts are equally resistant to generalization. The rule of thumb for monocots (one of the two great classes of flowering plants) is that floral parts are in threes or multiples of three, but aroids may have four to nine tepals and anything from four to 12 stamens – with numbers often varying within a species, and even in an individual from inflorescence to inflorescence.

Perhaps the most interesting aspect is how infinitely diverse the shape of the spathe-and-spadix unit can be. The shape of the spathe bears a close relationship to the way the species is pollinated and to the arrangement of flowers on the spadix. The simplest play little part in securing pollinators and are similar in appearance to small leaves – usually greenish and lanceolate, elliptic or oval. During flowering they are held parallel to the spadix or reflexed. This type is prevalent in the pothoid climbers and in *Anthurium*. The smallest and most insignificant spathes are probably those of *Orontium* inflorescences which appear spatheless.

From the plain and leaf-like it is quite a jump, visually at least, to the concave, hooded and funnel-shaped spathes. The most complex have a tubular lower portion which forms a chamber round the flowers and, according to the ploy used by the species, may serve as a trap, dining area or mating arena to detain pollinators during the consecutive opening of female and male florets. Such inflorescences provide all that is needed for an overnight stay: warmth and shelter, carefully adjusted humidity, ventilation, lighting and a supply of food, in order to keep the inmates suitably occupied, and alive, until they have served their purpose. A classic shape is the elegant chalice of the calla lily (*Zantedeschia* species), but extremes abound: *Helicodiceros* with hairy spathes the size, shape and texture of a pig's ear; *Amorphophallus titanum* with a massive fluted vase; *Arisarum proboscideum* and some *Arisaema* species with long-tailed hoods; *Urospatha sagittifolia* whose spathe is barely cupped at the base and narrows into a corkscrew twist at the apex; and *Cryptocoryne* species with delicate tubes, adjustable in length according to the depth of water so that the spathe entrance is not flooded.

The spadix is equally versatile and its numerous manifestations are the most bizarre features of the family and certainly the ones for which it is most notorious. Characteristically it is an erect cylindrical structure which may be anything from minuscule to over 1 m in length. Some spadices are highly reduced and scarcely recognizable as such. This happens when the number of flowers is very small, as in *Pistia* and *Ambrosina*. In *Cryptocoryne*, *Lagenandra*, *Ambrosina* and *Arisarum proboscideum*, the spadix is completely hidden within the spathe and careful surgery is required to remove one half of the spathe tube to reveal its intricate details. At the opposite extreme are those which protrude, often at some length, from the spathe: shiny, black and rigid in *Dracunculus vulgaris*; wrinkled or fungoid in several *Amorphophallus* species; whip-like in a number of arisaemas.

These phallic or tail-like sterile appendages of unisexual-flowered species are unique in the plant world. Their appearance is encapsulated in the names given by botanists; *Amorphophallus* translates as 'shapeless phallus' and *Anthurium* means 'tail flower'. They elicit the same inevitable responses of curiosity and amusement today as they did thousands of years ago, only now we may add some information about their structure and function to what we know of their folklore and symbolism. Bisexual spadices are less bizarre on the whole, though there is plenty of variety in shape, from coiled

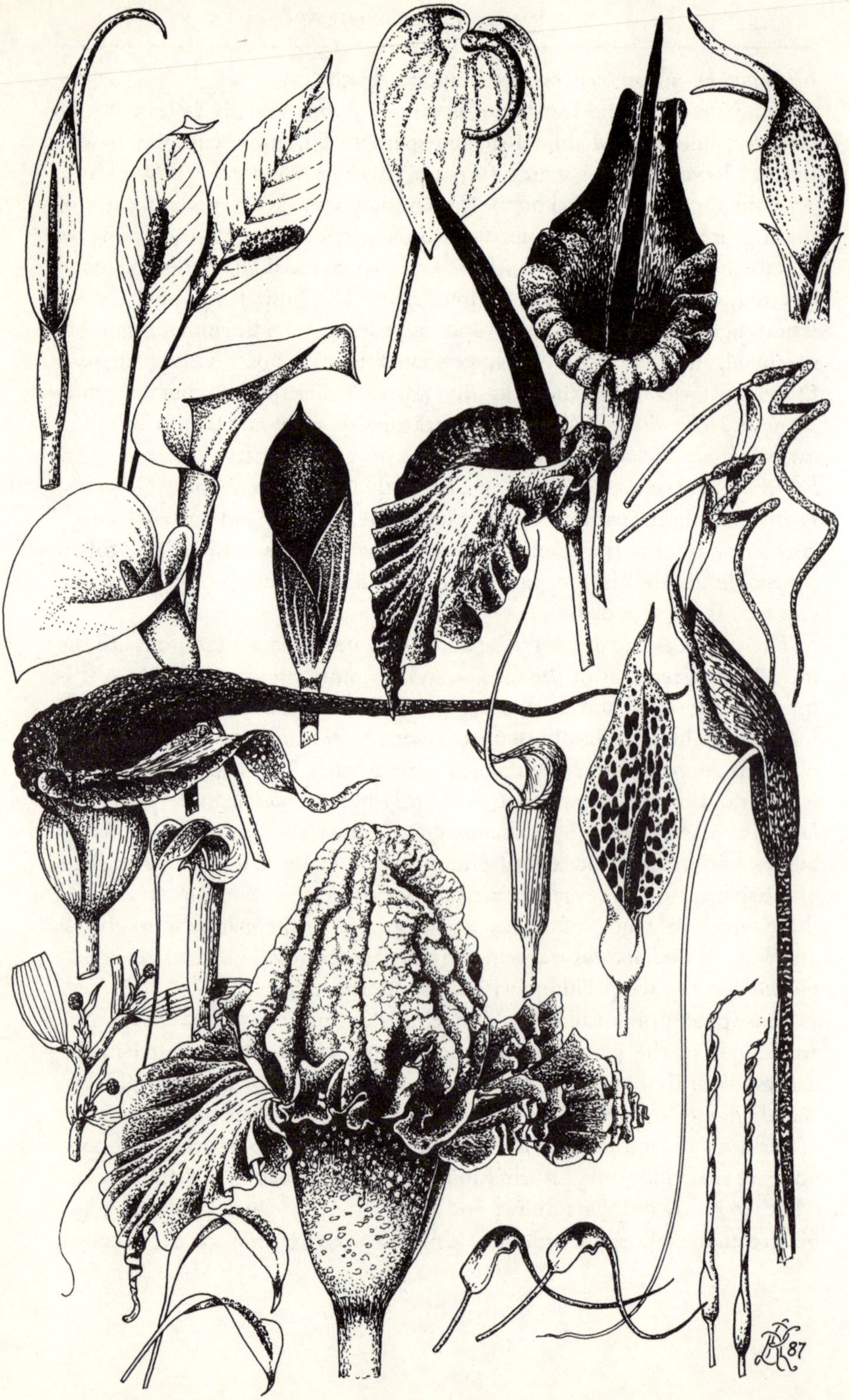

in *Anthurium wendlingeri* to globose in *Symplocarpus foetidus* and some species of *Pothos*. They are entirely covered in flowers which are often arranged in spirals, giving an overall geometric pattern to the surface.

—— Foliage ——

In spite of the diversity of spathe-and-spadix sizes, shapes, colours, textures – and odours – all but the most eccentric are quite easily recognizable as such, leaving little doubt that the plant in question is an aroid. Non-flowering aroids are a different matter, with foliage so variable that generalization is near-impossible. In fact, virtually every known leaf shape occurs in the family, and the beginner is bound to find it incomprehensible that species as different in appearance as *Acorus gramineus*, *Monstera deliciosa*, *Amorphophallus konjac*, *Pistia stratiotes*, and *Dieffenbachia seguine* could possibly be related. To complicate matters, venation patterns are just as variable. Though the ground plan of the primary lateral veins is generally parallel the end result is often netted (reticulate), as broad leaves, which are so common in the family, must develop a subsidiary network to reach all parts of the blade. This is unusual among monocots, which generally have simple, relatively narrow, parallel-veined leaves.

The simplest leaves are linear, lanceolate, oblong and oval. These occur plentifully in *Aglaonema*, *Dieffenbachia*, *Spathiphyllum* and *Anthurium*. In some genera these types show adaptation to extreme habitats. Aquatics such as *Cryptocoryne* and *Aridarum* mostly have willow-shaped leaves which sustain least damage in strong currents. At the other end of the scale, in arid conditions very narrow leaves reduce water loss and have been devised by several *Biarum* species which grow in hot dry Mediterranean

Aroid inflorescences (not exactly to scale)
The spathe-and-spadix inflorescence is the main distinguishing feature of aroids. It does, however, display great plasticity in size and shape and there are many bizarre variations on the theme.

a] *Colocasia esculenta*
b] *Spathiphyllum wallisii*
c] *Anthurium scherzerianum*
d] *Dracunculus vulgaris*
e] *Biarum davisii*
f] *Zantedeschia aethiopica*
g] *Dracontium polyphyllum*
h] *Anthurium wendlingeri*
i] *Amorphophallus cirrifer*
j] *Arisaema consanguineum*
k] *Arisaema tortuosum*
l] *Arum dioscoridis*
m] *Amorphophallus pendulus*
n] *Pothos scandens*
o] *Spathicarpa sagittifolia*
p] *Amorphophallus paeoniifolius*
q] *Arisarum proboscideum*
r] *Cryptocoryne retrospiralis*

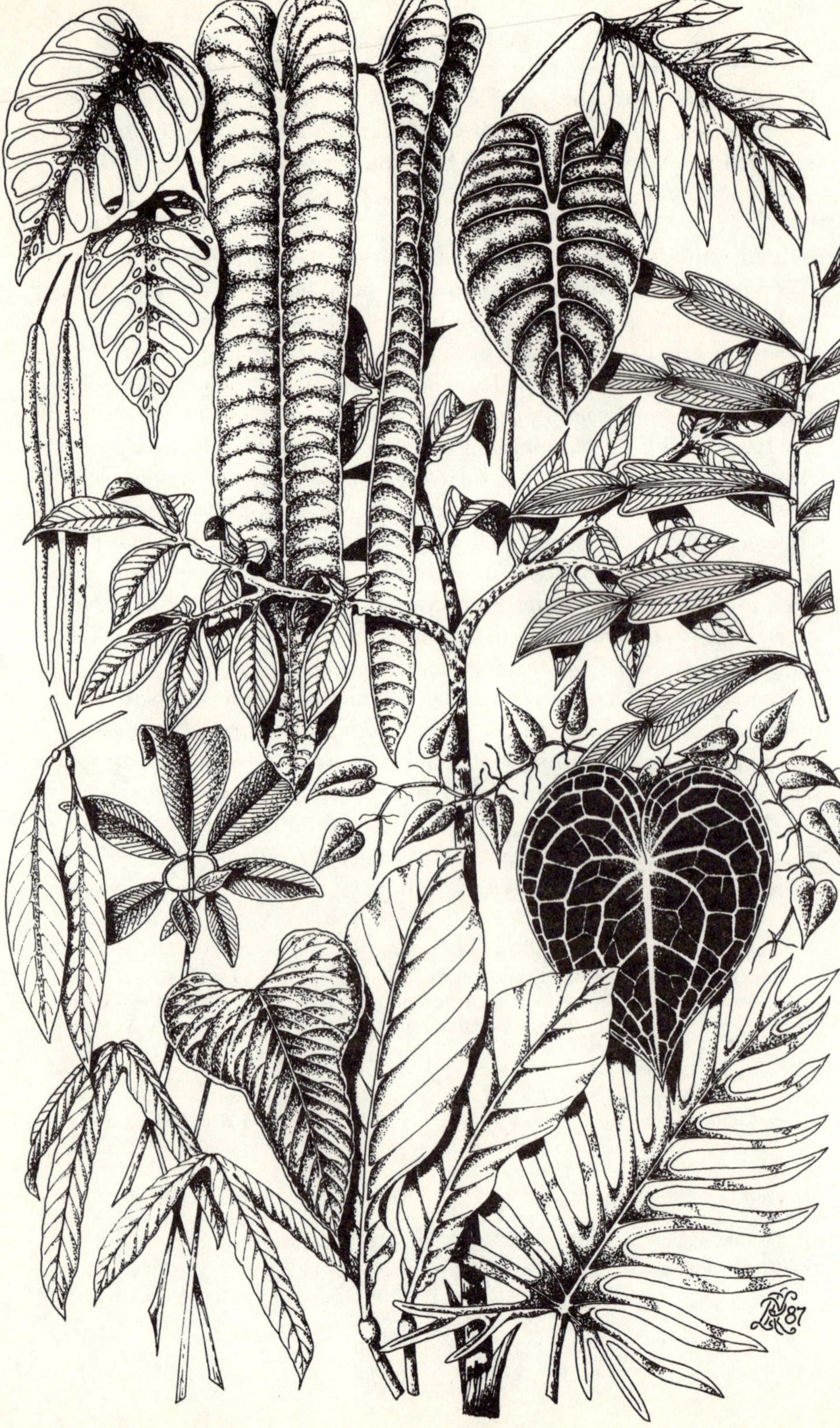

regions. Even simple leaves may vary enormously in character – dangling like straps in pendent anthuriums; held erect in large rosettes in bird's nest anthuriums; elegantly arched in spathiphyllums; or in reed-like tufts in *Acorus*.

Equally widespread are arrow-head shaped (sagittate), spear-head shaped (hastate) and heart-shaped (cordate) blades. Again, sizes and proportions vary greatly, from a fragile few centimetres in *Theriophonum minutum* to taller and considerably wider than a man in *Philodendron speciosum*. In some species the basic shape is elongated, giving an elegance unmatched elsewhere in the plant world, and for which the family is justly famous – the graceful oval-cordate of *Anthurium warocqueanum* and stunning oblong-cordate of *A. veitchii* being two of the best known. *Alocasia* boasts many of the finest elongated sagittate and hastate shapes, often made shield-like by the stalk being attached to the undersurface of the leaf instead of the base (peltate) and having the basal lobes joined for most of their length.

Compound leaves are plentiful too. The simplest have just three main divisions. Species with this pattern are often named after it: *Arisaema triphyllum*, *Typhonium trilobatum*, *Philodendron tripartitum*. Next in complexity are the palmate. These have five or more main divisions, sometimes with subdivisions and either lobed (as in *Anthurium pedatoradiatum*) or cut to the base into separate leaflets (as in *A. clavigerum* and *A. polyschistum*). Again, the name may reveal the leaf structure: *Anthurium polydactylum* (meaning having many fingers), *A. pentaphyllum* (five-leaved). A few species, such as *A. croatii* and *Arisaema exappendiculatum* have radiate blades with the leaflets symmetrically arranged around the stalk. Most species with compound leaves do not produce their most ornate foliage until

Aroid leaves
Almost every known leaf shape occurs in the family.

a] *Monstera obliqua*
b] *Anthurium veitchii*
c] *Alocasia cuprea*
d] *Alloschemone occidentalis*
e] *Anthurium pallidiflorum*
f] *Amorphophallus napalensis*
g] *Pothos scandens*
h] *Philodendron scandens*
i] *Holochlamys guineensis*
j] *Philodendron leal-costae*
k] *Anthurium clarinervium*
l] *Anthurium tilaranense*
m] *Anthurium signatum*
n] *Anthurium oblanceolatum*
o] *Philodendron elegans*

mature, so on the same plant may be found leaves of different ages with three, five, seven or more main lobes or leaflets.

The most complex and also the largest leaves in the family occur in the subfamily *Lasioideae*, especially in *Amorphophallus*, *Anchomanes* and *Dracontium*. They produce an umbrella of profusely dissected segments arising from three main divisions and many subdivisions. Watching one of these blades unfurl is even more absorbing than observing the opening of a flower. Those of *Anchomanes difformis* are folded very much like an umbrella and as tightly crumpled as a butterfly's wing in a chrysalis as they are raised up on stout prickly stalks to spread and stiffen several metres above the ground.

To add to this plethora of leaf shapes, aroid foliage often has the embellishment of wavy and lobed margins. Every degree of undulation occurs, from the near-rectangular indentations of *Alocasia sanderiana* to elaborate frills in some forms of *Philodendron bipinnatifidum*. Many species are cut close to the midrib, giving an array of almost pinnate leaves which resemble those of palms, as in *Philodendron elegans* (also known as *P. radiatum*). True pinnation, in which separate leaflets are formed, is rare in the family but can be seen in the African genera *Gonatopus* and *Zamioculcas*, and in the lower lobes of the Indian *Anaphyllum*. Serration is also uncommon, though found as a variable feature in a few arisaemas – *A. sikokianum* var. *serratum* (or *A. sazensoo* var. *magnidens*, as it used to be called) having deeply toothed margins.

A most distinctive araceous feature is the occurrence of natural holes in leaves, a process known as fenestration. It is found in many monsteroid climbers and in *Cercestis* and *Dracontioides*. Elsewhere in the plant world it is recorded only in the Madagascan lace plant (*Aponogeton madagascariensis*) and in *Pentagonia* (a genus of *Rubiaceae*),[2] though structurally they are quite different.

The colours and textures of aroid foliage are just as varied as the shapes. Variegation runs riot in the family, and in addition a high proportion of tropical rain forest species have fascinating glossy or velvety textures. Not only the leaf blades, but even the stalks and undersides of veins are highly ornamented – striped, mottled, bristled, frilled, prickled and all shades imaginable. Plant collectors have tended to select outstanding forms from wild populations and many of the plants seen in cultivation today are 'improved' through further selection and breeding, giving some of the most beautifully shaped and richly coloured of all ornamental plants – the gaudy reds, pinks and whites of *Caladium* hybrids being a prime example.

On occasion, collectors have been spoilt for choice, coming across a population in which every plant is differently marked. This happened in one small area of the Thai Peninsula where there were 22 distinct varieties of *Aglaonema nitidum* forma *curtisii*,[3] and along the banks of a stream in El Copé, Panama, where plants of a *Homalomena* species were anything from plain green to tricoloured.[4] Even more interesting, a search for the dark velvety *Anthurium dressleri* in Panama yielded not only the species in question but also, growing in the same area, several other aroids as well as *Calathea dressleri* and a species of *Melastomataceae* with similar-sized blackish velvety leaves.[5] This observation shows that there is probably much more at work in leaf coloration than mere genetic roulette. It is thought that velvety textures trap light and glossy surfaces protect against colonization by epiphytes (plants which grow on the surface of other plants, deriving their moisture and nutrients from the air) and algae: useful adaptations in gloomy humid rain forests. Whether variegation helps in the defence against herbivores is not certain, though it may be surmised that masquerading under different guises could fool an insect looking for the right leaf to lay eggs upon. In effect, the colours, textures and shapes we find so ornamental might well perform practical functions for the plant: an aspect of plant life that remains largely unexplored.

—— Growth ——

In aroids each new leaf emerges from the base of the preceding leaf, protected during development by a scale or cataphyll (a simplified leaf) or by a sheath (a slit in the stalk which envelops the new leaf before emergence). Sheaths can take up any proportion of the stalk, from the lower few centimetres (as in *Caladium*), to about half (*Monstera deliciosa*), or almost the whole length (*Dieffenbachia seguine*). Young cataphylls are often showy and a feature of some ornamentals such as *Philodendron erubescens*, in which they are an attractive pink. After the new leaf has emerged the cataphylls dry out, turning brown or fading and becoming papery or fibrous. Some are quickly deciduous, as in *Philodendron scandens*, but others are persistent. The nature of the cataphylls is important in taxonomy: in section *Calomystrium* of *Anthurium*, which includes *A. andraeanum*, the cataphylls persist and remain intact, whereas in section *Tetraspermium* they persist but weather into shreds – a characteristic of *A. laciniosum* which has a fringe around each node.

The normal juvenile pattern of growth is monopodial, in which the stem may put out branches but continues growing itself, after the manner of a Christmas tree. Some scandent aroids, such as *Pothos* and *Heteropsis* retain this pattern even when mature. The usual adult mode is sympodial, with successive branches taking turns as the main shoot. When the shoot comes to a full stop, through flowering or abortion of the bud at its apex, the developing lateral bud takes over as the main shoot. These patterns are difficult to detect when highly condensed, as they are in many rosulate species. Some scandent species switch patterns and revert to long monopodial flagella (leafless shoots) in low light conditions or after flowering, producing new sympodial shoots when they reach a better situation.

Aroid roots vary greatly in appearance, from the feathery water-borne ones of *Pistia* to the tough thongs and guy ropes of tropical climbers, but all are adventitious – which is to say they sprout from the stem (nearly always from the leaf nodes). They may be subterranean or aerial according to the habit of the species. Subterranean ones are sometimes wrinkled and contractile to adjust the level of the plant in the ground after soil disturbance through freezing, parching or flooding. The seedlings of tuberous species often have contractile roots to draw the young plants down into the soil after germination so that their tubers do not develop near the surface.

Sheaths and cataphylls
are important parts of aroid foliage. They protect each new leaf, and inflorescences too, and often remain as a distinctive feature after fulfilling their function.
a] ($\times \frac{1}{2}$) **Sheaths** are slits in the stalk (here seen in *Dieffenbachia*).
b] ($\times \frac{1}{2}$) **Cataphylls** (or scales as they are sometimes called) are simplified leaves (here seen in *Anthurium*). When the new leaf or inflorescence has emerged they may fall off, disintegrate on the plant, or dry and persist.

Climbers and epiphytes have aerial roots. Two types are formed: claspers which are insensitive to gravity and make for dark crevices to anchor the plant to the climbing surface; and danglers that respond to gravity and ignore light, sticking out from the plant – often at great length – to soak up rain, dew and atmospheric moisture. On a smooth surface, claspers are produced at right angles to the stem. Danglers branch profusely when they enter soil, whereas claspers, if planted in soil, fail to develop as their root hairs with microscopic suckers cannot grip particles. Some of the roots produced by epiphytic bird's nest anthuriums neither dangle nor clasp, but are relatively short and point upwards, penetrating the debris collected at the leaf bases and so tapping the nutrients in these arboreal compost heaps.

In many aroids the stem grows above the ground but quite commonly part of it remains underground in the form of a rhizome or tuber which has buds or 'eyes'. Rhizomatous growth, both horizontal and vertical, is often found in species associated with wet habitats, enabling the plant to consolidate itself in unstable substrates such as mud or the gravel of river beds. The horizontal rhizome system forms a grid which traps the substrate and the vertical type acts as a stake driven deeply into it. Not surprisingly, species using the latter technique can be very difficult to dig up. One would probably need a bulldozer to extricate a skunk cabbage from its mud! On the other hand, horizontal growers such as *Acorus* break off easily if undermined, and with purchase lost in one situation, float off downstream to river banks anew – turning disaster into an opportunity for further propagation.

As a rule, tubers in aroids are formed from the stem, not the roots. One of the few exceptions is the African genus *Stylochaeton*, in which the roots are swollen and fleshy. Whether aroid tubers are technically corms is debatable, but they are often referred to as such. Thickened stems, rhizomes and roots store nutrients and water, enabling the plant to undergo long periods of dormancy in dry or cold seasons. They also contain, in highly condensed form, the shoot which will renew growth after dormancy. Araceous tubers vary considerably in shape, from horizontal to vertical, and from the knobbly and asymmetrical ones of *Zantedeschia* to smooth and almost spherical in *Synandrospadix*. When it comes to size, variations are extreme: pea-sized in *Theriophonum sivaganganum* and heavyweights of 60–80 kg[6] in *Amorphophallus titanum*. Many produce miniature tubers on their surface or at the end of stolons which are sent out to some distance from the main tuber. These offsets are the size a seedling would reach after several years of growth, but of course, unlike seedlings, they are genetically identical to the parent tuber. Offsets in the form of bulbils may also occur on stems and leaves. *Amorphophallus bulbifer* bears large ones at the base of the leaflets, looking like strange brown galls. Tubers are renewed annually. As the plant enters dormancy the leaf and flower buds for the next season are already formed. When it starts into growth, the reserves stored in the tuber are used up and it shrivels. Then, as soon as the new foliage unfurls and starts to photosynthesize, the storage process begins again and a new tuber swells.

—— Habitats ——

Aroids have a definite liking for moisture and shelter. By far the majority are creepers, climbers and epiphytes of tropical rain forests – anything from insignificant members of the ground flora, such as West African *Nephthytis* to scrambling shrubs such as Asian *Pothos* and tree-like South American philodendrons of subgenus *Meconostigma*. Aroids in cooler, drier regions tend to be tuberous and seasonally dormant, escaping cold weather on the one hand and drought on the other by losing all aerial parts for several months of the year. Many of these species grow in woodland or scrub, in the shade and shelter of rocks or walls or in the vicinity of water, as they have not entirely lost the need for humidity and their fleshy leaves fare badly in strong winds and sunshine. In cold climates winter dormancy and renewed growth in spring is the norm. In warmer regions dormancy takes place in the summer or dry season and new growth begins when cooler, rainy weather returns.

A high proportion of aroids are aquatics and semi-aquatics – the latter including marginals, rheophytes (which live in fast-flowing water) and marsh species. Wetlands are present in most climatic zones from subarctic to tropical and vary greatly in ecology. Aroids are found in most of them, from tropical swamp forests, clear acid blackwater rivers and tidal areas to temperate ponds and wet woodland. Many of the most popular ornamental plants for aquaria are aroids, especially *Cryptocoryne* and the floating *Pistia*. A few aroids (notably *Aglaodorum griffithii*, *Cryptocoryne ciliata*, *Montrichardia arborescens* and swamp taro (*Cyrtosperma merkusii*) even inhabit brackish waters.

Only the sea and the extremes of desert, arctic and high alpine regions are without aroids.

Distribution

To date, the family *Araceae* has about 110 genera and more than 2500 species, more than half of which occur in the New World. And of these, over half belong to *Anthurium*, the largest genus by far, with over 700 species. Many genera are very small – 90 per cent of them account for only 15 per cent of the species.[7] The genera are contained in subtribes, tribes and subfamilies which cluster together related groups. For example, *Arum* is in subtribe *Arinae*, one of six subtribes belonging to the tribe *Areae*, which in turn comes under the subfamily *Aroideae*. Like all taxonomic groups, these categories are subject to change as understanding of the family grows, but the most recent account[8] puts the number of subfamilies at nine. They are arranged in order of evolutionary advancement, the first being *Acoroideae* which contains the problematic *Acorus* and *Gymnostachys*, followed by *Pothoideae*, *Monsteroideae*, *Calloideae* (containing only *Calla*), *Lasioideae*, *Philodendroideae*, *Colocasioideae* and *Aroideae*, ending with the highly idiosyncratic floating aquatic *Pistioideae* (which consists solely of *Pistia*).

There are two main centres of distribution for the family – tropical America and tropical Asia – with the number of endemic genera almost equally divided between them: 35 in the New World and 34 in the Old World. There are also 14 genera found only in Africa and four or five restricted to the Mediterranean region. Australia has only one endemic genus: the odd primitive monotypic *Gymnostachys*. When it comes to the actual number of genera, Asia is far richer than the Americas, with 75 compared to 46. As for species, tropical America is unbeatable, with some 1350 so far described.[9] The complex buckling of the Andes created innumerable niches, each forested peak becoming a virtual island around which separate populations evolved into a profusion of species.[10]

The two centres do share some genera: *Homalomena*, *Spathiphyllum*, *Schismatoglottis*, *Lysichiton* and *Arisaema* for example have species in both Asia and the Americas. Transoceanic links are uncommon but they are consistent with the drifting apart of the continents which halted the movement of species between east and west. Those marooned on the various land masses continued to evolve and began to spread across new bridges formed between North and South America on the one hand and between Africa, India and Eurasia on the other, giving the present pattern of distribution. Many must have died out over millennia of changes, especially in Africa which underwent extensive dry periods.

Family Relationships

If a family of plants is likened to a jigsaw then the species are the pieces. To take the analogy further, each family jigsaw has its own unique theme but fits into the giant jigsaw of plant life as a whole. Putting the pieces together is done by the meticulous work of taxonomists. The task begins with observation and description, then comparison, and finally with classification – the actual putting of the piece in its place.

As anyone who has ever done a jigsaw puzzle will know, some pieces fit quite easily next to others with similar characteristics. *Arum* and *Typhonium*, for example, look so alike that anyone would guess they go side by side. Others are far more tricky, showing few clear affinities and have to be tried first in one place and then in another. A good example is the genus *Scaphispatha*. The 19th century taxonomist Heinrich Schott, whose classification was based mainly on floral characters, was not sure where it fitted and so put it, aptly enough, into 'subtribu Problematicae'.[11] The next move was made by Adolf Engler, who began publishing works[12] on the family about 20 years after Schott's final monograph of 1860. His approach took into account both floral and vegetative structure. He tried it in subfamily *Aroideae*, as it shows reductions which indicate it is an advanced genus. There it remained until 1980 when Josef Bogner (who published a revised classification in 1978)[13] moved it (on the basis of similar male flowers and the same chromosome number) to the *Colocasioideae* – tentatively![14]

OPPOSITE]
Alocasia × *amazonica* A backlit alocasia leaf showing the peltate structure (the stalk being attached to the middle of the undersurface) and characteristic venation in which the primary lateral veins are prominent on the undersurface and the veinlets curve towards the margin in a parallel formation.

OVERLEAF]
ABOVE LEFT *Scindapsus borneensis* (Borneo) In this genus the developing seeds are protected by a layer of caps containing needle-like trichosclereids which deters the unwary from sampling unripe fruit. When mature, the layer breaks away, leaving the berries exposed for dispersal. Similar mechanisms are also found in *Monstera* and *Stenospermation*.
ABOVE RIGHT *Biarum tenuifolium* (Mediterranean) A tuberous aroid which produces roots at the base of the shoot and offsets around the tuber. The yellowish bulges are the beginnings of contractile roots which pull the tuber down into the ground.
BELOW *Spathicarpa sagittifolia* (Brazil, Bolivia and Paraguay) The small tropical American genus *Spathicarpa* has the spadix fused to the spathe. Male and female florets are in rows: the two outer ones being female and the inner ones male. The male flowers can be seen extruding pollen.

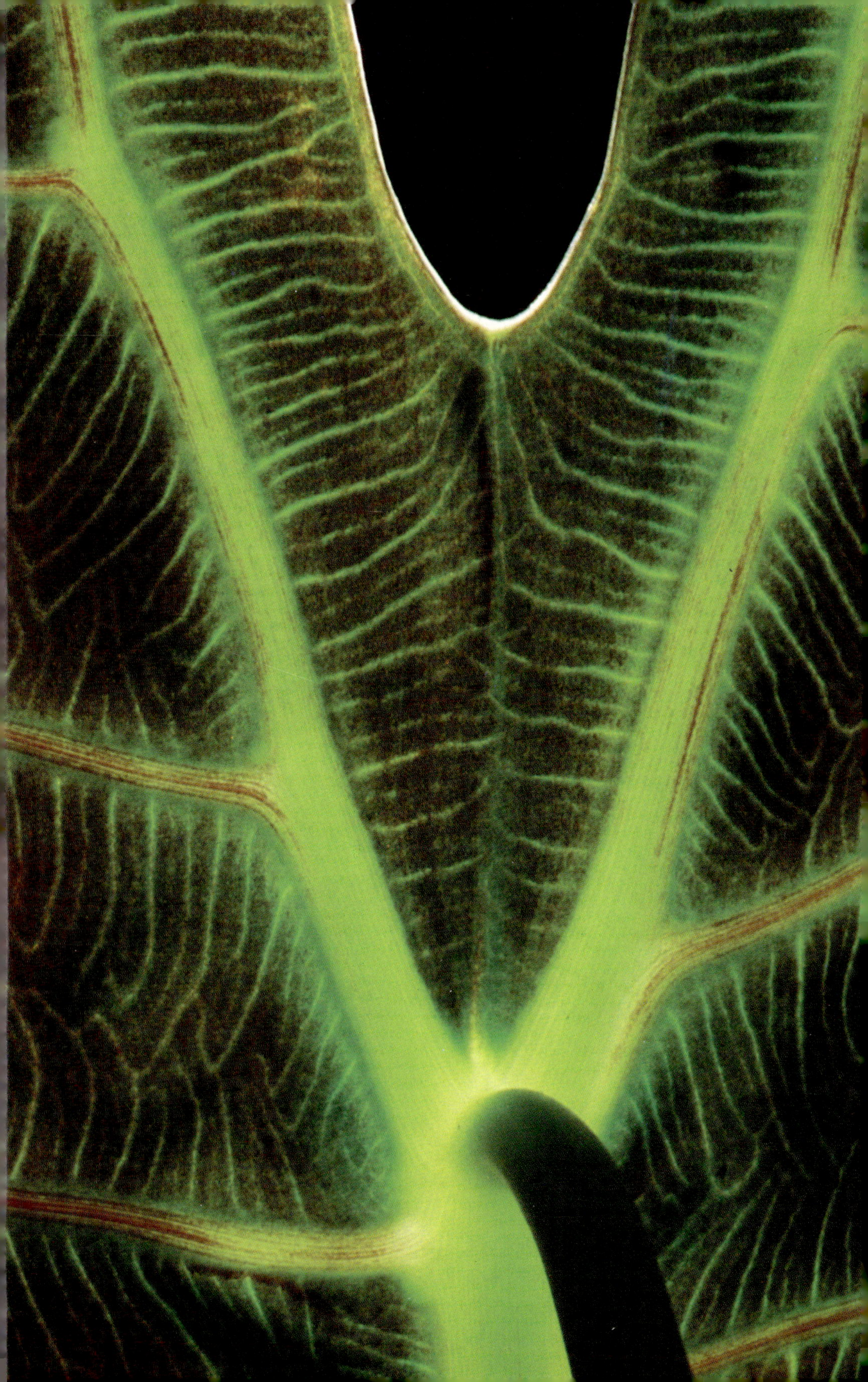

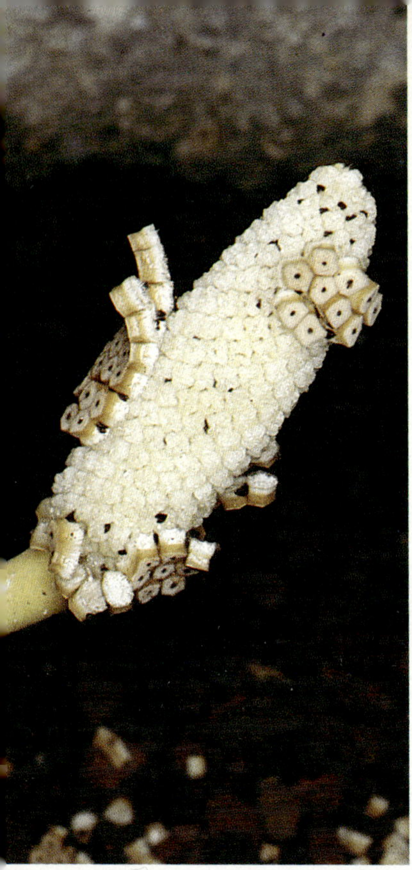
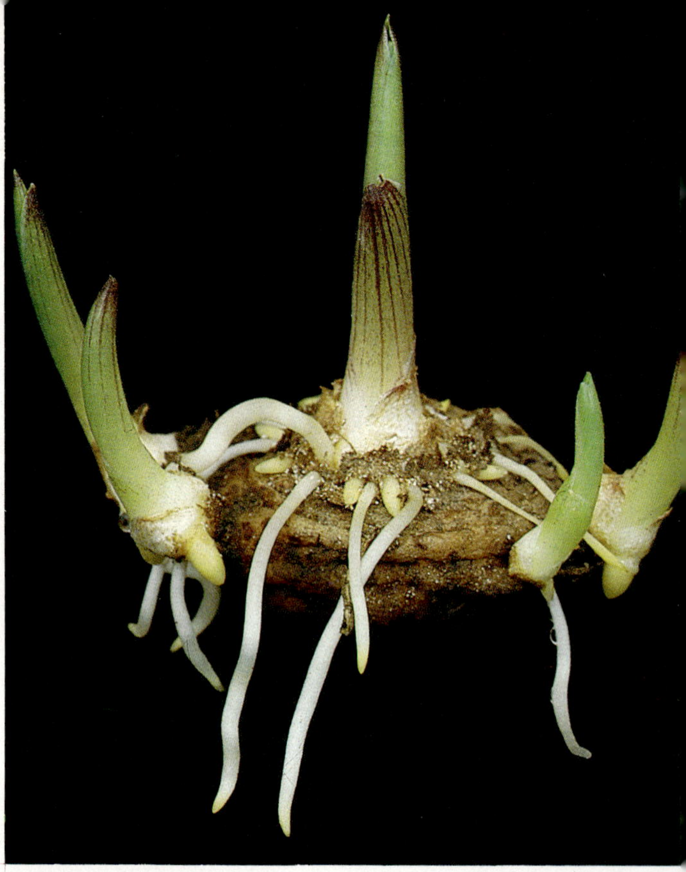
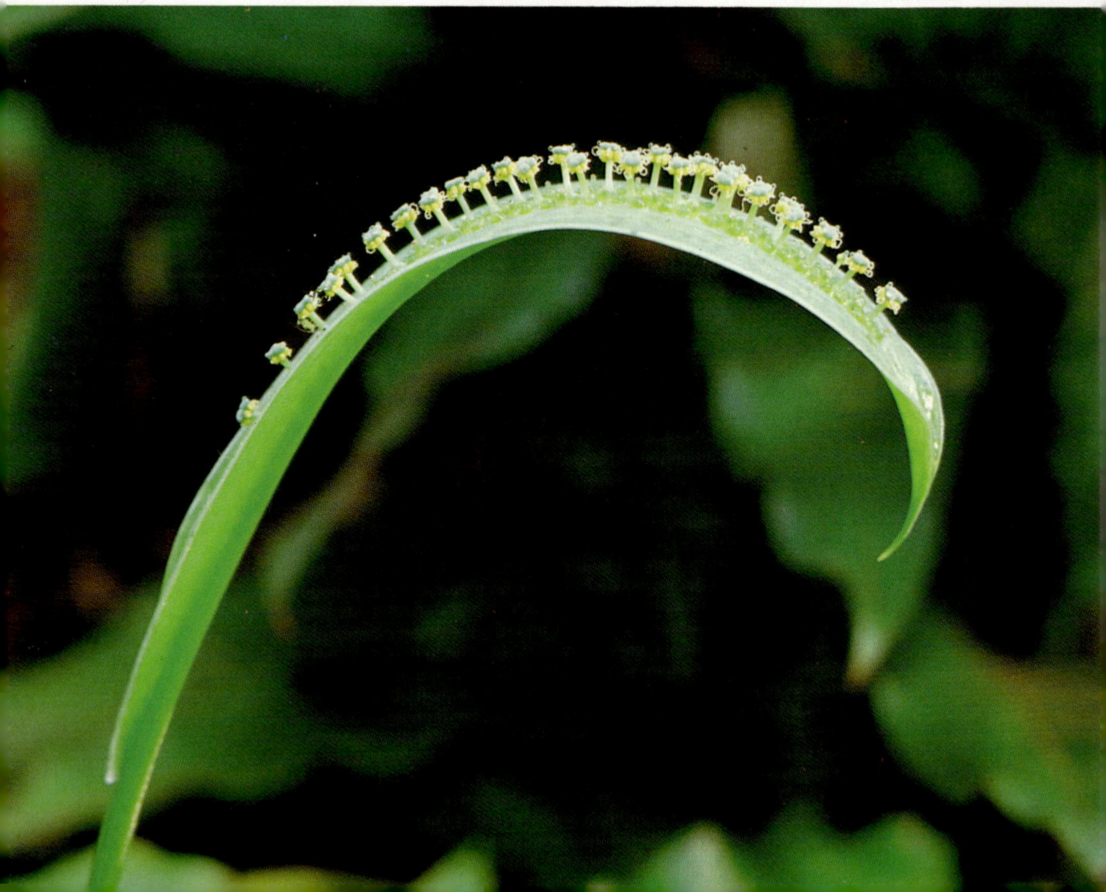

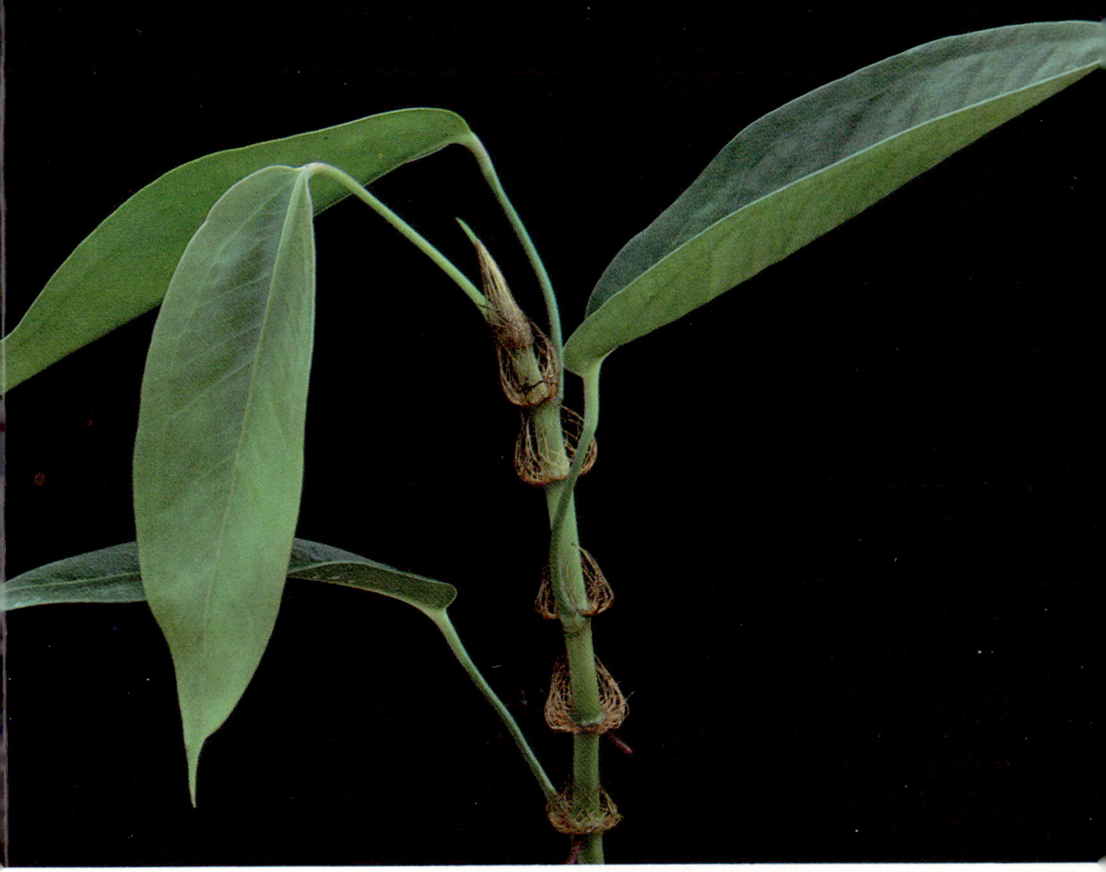
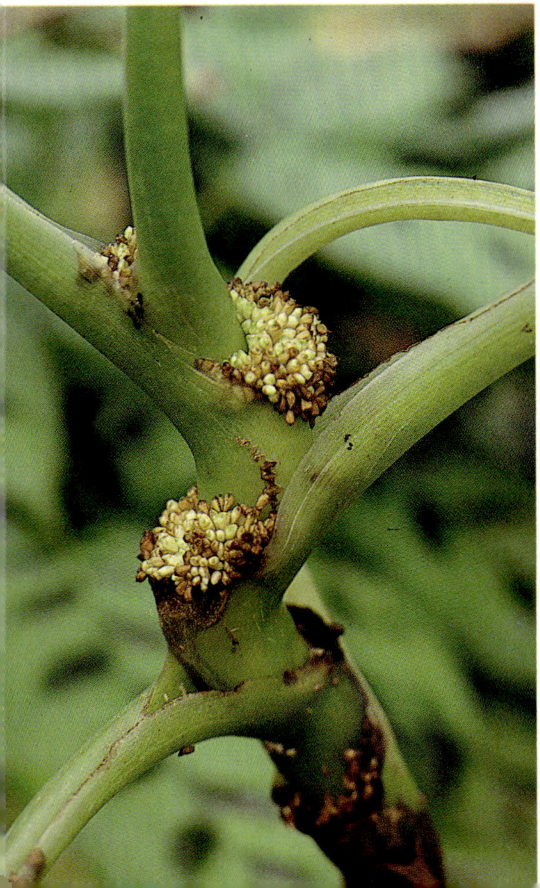
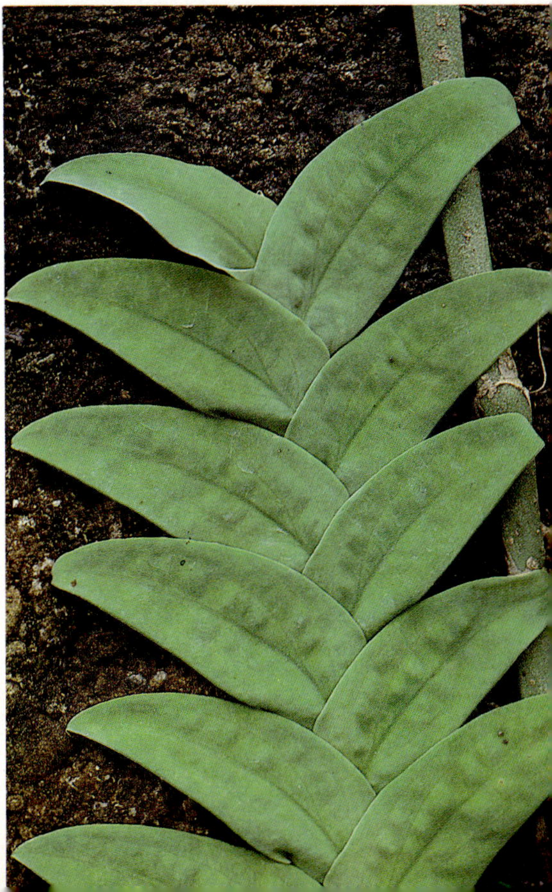

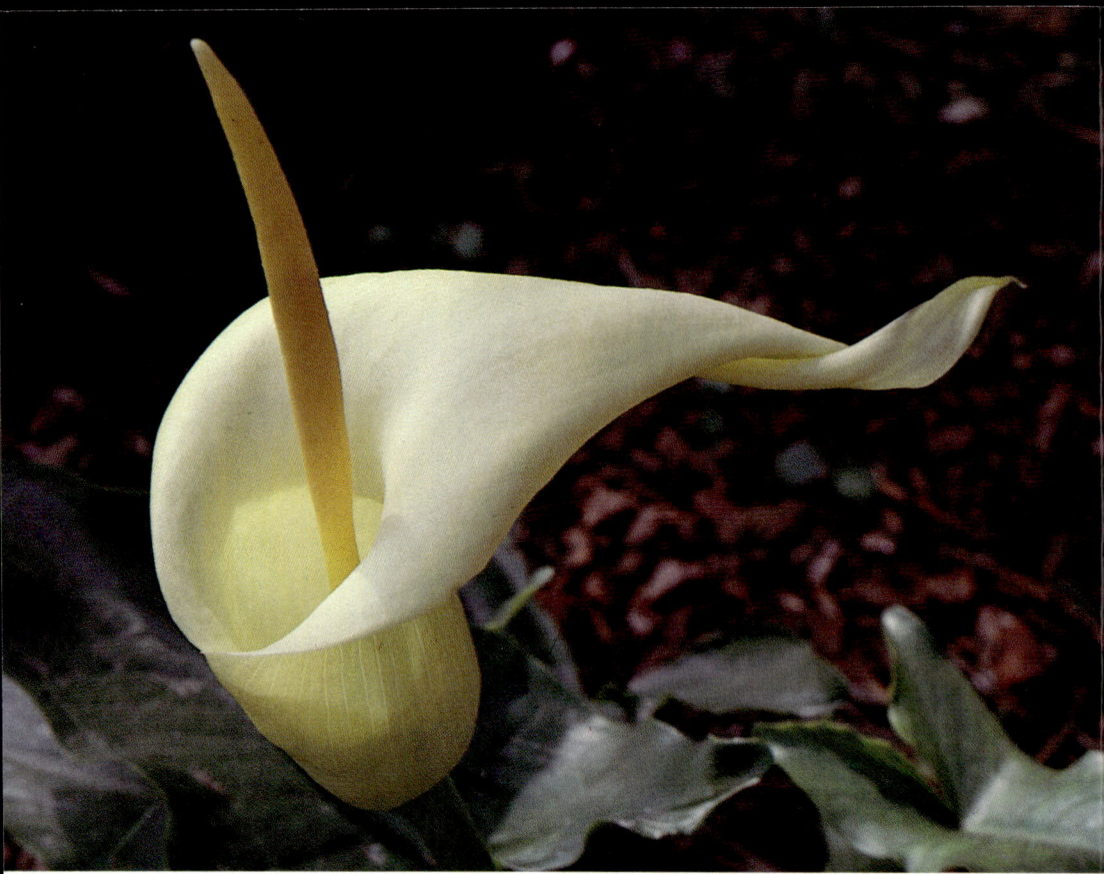

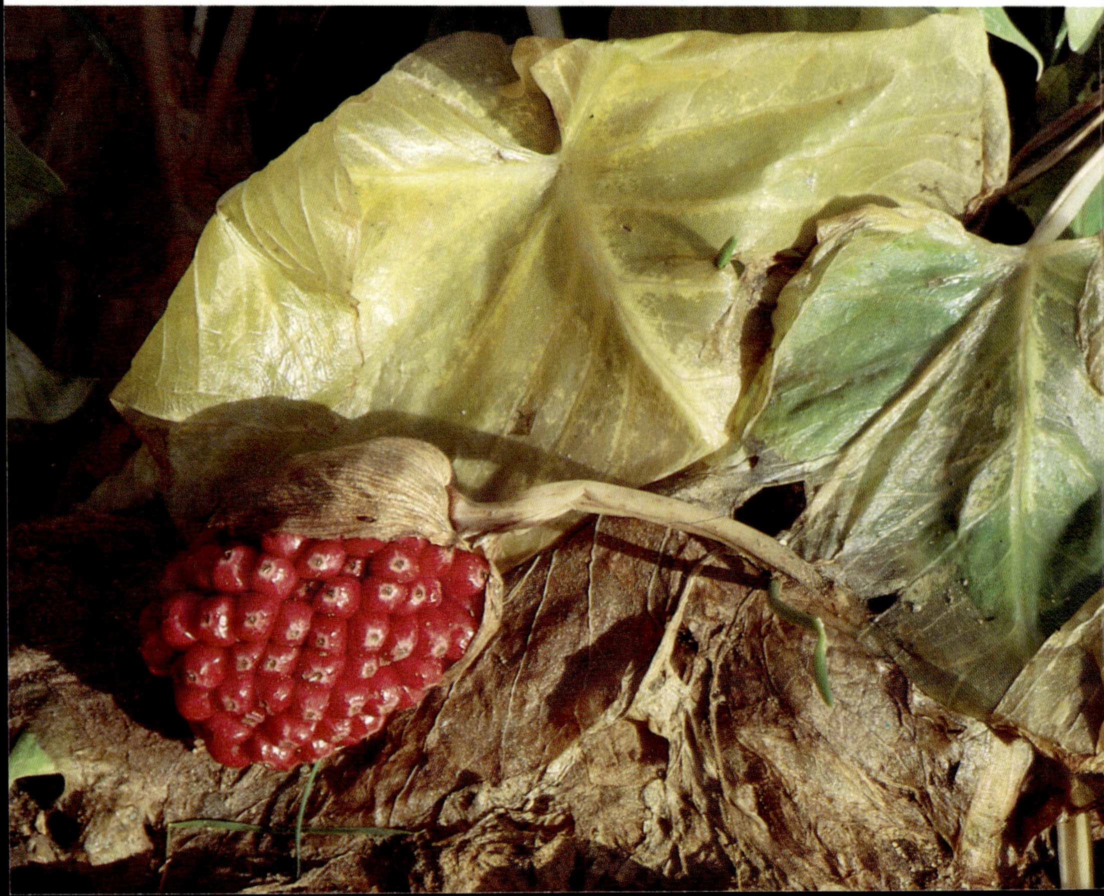

In the early days, classification was largely intuitive and based on the most obvious features of the plant. Much more detailed descriptions are possible now and it is usual to take into account not just outward appearances but pollen structure, chromosome numbers, embryology and chemical constituents. Nevertheless, appearances and accurate descriptions of easily observable features still make up most of the taxonomic information given on a species and can be undertaken with nothing more than the senses, which is encouraging to amateur botanists and horticulturists who do not have techniques to study the more minute and abstruse details. And, as a final consolation, working with plants, as with animals, is greatly dependent on the sixth sense, intuition. The work of Schott and Engler, which was done with little technological help, has proved fundamentally right: the aroid bible with which subsequent botanists basically agree, only begging to differ when their research casts new light on the most puzzling pieces of the jigsaw.

Araceae is at present undergoing quite extensive revision. Hence new species and even genera are still being described, while others are occasionally 'sunk' as no longer valid. As aroids are pre-eminently tropical rain forest plants, work on the family has taken on considerable urgency in recent years, fuelled by the desperate crisis over the rain forests and their little known or undescribed species which are being extinguished daily. It is no easy task. There are too few botanists working in the field and not enough money to fund the research needed. On top of these ubiquitous

OVERLEAF]
ABOVE *Anthurium laciniosum* (Ecuador) In many aroids the new leaf is protected by a scale (cataphyll) which later falls or withers. In this species the cataphyll remains as a distinctive feature.
BELOW LEFT *Xanthosoma viviparum* (Ecuador and Peru) Hundreds of minute bulbils are formed in the axils.
BELOW RIGHT *Rhaphidophora* species A 'shingle plant': the juvenile phase of a robust climber with completely different mature leaves. This is an efficient way of becoming established on a host tree before conditions are suitable for the production of adult – in this case, deeply divided and long-stalked – foliage.

OPPOSITE]
ABOVE *Arum creticum* (Crete, Karpathos, Samos and eastern Turkey) An unusual arum with large bright green leaves all winter and fragrant primrose yellow inflorescences in early spring.
BELOW *Typhonium giraldii* (China) Used in Chinese herbal medicine to relieve pain and spasms, this hardy species has bright purple berries in late summer.

problems, there are the sheer physical difficulties of observation and collection in these exceedingly complex and hazardous habitats. And it is not only wild species which deserve attention. The popularity of aroids in cultivation has led to countless cultivars and hybrids, mostly with scant or dubious records of their origins: a taxonomic Tower of Babel which again needs a wealth of expertise to unravel. And not least in importance are the edible aroids. Thousands of varieties are grown, few receive any systematic investigation, and, as social and agricultural practices change, many are in danger of being lost.

—— Origins ——

Solving taxonomic puzzles may be helped by information from fossil remains or by comparison with closely related families. Tracing the ancestry of any family of flowering plants is fraught with difficulty as flowers, which provide the most pieces of information, are delicate and not well represented in fossil remains. In addition, over the millennia species come and go and their extinction may leave yawning gaps in the family so that in places the family tree is reduced to a sole remaining twig. In such cases we can only theorize about the closest relatives and paths of development.

A few aroid fossils have been found but their significance is hard to determine. The spadix is often quite tough and there have been several finds of spadix-like remains. Some were later identified as water lily rhizomes but one was named *Acorus heeri* and others included fragments possibly belonging to ancestors of *Lysichiton* and *Orontium*.[15] The most notable find was made in the early 1970s in clay pits in west Tennessee, an area which in the Middle Eocene period (45 million years ago) was periodically flooded. Here the leaves of an aroid were deposited, probably after sinking gently beneath standing water, as they were undamaged and facing the same way. They were well preserved, although no apex was found. Comparison was made with present-day species such as *Philodendron speciosum* and its kin and the similarities were such that the extinct species was named *P. limnestis* (meaning 'marsh plant').[16] However, opinion is not unanimous on this point and it has also been remarked that they resemble those of North American *Peltandra* and Madagascan *Typhonodorum*.[17]

The very beginnings of the family are put in the Cretaceous period (which began 136 million years ago and lasted 65 million years) — long before any known fossils — when the subclass *Arecidae* is thought to have become differentiated into the classes *Arales* (duckweeds and aroids), *Arecales* (palms), *Cyclanthes* (Panama hat plants) and *Pandanales* (screwpines).

So which groups today show the closest affinity to aroids? From the earliest days of classification, *Lemnaceae* (the duckweeds) have been regarded as the next-of-kin, but this has recently been questioned.[18] Engler put *Araceae* and *Lemnaceae* together in *Spathiflorae* which in his classification was placed between the *Synanthae* (*Cyclanthales*) and *Pandanales* (though not implying that aroids are derived from either of those groups).[19] The inflorescences are superficially similar but the flower structure is quite different. In addition, the *Synanthae*, like the palms, have plicate leaves which are not found in *Araceae*.

A quite different approach was taken by Hutchinson, who in 1934 suggested that aroids are lily-like.[20] He went as far as to propose that the *Arales* were derived from the tribe *Aspidistreae* of *Liliaceae*: a group of rhizomatous Asian genera with dense spikes of small bracteate flowers. In the foreword to Hutchinson's controversial classification, Arthur Hill wrote: 'It is pleasant to find that *Aspidistra*, beloved of Bayswater landladies, has at last received the scientific distinction which apparently it deserves. For it bears on its shoulders the *Arum* family, which is considered to be the culmination of the phylogenetic line *Liliales-Aspidistreae-Arales: Per Aspidistra ad Astra*!' – a sentiment which was not, however, to be widely shared. Another genus of the *Liliales* has also been compared to the aroids. *Dioscoreaceae* (yams) are generally tuberous shrubby climbers with mostly heart-shaped reticulate-veined leaves and spikes of inconspicuous flowers. The differences are mainly seen in the venation pattern and the absence (in aroids) of a bract beneath each flower.[21]

The relationship with the lily clan has been largely eclipsed by other findings. Affinities have been noticed between aroids and the *Typhales* (reedmaces), especially between *Acorus* and *Sparganium* (bur-reeds), which are attacked by the same rust infection (*Uromyces sparganii*).[22] Great interest was also aroused by the suggestion that *Arales* and the dicotyledon *Piperales*, familiar as the condiment pepper (*Piper*) and the houseplant *Peperomia*, may have a common ancestor.[23] If this were the case, it would blur the separation between monocots and dicots which has long been maintained as a clear division within the class of flowering plants. More recently, another dicot group, the aquatic *Nymphaeales*, which includes the water lilies, has also been considered as possibly sharing a common ancestor with the aroids,[24] and convincing links between aroids and the *Alismatiflorae* have been established. Both have spathes and spikes of flowers, and similarities in details of floral and vegetative structure, embryology and chemistry.[25] *Alismatiflorae* mainly consists of aquatics and amphibians and includes not only such well-known aquarium plants as *Vallisneria*, *Elodea* (Canadian pond weed) and *Echinodorus* (Amazon sword plant) but also the only truly marine flowering plants – the commonest being *Zosteraceae*, the eel grasses.

And so the plot thickens. The theme is fairly clear, though some of the details and variations are baffling. What we have today is a jigsaw puzzle with some pieces possibly in the wrong place and many undoubtedly missing, lost through the natural extinctions of geologic time and accelerated now through destruction of habitats. Bit by bit it takes shape as taxonomists examine the evidence and reposition the misfits. A great deal remains to be discovered about the evolution and classification of *Araceae* but there emerges a recognizable picture nevertheless: shaped as differently as *Pistia*, *Monstera* and *Amorphophallus* they are unmistakably aroids.

2
Of Tails and Traps and the Underworld

Mechanisms of reproduction

One thing above all makes aroids stars of the botanical world and that is their strange and fascinating sex lives. Few accounts of reproduction in flowering plants fail to include the not-so-tender traps of *Arum maculatum* and an illustration of the cut-away section of the spathe tube accompanies nearly every description. This may be the most studied and commonly given example but *A. maculatum* is not alone in coming up with such a clever device. Hundreds of aroid species have devised the same sort of thing, though each has its own angle on the design, its unique variation on the theme, the majority of which have not yet been studied in any detail. These ruthless floral confidence tricks inevitably get the most publicity but are not the only reproduction techniques used by aroids. Many species favour a more open straightforward relationship with their pollinators (the huge genus *Anthurium* with over 700 members being the obvious example). Some even dispense with cross-fertilization and form fruits without sex – the 'virgin births' of the plant world – or more or less give up on sex and resort instead to non-sexual ways of reproducing.

—— Sexual chemistry ——

Anyone making the acquaintance of flowering aroids is in for shocks and delights, often in quick succession, and this is at no time more likely than when sampling their smells. Human beings may not be designed to pollinate flowers but the scents put out to attract pollinators certainly exert a powerful effect on us. The chemistry responsible for an odour is complex enough but the nature of a scent virtually defies description and may be experienced differently from one person to the next. In botanical works size, shape, texture and colour can be accurately defined but smell is an uncharted property which founders on the rocks of our sensitivity, prejudices and memories.

Some aroid inflorescences seem to have little or no smell, but those that do usually have one that is memorable, whether nauseating or blissful. Compare the repugnant fascination of *Amorphophallus konjac* in full bloom with the charm of *Anthurium amnicola*. The first was described by an aroid lover who came across it unexpectedly at a spring flower show: 'Amidst all of the tulips and narcissus and hyacinths, we detected the scent of pure rot. Following our noses, we came across two huge, liver-colored inflorescences atop five-foot stalks which arose from naked corms the size of small pumpkins.'[1] The second was found by a botanist in Panama, equally unexpectedly, when he fell down a cliff into a stream and landed on a huge mat of 'numerous deep lavender blossoms' and 'was instantly aware of an intense, minty fragrance'.[2] Sometimes even the same inflorescence presents contrasting scents. One botanist discovered that, after cutting away the enveloping lower spathe of a *Biarum* inflorescence 'the admittedly pungent aroma of the black terminal appendix turns unexpectedly to a sweet, gentle fragrance when the bolder nose investigates the more basal regions of the spadix.'[3]

These episodes give an idea of how different aroid scents can be and what adventures await prospective connoisseurs of the family who miss no opportunity of thoroughly, if cautiously, exploring the odour of every

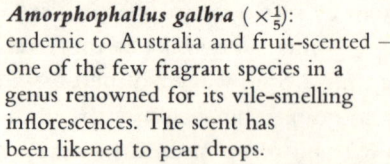

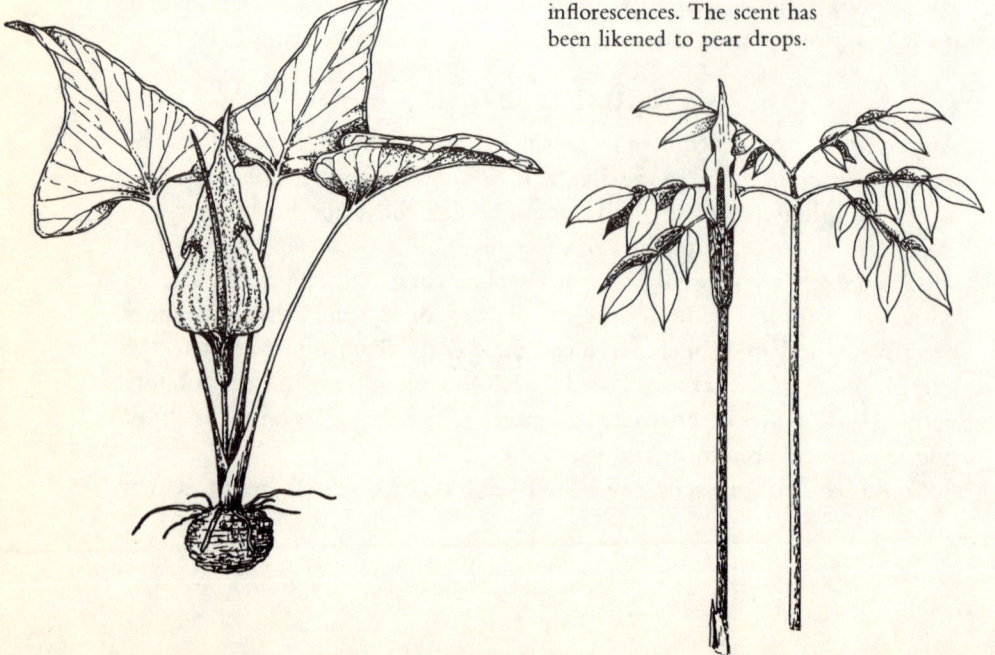

Typhonium roxburghii ($\times \frac{1}{2}$): an odour reminiscent of 'mild carrion, Stockholm tar and molasses'.

Amorphophallus galbra ($\times \frac{1}{5}$): endemic to Australia and fruit-scented – one of the few fragrant species in a genus renowned for its vile-smelling inflorescences. The scent has been likened to pear drops.

inflorescence encountered. Indeed, this particular aspect of araceous diversity has to be experienced first hand. One naturalist obviously did just this and in one of the very few papers which goes into aroid scents in any detail, he describes the odour of *Typhonium roxburghii* as being like 'mild carrion, Stockholm tar and molasses', *Spathiphyllum cannifolium* as smelling wonderfully of stocks (*Matthiola* species), *Alocasia odora* as reminiscent of violets and an *Amorphophallus* species (the Australian endemic *A. galbra*) as very fragrant among a genus renowned for its vile smells. His experiences of sniffings aroid blooms led him to conclude:

> I was impressed by the diversity of odours manifest in this large family – from indescribable nauseating stenches to transportingly beautiful perfumes. Probably no plant family on earth exhibits the whole gamut of flower scent. In the highly advanced *Orchidaceae*, numbering some 20,000 species, we certainly find heavy, spicy, vanilla, fruity, lemony, musky, and animal-like smells. At the other extreme of monocotyledons stands the comparatively simple, primitive *Araceae* with even greater contrasts in odour.[4]

Most students of the family would not agree with the diagnosis that aroids are 'primitive' but all must be unanimous about the variety of scents. The majority would also concur that by far the worst are produced by *Amorphophallus* species. In sheer volume of odour-producing tissue, *A. titanum* or *bunga bangkai*, the corpse flower, must be the very worst. Sir Joseph Hooker described it as 'a mixture of rotting fish and burnt sugar',[5] which turns your stomach over and makes your eyes run. Others come close. An Australian soldier wandering through the Southeast Asian jungle in 1945 found a *punga pung* (*A. paeoniifolius*) and wrote:

> ... my nostrils were shockingly assailed by the most nauseating stench I have ever encountered. The rain forest, dark, wet, cavernous and vine-curtained, abounding in pinkish, jelly-like, luminescent toadstool growths, giant trees supported by huge flying buttresses, pulpy soil and rotting breadfruits, gave no clue to the origin of the smell. I looked into the gloomy recesses formed by the buttressed base of an *Octomeles sumatrana* and saw.... The bloom, marbled in reddish magenta, yellow and green, not unlike a piece of decomposing liver, both in appearance and odour.[6]

He was nevertheless so taken with his find that he transplanted it to his garden. The disturbance put it out of its stride for a few days but 'on the third evening it polluted the pure equatorial air for a distance of 50 yards'. The smell was obviously produced at the time of day when its pollinators, if not its admirers, are most active and in revulsion he hurled it into the roaring waters of a nearby river, accompanied by some choice Australian epithets.

In contrast, many of the 320 or more species of *Philodendron* are deliciously fragrant, as are a number of anthuriums, especially in the section *Calomystrium* which may be recognized by its pale waxy spathes. Spathiphyllums are mostly scented too and just one or two inflorescences can perfume a whole greenhouse. Some have a sweet yet medicinal odour which is very intriguing. Again, many turn their fragrances on and off each day during the flowering period with great precision so that the right pollinators are attracted and no energy is wasted in producing chemicals and releasing scents when the objects of the exercise are inactive.

—— Red hot pokers ——

It was the French naturalist Lamarck who in 1778 first reported that aroid inflorescences produce heat and this has subsequently become one of the hallmarks of the family. The poker-like spadices of many aroids (especially in the subfamilies *Aroideae, Lasioideae* and *Colocasioideae*) become distinctly warm to the touch. They may not glow red, but if a spadix is painted with a film of liquid crystals as soon as the spathe opens, even more exciting colour changes can be seen as it turns from gold to green to blue with the rise in temperature.[7]

The raised temperature does not in itself attract pollinators. In one experiment an artificially heated but scentless spadix was substituted in *Arum nigrum* inflorescences and did not interest insects at all.[8] Rather, its main purpose is to volatilize and disperse the odours. In most foetid aroids these are compounds of ammonia, amines and amino acids, together with the substances skatole and indole (which are rarely found in higher plants).[9] As a side-effect, the heat creates a micro-climate and this may make insects more active once inside, so that they are more likely to come into contact with the floral sex organs and carry out the transfer of pollen. In some cases the warmth and humidity (and possibly the odour) encourages insects to use the inflorescence as a mating area. This has been observed in the fragrant South American *Xanthosoma sagittifolium* which is pollinated by quite large *Cyclocephala* beetles.[10] In some *Dieffenbachia* species and in

Alocasia puber (formerly known as *A. pubera*) and *A. maquilingensis*, the pollinators not only mate within the spathe tube, but lay eggs. On hatching, the larvae feed on decaying female flowers and pupate around the developing fruits.[11]

In the eastern skunk cabbage (*Symplocarpus foetidus*), the warmth from the spadix dissipates foul-smelling volatile amine, indole and skatole compounds to attract flies and beetles (and possibly bees).[12] A secondary effect of this heat output is that its fist-sized ground-level inflorescences, which are not frost-proof, can melt their way through ice and snow by radiating some of the heat produced by the spadix. Regardless of near-freezing air temperatures, it can raise its tissues 25°C above the surroundings.[13] This is a remarkable feat, but even more astonishing is the fact that this high respiration rate can be maintained for around two weeks and can be regulated according to fluctuations in the ambient temperature. The colder the weather, the higher the respiration rate, and a 10°C drop in the air temperature almost doubles the oxygen consumption of the spadix. Interestingly, the minimum weight for a heat-regulating organism has been put at 2.5 g, which is the weight of the smallest spadices, and their oxygen consumption is similar to that of small mammals of the same size.[14]

The source of energy needed for *Symplocarpus* to produce its tropical blooming in the frozen American north is provided by massive stores of carbohydrate in the stout vertical rhizome. If the inflorescences are cut off from their generator, they soon cool down.[15]

Similar mechanisms may operate in tuberous species, but carbohydrate is not the only fuel burnt in aroidean floral furnaces. Recent investigations into *Philodendron selloum* (an arborescent or tree-like Brazilian species now considered synonymous with *P. bipinnatifidum*) have revealed that lipids

Philodendron bipinnatifidum ($\times \frac{2}{5}$): the spadix heats up to 35°C by oxidizing lipids and has an oxygen consumption close to that of a flying hummingbird. The fruity scent attracts large dusk-flying scarab beetles.

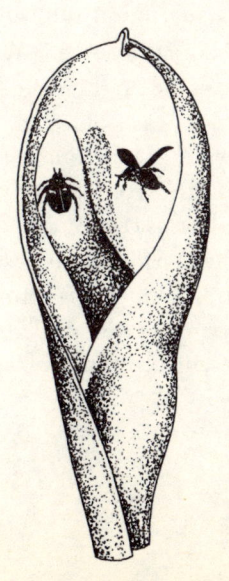

may also be oxidized directly for heat generation.[16] Lipids are fatty or waxy substances that can be mobilized and broken down very rapidly to meet the cells' demand for energy. On the whole, plant tissues have far lower energy requirements than those of animals and so lipid metabolism is not associated with normal functions in plants (it was once observed as a brief response to wounding in potatoes). This remarkable discovery demonstrated not only that the spadix of *P. bipinnatifidum* utilizes lipids for heat production but also has an oxygen consumption approaching that of a flying hummingbird,[17] whose muscles have extremely high requirements to power a wing beat of 90 times a second.

In *Philodendron bipinnatifidum*, lipids are contained in the sterile male florets which are situated between the male and female zones. Flowering lasts for two days and the production of heat and odour begins as soon as the spathe starts to open. Most of the time the scent is fruity and pleasant and the spadix is only about 10°C above the surrounding air. At about seven o'clock in the evening though, things really hot up. The temperature of the spadix rises and emits a pungent spicy odour 'like a mixture of black pepper, cinnamon, vanilla, and a resinous component'.[18] It usually reaches around 35°C but has been recorded as high as 46°C. In one experiment it was found that the spadix could still maintain these temperatures when the surrounding air was near freezing.[19] The peak lasts for 20–40 minutes and then cools down around 9 p.m. The warm scent is switched on and off each evening by light-sensitive hormones and permeates the cool evening air just as the pollinators – large dusk-flying scarab beetles such as *Erioscelis emarginata* and *Cyclocephala* species – become active. The inflorescences are a big attraction and some have been seen with nearly 200 beetles covering the spadix and filling the chamber in an orgy of mating and feasting on the nutritious sterile flowers and secretions. In the *mêlée* the beetles are smeared with resin that oozes from the spathe, especially at the constriction which they have to squeeze past, and so when the male florets release pollen in the final stage of flowering, it sticks easily to their hard shiny bodies.

Heat is always produced by the spadix, and usually by sterile flowers, but the odour it disperses may arise from either the spadix or the spathe. Detecting the origin of the odour is complicated by the fact that as a result of dispersion it adheres to other surfaces. Analysis has shown that indole, a chemical which makes smells last longer, is found in the appendix (the tip of the spadix) of *Sauromatum venosum* and *Dracunculus vulgaris*, but in *Lysichiton americanus* it is present only in the spathes.[20]

Tails

In *Lysichiton* the broad spathe both produces and disperses odours. More commonly the spathe or spadix ends in a long thin point, or even in a tassle or fringe, to perform these functions. Such elaborations may either be the source of the smell or may pick up the volatile odour produced elsewhere. Being very sensitive to air movement, they also serve to waft the alluring molecules into the surrounding countryside. The smell is certainly the initial attractant but moving parts also catch the eyes of insect pollinators and enable them to home in on the target, their faceted eyes giving poor definition but acute detection of movement at a distance. Dangling threads and quivering wands may also provide easily located landing places and be positioned so as to guide the insects into the floral chamber, which in many cases is out of sight beneath the foliage (as in *Pinellia ternata*, *Arisarum proboscideum* and a number of arisaemas, for example), or even hidden underground (in some *Eminium* and *Stylochaeton* species).

Shades of deception

The irresistible smell which first draws the insects to the open inflorescence is just one of a series of confidence tricks. It may suggest carrion, excrement, fungi, over-ripe fermenting fruit or decaying vegetation but a number of other deceits are needed to complete the illusion. The combination of heated spadix and concave spathe provide the warmth, humidity and shelter which the insects expect from feeding or breeding grounds. Colour and texture are also important and species which mimic decaying matter often have brownish, greenish, or maroon coloration and various devices such as twisted appendages to imitate shrivelling tissues and leathery, hairy or warty surfaces which look like skin and hair or mouldy

Far right: ($\times 1$) **Pterostylis coccinea**
Right: ($\times \frac{1}{2}$) **Arisaema hunanense**
An example of convergent evolution in which an orchid and an aroid have come up with remarkably similar designs for trapping flies.

surfaces. All these features make up what is recognized as the 'trap-flower syndrome'. Unlike most other pollination devices, the insects are not part of a reciprocal arrangement whereby they receive rewards – usually food in the form of nectar and pollen – for carrying out the role of go-between. Trap-flower artifices deceive the insects (mostly flies and beetles) into visiting the bloom, retain their interest by false pretences and in many cases imprison them too. In addition, the most sophisticated have areas of refractive tissue which take advantage of an insect's reflex to fly towards the light and thus direct the captive's movements during detention.

The trap-flower syndrome is an excellent example of convergent evolution as several different families of plants have independently developed similar tools for the same job. *Aristolochia* has evil-smelling saxophone-shaped flowers which are mostly purplish and mottled. In the same family (*Aristolochiaceae*) is *Asarum* with dingy-coloured triangular ground-level flowers, often with long-tailed petals. The succulent *Ceropegia* has intricate fly-catching lanterns and the closely related stapeliads (*Asclepiadaceae*) have vile-smelling flowers which look like wrinkled and hairy starfish. The orchids – tricksters *par excellence* – display a similar deviousness: twisted warty and bristled sepals and pouched lips in *Paphiopedilum* and its kin; tails and even fungus-mimicking lips in *Masdevallia*; mobile hairy lips, trailing appendages, and fairly disgusting smells in *Bulbophyllum*; and both tails and light windows in *Pterostylis*.[21]

—— Death traps ——

Aroid trap-flowers also bear a strong resemblance to several insectivorous pitcher plants such as *Darlingtonia*, *Sarracenia*, and *Nepenthes*. Their traps are superficially similar with hooded tubular chambers, light windows and greenish-maroon coloration, but differ significantly in being trap-leaves, not trap-flowers (their flowers being produced quite separately). Nevertheless, a number of observers have suspected that, like carnivorous pitcher plants, certain aroid inflorescences kill and devour the insects they waylay. The *punga pung* was described as insectivorous because the 'flies, beetles, bugs, worms, etc. on entering the tube, became entangled in the thick soupy syrups at its base.'[22] It cannot be denied that trap-flowers do prove fatal to some of their prisoners. *Sauromatum venosum* inflorescences have been recorded with as many as 70 large and over 120 small dead flies in the chamber,[23] and one inflorescence of *Arum maculatum* was estimated to contain 4000 tiny *Psychoda* flies, some of which must surely have expired from overcrowding. A number of *Arum*, *Helicodiceros*, *Dracunculus* and *Arisaema* inflorescences that I have cut open have also contained corpses. More macabre still is the spectacle of insects being so duped by an inflorescence that they proceed to lay eggs in what they mistake as dung or flesh, on which the hatching maggots starve to death. (*Helicodiceros* is especially guilty in this respect.)

It is certainly possible that the spathe – a modified leaf after all – may absorb some nutrients from the decaying remains of these casualties but this must surely be incidental. Murdering pollinators is obviously self-defeating and in the main they must leave prison none the worse for the experience, ready and able to be taken in time after time if cross-pollination is to be achieved. However, certain inflorescences seem to attract far more insects than necessary for the purposes of pollination. Could it be that some suspicion is justified and that some aroids are getting a bit extra from their victims?

—— Food traps ——

Insects deceived into behaving as if an inflorescence was the perfect place to feed, mate, or lay eggs have to be detained long enough for cross-pollination to take place. This means that they must first be enticed close to the female flowers when they are receptive to any pollen the insects are already carrying (in most species this phase lasts a day or so) and finally come into contact with the male flowers to pick up another load of pollen (which only happens after female receptivity). The trouble is that insects do not tend to hang around for that length of time, and so they have to be persuaded, or coerced, into doing so. As far as some of the large clumsy beetles are concerned, this can be done by giving them something to get their mandibles into. Keeping beetles immobile also avoids the damage their sharp tarsal claws and tibial spines might do if they blundered aimlessly about the flowers.

Amorphophallus variabilis is pollinated by *Nitidulidae* flower beetles. They arrive in response to the sickly sweet durian-like smell which is turned on promptly at 4.30 p.m. and off three hours later for several days. The beetles settle in the chamber which is lined with thick yellow pimples of oil and starch. Scarcely moving, they champ away until the feast is over, by which time the male flowers have sprinkled them with pollen and, well-fuelled, they take off for the next floral restaurant.[24]

The problem with providing excellent dining facilities is that some guests overstay their welcome. From the plant's point of view, this is as soon as both female and male flowers are finished, after which the continued presence of gluttonous beetles might endanger the newly fertilized flowers. *Philodendron*, which feeds its visitors on pollen, exudates, sterile flowers and even the fleshy stigmas, has come up with some answers to this problem which are so ruthless as to give the impression that it too has carnivorous tendencies. *P. acuminatissimum* attempts to pressurize lingerers into leaving by tightening the spathe around the flowers so that the beetles are crushed unless they make their escape in time or gnaw through the spathe to safety. Some other species merely turn their spathes so that they fill up with water at the next tropical downpour and drown persistent hangers-on. *Xanthosoma* appears to have arrived at the same solution. Most species lose the upper part of the spathe after flowering, but the base persists to form a cup which at first fills with rain to flood out the last remaining pollinators and then adopts a pendent position to protect the developing fruits.[25]

Pitfalls and windows

Less cooperative insects such as flies, which are much more flighty and less likely to settle for hours on end, must be forcibly detained. A number of different mechanisms have been devised to ensure that they first land among the receptive female flowers and later clamber past the pollen-shedding male florets in their attempts to escape.

Arum nigrum has perfected a pitfall for the assortment of dung-loving insects it attracts. After smelling the bait they land on the spathe. The outer surface has a normal cuticle which the flies walk over with ease. The inside is quite different – a treacherous oiled slide of cells which project downwards. The edge of the spathe forms a transitional zone on which the insects tread warily, pausing to clean the oil from their feet but sooner or later taking an irrevocable step and falling into the chamber. Some are lucky and get airborne again if they are caught on the ring of filaments formed by the sterile florets. This is designed to prevent entry of large insects that might damage the delicate flowers (or smaller pollinators) if they fell into the chamber.

Down in the dungeon the insects can only get a foothold around the female flowers which are ready to receive any pollen they may be carrying. In an opaque chamber, the flies would probably settle quietly, but here they make agitated attempts to escape towards translucent markings in the base of the spathe, the real light of day being obscured by the hood of the spathe. These windows of bright stripes or dapples which are so common in fly-trapping aroids (either at the base or further up the spathe), are made up of refractive cells which make them appear much brighter than the surroundings; an effect further accentuated by contrasting dark patterns and borders. They therefore provide a light source which stimulates the insects to fly, ensuring that they repeatedly come in contact with the flowers in the chamber and carry out pollination.

For the 24 hours of imprisonment the insects are kept alive by sugary secretions from the withered stigmas and by the finely balanced humidity and ventilation provided by the structure of the spathe. The next morning the male phase begins. The filaments shrivel, the slippery lining breaks down, and pollen showers on to the insects as they climb out.[26]

⎯⎯ Cliffhangers ⎯⎯

There are even more ways of detaining pollinators than by well-stocked larders and unclimbable prison walls. In the giant *Amorphophallus titanum* the spadix has two distinct sections. The pollinators are mostly large sylphid beetles *Diamesus osculans* which drop in on the spathe from far and wide in response to the stench of decay emanating from the gargantuan spadix. They can move freely over the lower part of the spadix which bears the zones of male and female flowers but are unable to negotiate the overhanging ledge which marks the beginning of the thick sterile appendage. At each attempt they fall back to the bottom of the spathe and in doing so walk over more flowers, at first distributing pollen from another inflorescence and finally picking up a new load to carry off to the next. They escape when the spathe collapses around the spadix and bridges the overhang.[27]

⎯⎯ Constrictors ⎯⎯

Several aroids have spathes that tighten to imprison pollinators. This has been described in several species of *Colocasia*,[28] *Alocasia*,[29] and in *Typhonium trilobatum*.[30] The basic mechanism is similar in all cases but the timing, odour and pollinators may be quite different. In *C. esculenta* (formerly *C. antiquorum*) the odour is strong and unpleasant, attracting flies; *A. odora* is sweetly scented; and *T. trilobatum* has violet pollen and a revolting smell that appeals to small dung and carrion beetles and flies. All have a spathe which narrows between the male and female zones, giving an egg-timer shape.

During the first stage of flowering, the spathe is open and admits insects

Alocasia odora ($\times \frac{1}{2}$):
a sweetly scented species which traps insects inside the tube by tightening up the hourglass constriction. During overnight imprisonment they pollinate the receptive female flowers. The following morning the male flowers release pollen, the constriction loosens and the insects crawl out through the layer of pollen which has collected around the exit.

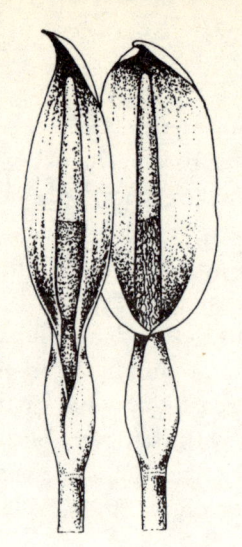

to the chamber where the female flowers are ripe for pollinating. Excited by the floral aroma the insects crawl over the flowers. Any pollen carried from another inflorescence jumps on to the stigmas through electrostatic attraction and fertilization takes place.[31] Then, with a full house of insects whose job is only half done, the spathe constriction tightens and shuts them in the chamber for the night. The next morning it is the turn of the male flowers to ripen. They release pollen which falls on to the shelf formed by the widening of the spathe limb above the constriction. Finally the constriction loosens and the insects emerge to wade through the deep layer of pollen, feeding on it and obtaining a generous coating before they fly off.

——— One-way systems ———

The genus *Arisaema* differs from others in *Araceae* in having a number of species which are sexually versatile, with individual plants varying the gender of their inflorescence from year to year. This means that in a given population there may be inflorescences bearing all-male, all-female and monoecious (both male and female) flowers. Few species have been studied in any detail, but some of those observed show a modification to the spathe design which safeguards against imprisoning insects that have crawled or fallen into the chamber. After all, the single-sex spadix has only one phase and consequently there is no advantage in detaining pollinators after they have completed the floral obstacle course once. Typically, male inflorescences have a 1–2 mm gap at the base where the edges of the spathe overlap to form the tube. Insects therefore enter at the top of the spathe and leave at the bottom after wading through fallen pollen: an efficient one-way system.[32]

Open House

The generalization is often made that aroids are pollinated by beetles and flies which are deceived into visiting the inflorescences by the drab colours and foul smells reminiscent of home – the dung heap or refuse tip. This may be true of many, but if the pollination biology of a wider range of species were studied, especially in tropical forests, it might well turn out that surprising numbers attract other creatures. Exploration of the rain forest canopy, where most epiphytes and climbers flower, is in its infancy but aerial walkways and climbing techniques are now enabling scientists to start work on the complexities of relationships at this level – not least of which is how pollination takes place.

It is already known that a number of *Anthurium* species have inflorescences which produce aromatic substances such as waxes or resins. These are collected by various bees and wasps as scent attractants in courtship or as waterproofing materials for their nests (a rather important component in the wet tropics). Territorial *Eulaema* bees have been seen visiting anthuriums for this purpose.[33] Bees are also recorded as the pollinators of several *Monstera* species, including *M. deliciosa*.[34] Monsteras generally have pale inflorescences which show up well in deep shade, and some give out a faint sweet scent. The sterile flowers at the base of the spadix seem to function as nectaries[35] and ooze drops of syrup which the bees gather into receptacles on their legs.

Iridescent *Euglossa* bees, which fly high in the canopy, build tunnel nests of bark chippings sealed together with resin. They are further candidates for the pollination of some rain forest aroids.[36] This group of insects has been called a 'mobile link' species. They pollinate plants such as aroids that are major food sources for other insects, which in turn pollinate other plants that provide the buds, leaves and fruits eaten by numerous other creatures. A 'mobile link' is not necessarily a common species but is proficient and wide-ranging – a *Euglossa* bee covers over 10 miles in an hour – and its industrious fumblings are a key strand in the food chain. Its demise would therefore cause the intricate web to unravel, setting off the decline of seemingly unconnected species.[37]

Even if the pollinator of a particular species is unknown, a prediction can often be made about the kind of pollinator if the pollen is examined under a scanning electron microscope. Pollen grains are the fingerprints of the plant world, with every species of flowering plant producing its own unique type. Aroids are extremely diverse in their pollen structure, which in itself suggests a wide variety of pollinators. The grains may vary in shape, size, surface texture and in consistency, being either starchy or starchless and lipid-rich.[38]

As one might expect, there is a strong correlation between pollen size and the size of pollinator destined to bear it, though in some cases (*Xanthosoma* for example), small grains are produced in clusters of four (tetrads) for large beetles, such as scarabs of the family *Rutelinae*, to carry away. On the whole, spiny pollen is best suited to flies and bees as it catches easily in their bristles and hairs. Shiny beetles are more likely to be served with smoother types which are glued on by a previous application of sticky syrup or resin. Starchless pollen is ideal for bees who cannot digest starch, though they sometimes pollinate species with starchy pollen if they come to collect another substance (resinous gum in the case of *Monstera deliciosa*)[39] and become powdered in the process.

With these requirements in mind, it is likely that the heavily fragrant inflorescences of *Montrichardia arborescens*, which have large starchless pollen,[40] and some species of *Syngonium* with a sweet scent and spiny pollen are probably bee-pollinated.[41] Another aroid which may be pollinated by bees is *Orontium*, again with large starchless grains.[42] Surprisingly, no observations of pollination in this common temperate species have been recorded. More certain is *Spathiphyllum* which has the complete bee-pollination syndrome: pale open inflorescences, sweet scents and starchless pollen.[43]

The largest genus of aroids is *Anthurium*. Characteristically the inflorescences are open and held clear of the foliage, and during the female phase the bisexual spadix drips with sweet fluid from the stigmas. It has been suggested that this may be fed upon by hovering pollinators such as bees and even hummingbirds.[44] A number of other pollinators, including flies and wasps,[45] have also been recorded though and it is unlikely that generalization is possible. The pollen is spiny and would therefore suit bees, wasps, flies – and theoretically birds too.

Unfortunately, learning about the pollination biology of a species is often extremely complicated. Few botanists are entomologists or ornithologists and vice versa, which makes identification of the characters in the drama a difficult process. And to confuse the issue, many visitors to inflorescences – slugs, snails, ants, pupating caterpillars, robbing, parasitizing and carnivorous insects, together with spiders preying on all and sundry

– have nothing to do with pollination. In addition, chance observations may be misleading as may those of plants in cultivation. There really is no alternative, whether the species in question grows at the roadside, the middle of a river or in the rain forest canopy, than to be there recording every event during flowering. Any enthusiast can do this if patient and methodical, able to take recognizable photographs and make good specimens of both plant and pollinator for subsequent identification. Indeed, pollination and seed dispersal are the areas where non-scientists may add significantly to knowledge of the family.

Self-sufficiency

Darwin concluded that 'Nature tells us in the most emphatic manner that she abhors perpetual self-fertilization'. On the whole, aroids support this observation, being predominantly protogynous with male flowers ripening only after female receptivity has ended. Nevertheless, there are exceptions to this rule and a number of species are known to produce fruits without cross-pollination though it has rarely been established whether they are self-fertilizing or apomicts, the latter being the 'virgin births' of the plant world in which no fertilization takes place at all.

One group which shows these tendencies more than most is that of epiphytes living in association with ants. The strap-leaved *Anthurium punctatum* in western Ecuador is an example.[46] Ants feed on sugary liquids and are notoriously aggressive. A number of epiphytic aroids have nectaries situated away from the inflorescences on cataphylls, stalks and leaves to secure the presence of ants to defend them. But it may be that this strategy not only deters would-be herbivores but potential pollinators as well and so such species may have had to evolve ways of flowering and fruiting independent of insects.

Another possible reason that ant-garden aroids appear self-sufficient for fruiting is that ants secrete myrmiacin, an antibiotic substance designed to protect them from disease-causing moulds and bacteria in their soil-based nests. Unfortunately it also prevents pollen tube development so that, even if a pollen grain comes in contact with the stigma, the male cells cannot be transferred to the ovule for fertilization to take place.[47] This is probably the main reason also that the ants themselves are not pollinators, even though they are often present in large numbers.

Self-pollination produces offspring genetically identical to the mother plant. In apomicts the situation is more complex and is usually associated with great genetic variability and hence evolutionary potential. One of the best examples in *Araceae* is *Anthurium scandens*. Not only does it have an exceptionally wide range, from southern Mexico right through central and South America to the West Indies, but all its parts are extremely variable in size, shape and colour. Species of this kind are often referred to as a 'species complex' because many distinct forms may be recognized. All attempts to use *A. scandens* in breeding have failed, which implies that it has been isolated from sexual contact with other anthuriums for a very long period.[48]

—— Fruiting ——

After the sexual activity of flowering, when all manner of display is used to gain the attention of pollinators, the plant enters the secretive phase of fruit development: the equivalent of pregnancy in animals. The immature berries have to be protected until they are ready for dispersal which, to continue the analogy, is like labour. If successful it results in germination – the birth of a new plant.

Protection is done in a variety of ways. The flowers in *Anthurium* have a perianth and this serves to protect the fruits until they are sufficiently developed to erupt through the covering. Quite different methods are seen in genera with unisexual flowers. In *Philodendron*, *Dieffenbachia* and *Syngonium*, for example, the spathe closes tightly around the flowers after the opening period and remains intact to protect the berries until they are ripe. This is taken a stage further in *Syngonium*: the spathe turning green in the protective phase and later changing to red to attract dispersers.[49] Where the spathe withers after flowering (as in *Arum*), the fruits are generally camouflaged green during their development and may further be protected by poisons. If the spathe is shed immediately after flowering – as is the case in *Monstera* – needle-like trichosclereids in the stylar portion of the ovary may be employed to deter the unwary from sampling unripe fruit.

When the seeds are fully formed, the plant turns again to advertising its needs: this time for dispersal agents. Like us, the majority of birds and animals are attracted by bright colours and many aroids respond to this by changing the cryptic shades of unripe fruit to flamboyant red, orange, yellow or purple. Taste is also catered for, usually by succulence, sweetening and a loss of the mechanical and chemical barriers to consumption (though of course there is no accounting for taste and many aroid fruits enjoyed by other animals remain poisonous to humans).

How the seeds of aroids are dispersed is even less known than the details of pollination. Indeed, the seeds of many species (and of some genera such as *Alloschemone*) are unknown and those which are familiar may be so largely as a result of chance collection in the wild or as a result of cultivation. What actually carries off the berries and deposits the seeds in a fit state for germination for the most part remains a mystery that only patient observations in the field will reveal.

It is presumed that the bright colours of many aroid seeds attract birds but hard facts about dispersal are usually lacking. There are however detailed reports of *Alocasia macrorrhiza* fruits being eaten by Lewin's honeyeater (*Meliphaga lewinii*) and the regent bowerbird (*Sericulus chrysocephalus*) in Australia,[50] and sightings of blackbirds (*Turdus merula*) eating the scarlet berries of *Arum maculatum* are not uncommon. Unfortunately, feeding in itself does not prove dispersal, as seeds must remain undamaged after consumption in order to germinate. Certain birds crush seeds before swallowing or grind them in the gizzard and thus destroy them.

Ripe *Anthurium* fruits are squeezed from the tepals to dangle on threads. The seeds have a very sticky coating which suggests that birds feed on the pulp but wipe the seeds off on branches when cleaning their beaks, leaving the seeds (of epiphytes at least) well-placed for germination. The black-and-white manakin (*Manacus manacus*) is known to feed on the berries of *A. dolichostachyum* in western Ecuador.[51] Manakins feed in wandering bands, mainly on small fruits which they pick in flight, for which dangling *Anthurium* berries would prove most convenient.

Though birds seem to be the main dispersers of aroids a number of others are known or suspected. Evidently many aquatic aroids are dispersed by water. Those of *Urospatha* are corky to aid buoyancy and in *Cryptocoryne ciliata*, *Aglaodorum griffithii* and *Typhonodorum lindleyanum*, the seeds germinate before leaving the plant to give them the best chance of becoming established in moving water.

Mammals must also be responsible for eating berries on occasions. Palm civets were seen feeding on the ripe green odoriferous fruits of *Colocasia esculenta* in the Philippines[52] and the orange musty-smelling berries of *Heteropsis integerrima* are positioned at the end of hanging branches, which suggests bat dispersal. The local names of *Philodendron bipinnatifidum* – *fruto de macaco* (monkey fruit) and *banana de morcego* (bat banana) – indicate that both monkeys and bats probably feed on its juicy fruits.[53]

Insects undoubtedly carry off some seeds too. *Philodendron* seeds are some of the smallest in the family and may stick to beetles and wasps that feast on the fermenting fruits. *Pistia* also has small seeds. They are water-dispersed but may travel further afield on the feet of wading birds that walk across the rosettes. Ants are attracted to seeds bearing fatty protrusions (elaiosomes) which they consume without damaging the seed. The Mediterranean *Arisarum vulgare*[54] and *Ambrosina bassii*[55] have these adaptations, as apparently does *Spathiphyllum floribundum* which grows in quite different habitats, mostly along waterways in Colombia.[56]

—— The last resort ——

It goes without saying that reproductive success is crucial to the survival of the species. For plants, sexual reproduction is dependent on the vagaries of go-between — the elements, insects and other creatures — and is therefore a more risky business than for animals that can go off in search of a partner and exert a certain amount of choice in the matter. As a result, many plants have a back-up system of non-sexual reproduction and duplicate themselves in a variety of ingenious ways — bulbils, offset tubers, branching rhizomes, stolons and runners — all of which aroids are adept at producing. Bulbils are much like seeds in that they must be dispersed. Some just fall near the parent but the tiny ones of *Xanthosoma viviparum* may be scattered by rain and the hooked burrs of *Remusatia vivipara* are undoubtedly picked up on fur and feathers. With the possible exception of bulbils, most of these specialized organs develop on or under the ground (or close to the substrate in epiphytes) and are thus protected against extreme weather, trampling and herbivores. They therefore not only make up for possible breeding failure but are also an insurance against serious damage or death in the mother plant.

The drawback of these last resorts is that the offspring are genetically identical to the parent. Of course, if the parent is doing well, it has obviously got what it takes for that particular situation and so more of the same may be the best strategy; so much so that *Acorus calamus* and several *Cryptocoryne* species rely mainly on underground spread and seem to have almost given up on fruiting. But the greatest advantage can be summed up as 'safety in numbers'. Then, when conditions are right, there will be more individuals in the area to pool their genes by advertising floral delights and fruitful nourishment to their insatiable consumers.

3
Woodlanders

Species of temperate woodland and higher altitudes of the tropics and subtropics

If you go down to the woods today, you may find green dragons, mouse plants, cuckoo pints, a one-cornered lotus, Jack-in-the-pulpit or, to put it more prosaically, *Arisaema dracontium*, *Arisarum proboscideum*, *Arum maculatum*, *Typhonium giraldii* or *Arisaema triphyllum*. These are just a few of the many aroids that grow in temperate forests and thickets. Species in this category are often hard to find, even when common, as their divided foliage and green or brownish inflorescences are inconspicuous in the undergrowth and they disappear completely for several months of annual dormancy. If you can locate them though, 'you are sure of a big surprise', as the nursery rhyme says, for they are not only exceedingly curious but are also some of the easiest aroids to observe, growing as they do at ground level in habitats which are easily accessible compared with the wetlands and rain forests favoured by most of their relatives.

—— Woods, trees and aroids ——

There are many kinds of woodland. In the far north and on high mountains evergreen coniferous trees predominate. More temperate regions have forests composed mainly of deciduous broad-leaved species, though conifers are often present too. Areas with a drier Mediterranean-type climate tend instead to have both conifers and broad-leaved evergreen or leathery-leaved (sclerophyllous) trees and shrubs. And in tropical and subtropical zones which have lengthy dry seasons, there is open savannah woodland: a mixture of grass and trees such as acacias. Lastly, in the tropics where rainfall is more evenly distributed and often heavy throughout the year, tropical rain forest dwarfs all else. The limit for continuous forest is about 3000 m above sea level (nearer 4000 m in the tropics) and within 3200 km of the North Pole. Beyond this, trees are sparse and stunted and soon give way to alpine or tundra vegetation which is better able to withstand prolonged severe cold. Aroids are found in all types of forest, and even beyond the tree line.

The aroids described in this chapter belong mainly to temperate woodlands and higher altitude (montane) forest in subtropical and tropical areas. Those native to drier regions come under chapter 5, and the rain forest species are so numerous that they take up chapters 6 and 7.

At one time the land was extensively forested. In human history there has been a Stone Age, an Iron Age, and now a Nuclear Age but throughout every era the exploitation of wood for fuel and construction has outstripped that of any other natural resource; a demand which in many parts of the world exceeds natural regeneration rates. In addition, these losses are compounded by the destruction of woodland to make way for agriculture and development. As a result, vast areas are now almost completely deforested, and the felling continues apace. Forests now cover less than two-fifths of the earth's surface.

A forest is, by definition, a sizeable area of trees but among them grows up an assortment of shrubs, herbaceous plants, ferns, mosses and lichens that together create a complex habitat of what would otherwise be a mere plantation. Trees are the great protectors, shading and sheltering everything beneath them. They act as a sunshade, windbreak and umbrella so that on the forest floor stressful fluctuations in light, temperature and moisture are greatly reduced. When the trees are felled, most of the other woodland plants die of exposure. It would seem however that, at least in temperate and Mediterranean-type regions, the aroids among them often survive – possibly because they can escape the greatest extremes by dormancy in underground tubers and also because vestiges of shelter, such as rocks and shrubs, remain here and there. Thus many species that were once true woodlanders can now be found in grassland and scrub and along hedgerows and field edges. It is only too easy to find examples. In Ethiopia, a country drastically deforested in recent years, several *Arisaema* species are recorded as being very probably formerly woodland plants. *A. addisababense* grows where once olive and juniper forest flourished, sheltering now under shrubs of Natal plum (*Carissa edulis*). A newly described species, *A. mooneyanum*, is also thought to have originated in mountain forests but is now colonizing the resulting grasslands successfully,[1] a process aided by the fact that it increases mainly by offsets and is not wholly dependent on the pollinators and seed dispersers that may well have disappeared with the trees.

The genus *Arisaema*

Arisaema is by far the largest genus of woodlanders in *Araceae*. With some 150 species it greatly outnumbers all other predominantly tuberous genera (second comes *Amorphophallus* with about 100 species) and is in fact the third largest genus in the family after *Anthurium* and *Philodendron*. Unlike those huge genera though, which are composed solely of New World species, *Arisaema* has an extraordinarily wide range, from Mexico and through eastern North America, eastern Asia, and eastern and central Africa. Its northernmost station is Sakhalin (a Soviet island north of Japan) at 51°N and its southern extreme is Tanzania at 7°S. The centre of distribution is southwest China, in Yunnan and Sichuan (Szechwan) provinces. Japan is also rich in species (with 25 according to a recent revision),[2] as are the Himalayas.

Characteristics

The genus *Arisaema* was first described by Martius in 1831. Since then its elegant divided leaves, striped spathes and often strangely shaped spadices have fascinated plant lovers and received the attentions of avid collectors. Characteristically, arisaemas are tuberous and have an annual dormant period. Most flower in the spring or early summer just before or with the new foliage. Some of the tropical ones are, however, rhizomatous and usually evergreen and flower at various times of the year. One of these is *A. filiforme* which is found in the Malay Peninsula, Borneo, Java and Sumatra. It has one or two glossy, sometimes grey-variegated leaves with three to five distinctly stalked leaflets and a clear green to purplish-brown spathe and long dangling spadix. The strangest is *A. rhizomatum* from western China and northeast India, which not only has a rhizome and flowers in the autumn but also has a peculiar spadix that ends in a tuft of bristles.

Many species have only one leaf per tuber in each growing season. Most others have just two. This makes them exceptionally vulnerable to damage, though this is no doubt minimized by the fact that they are too poisonous for most herbivores. The leaves are nearly always compound – which may be another device to minimize damage, as the loss of one or two leaflets is survivable. The more complex kinds have anything from five to over 20 leaflets. Juvenile plants take several years to reach their full complement, and usually have simple leaves to start with.

Left: **Arisaema rhizomatum**
($\times \frac{1}{5}$): arisaemas are mostly tuberous but a few have rhizomes. This tends to occur in tropical species which are evergreen and have no need for annual dormancy.

Right: **Arisaema fimbriatum**
($\times \frac{2}{5}$): a species from the Malay Peninsula with a maroon tasselled appendix. Fringes, tails, warts, wrinkles and other elaborations are common devices in fly-trapping aroids, giving a greater surface area for the dispersal of odours.

Arisaemas generally have very attractive foliage. The stalks of some species have well-defined sheaths which overlap to form a pseudostem that may be quite tall, some 1.5 m in *A. tortuosum*. Like the stalks, the pseudostem is often beautifully mottled. Particularly graceful are those whose leaflets taper to long points. In *A. taiwanense*, a species described only in 1985,[3] the tails reach 20 cm in length. As arisaemas generally come into leaf during heavy spring rains or the monsoon season, these are presumably drip-tips to ensure rapid run-off of rain. As a further protection, the leaves of many species are characteristically held like umbrellas over the inflorescences. Several species have variegated leaves: *A. undulatifolium* and *A. yamatense* are fine examples. Some plants of these species also display another interesting, but variable feature of *Arisaema* foliage: serrated margins.

When it comes to spadix design, *Arisaema* sports a bewildering variety and some of the most bizarre in the family. They range from conventional cylindrical and club shapes to the spiked, hooked, hairy and thread-like and may emerge from the spathe either erect, horizontal, or drooping. *A. fimbriatum*, a species introduced from the Philippines, caused quite a stir when it first flowered in England in 1884 at the Chelsea nursery of William Bull,[4] for its spadix ends in a thin dangling tassel of maroon filaments. *A. album*, a white-stemmed Indian species, also has a feathered appendix. At the other extreme, *A. exappendiculatum* looks as if it is without a spadix. It does of course have one but lacks an appendix so that nothing shows above the mouth of the spathe.

In some species, spathes are very odd too. Some have the tip of the spathe, rather than that of the spadix, drawn out into a long tail which hangs down in much the same way and may well have a similar role to play in attracting pollinators. That of *A. consanguineum*, a widespread and variable species found from the Himalayas to Taiwan and Thailand, reaches 18 cm. Its spadix, on the other hand, is short and blunt, as it usually is in species with spathe tails. Quite unique in the family are the arisaemas with lobed spathes. Again, it is a variable feature but in the best examples, found in some specimens of *A. auriculatum*, *A. biauriculatum* and *A. limbatum*, it is almost as if the spathe is divided into three distinct lobes.

Whiplash arisaemas

One group of arisaemas is characterized by leaves with three leaflets and long thread-like extensions to the spadix which either trail along the ground at some length or are tossed back among the foliage. A number of these species are seen occasionally in cultivation: *A. propinquum* (often referred to as *A. wallichianum*), which has been recorded at over 3500 m in Kashmir; *A. thunbergii* (also known as *A. urashima*), a Japanese species; and the more readily available *A. speciosum* from China and Nepal, whose leaflets are finely edged with bright red. Its spathes are striped in maroon and translucent white, and the maroon appendix thread reaches up to 80 cm.

Watching the unfolding of leaf and inflorescence is quite fascinating. The leaf emerges first and, as if dragging the inflorescence after it, the bud follows, attached to the leaf by the appendix thread which is packed inside the central leaflet. The spathe hood is tightly closed and points upwards with the thread protruding from its tip and pulled taut by the growing leaf. As the leaflet unfurls the thread can be seen neatly compacted into many 'S' shapes lying along the midrib. When the leaf finally opens, the thread loosens and frees itself, though it often remains caught up on the leaf.

Occasionally the growing leaf exerts too much tension and the thread breaks – an especially sad loss on cultivated plants, which benefit from a helping hand.

Frank Kingdon-Ward came across colonies of whiplash arisaemas in northern Burma. They were growing in rich leafmould in a dripping wet forest of oak, chestnut, holly, magnolia, cherry, birch and rhododendron at about 2000 m. He described his find as 'a deluxe model of a cuckoo pint' and was puzzled by how the threads got into their odd positions among the leaves, not having watched one 'giving birth'. His plants, whatever the species, must certainly have been impressive. The central leaflets were over 45 cm long and the threads trailed down for an astonishing 1.5 m. He returned in the autumn, hoping to get some seed to send back to England for cultivation but was disappointed to find none set.[5]

All the whiplash arisaemas are handsome plants but two are quite outstanding. The first is *A. costatum*, a species native to semi-deciduous forest in Nepal and southern Tibet. It has unique parallel-veined leaflets which give a strongly ribbed appearance to the leaves, but although '*costatum*' means ribbed, it does not refer to this feature but to ridges on the inside of the spathe tube. The inflorescences of this species are darker maroon and brighter white than those of *A. speciosum*. It is reasonably hardy and a fine clump bearing many leaves, each about 50 cm tall and as much across, grows outdoors at Kew. The second is *A. griffithii*, to my mind one of the most extraordinary of all aroids. Its reptilian inflorescence is borne quite close to the ground in May or June in the forests of Nepal and

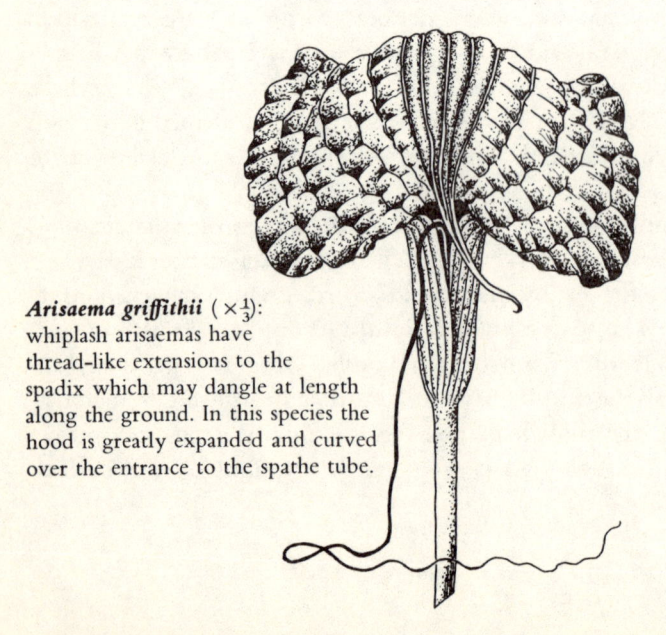

Arisaema griffithii ($\times \frac{1}{3}$): whiplash arisaemas have thread-like extensions to the spadix which may dangle at length along the ground. In this species the hood is greatly expanded and curved over the entrance to the spathe tube.

Bhutan. Usually there are two leaves on plain dark stalks which are held at about 45° on either side of the inflorescence, each with three rather stiff wrinkled leaflets. The spathe is blackish-purple, ribbed and netted with green veins and has a greatly expanded hood, about 15 cm across, that curls tightly over the entrance. From under it emerges 20–80 cm of purple appendix thread.

The extraordinary appendages of the whiplash arisaemas serve as osmophores which disperse odours at some distance from the actual flowers. Functionally, they are more like fishing lines than whiplashes, trailing down to the leafmould of the forest floor as a bait to lure insects up into the spathe. Though few have been studied in any detail, the pollinators are thought to be mainly small flies and gnats (*Nematocera* species) which are attracted to smells reminiscent of the vegetation or wet places in which they breed. Arisaemas are not particularly foul smelling and one has to sniff quite close to the inflorescence to detect the odour, which often is like mushrooms, compost or pond water.

—— Changing sex on the forest floor ——

In *Arisaema*, scents and floral structure are only part of the reproductive strategy. The feature which makes the genus outstanding in the family and unusual within flowering plants as a whole, is that individual plants in many species can change sex from year to year. Usually in plants, as in most other living things, an individual's sex is determined genetically and remains constant throughout life. In most aroids flowers of both sexes occur on the same spadix, those with unisexual flowers having a zone of female flowers at the base of the spadix and a male zone above them. *Arisaema* has unisexual flowers and in a number of species the spadix bears flowers of one sex only. This in itself is not particularly uncommon in flowering plants. As any gardener knows, if you want certain shrubs to bear berries you must buy both male and female plants. However, in *Arisaema*, such a selection would prove fruitless in the long term as your chosen males and females will probably change sex each year, leaving you on occasions with all males or all females.

Several studies of this phenomenon have shown that the situation is surprisingly complex. Arisaemas do not appear to have sex chromosomes and they change sex as a part of their survival strategy. The species most studied is the North American *A. triphyllum*, commonly known as Jack-in-the-pulpit. It forms colonies in a variety of woodlands from southern

Florida to Maine and New Brunswick. Like most forest plants, it is long lived and during a lifetime of anything up to 25 years will experience many changes. Even the most ancient forests have their ups and downs, especially in temperate regions where unusually hot dry summers, severe winters, late springs, storms or fires may topple the old and weak and, in doing so, make space for the youngsters. No two years are alike and this is reflected in the fact that a Jack-in-the-pulpit plant may vary as much as 200 per cent in size from one year to the next, according to growth conditions.[6]

A plant of *A. triphyllum* starts life in one of two ways: as an offset produced as a bud on the parent tuber; or as a seed. An offset has a head start, being the size a seedling would reach after several seasons' growth but is genetically identical to the one parent and grows in close proximity. It sends up a shoot when separated from the parent, which happens when the parent tuber sheds its old skin or when there is soil disturbance. A seedling is much more vulnerable, possibly germinating on a completely different site, but has a new combination of genes which might turn out to be just what is needed to found a new and vigorous dynasty. In order to gather strength for emerging into the unknown, it may spend its first year entirely underground, producing only a root and a tiny corm. Thereafter a juvenile plant concentrates on growing, putting out just one leaf to start with, then two, and from year to year increasing (or decreasing in a bad year) the size of leaflets.

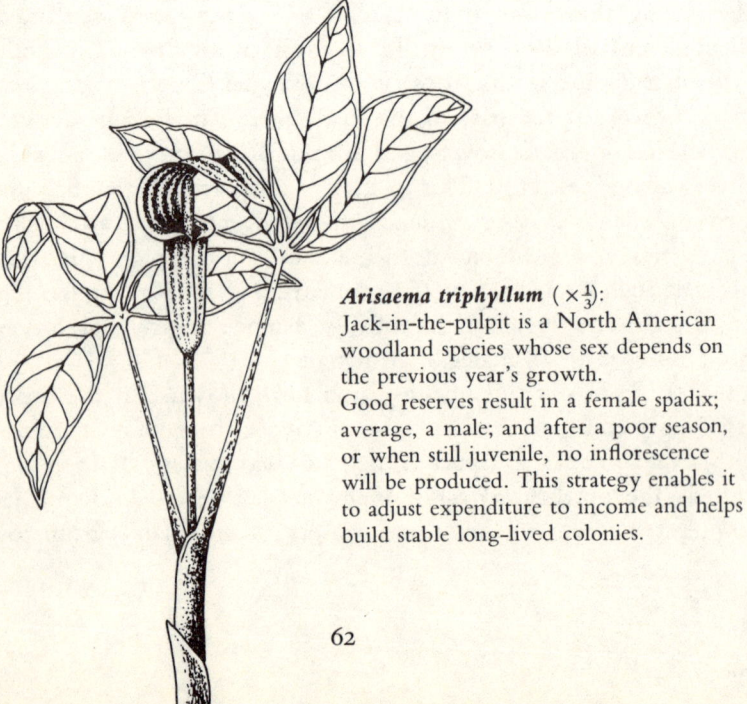

Arisaema triphyllum ($\times \frac{1}{3}$):
Jack-in-the-pulpit is a North American woodland species whose sex depends on the previous year's growth.
Good reserves result in a female spadix; average, a male; and after a poor season, or when still juvenile, no inflorescence will be produced. This strategy enables it to adjust expenditure to income and helps build stable long-lived colonies.

When it gets big enough, it flowers and produces a male spadix if only average-sized, or a female spadix if its reserves are sufficient – more energy being needed for the larger female flowers and for seed development. Its status is always determined by the previous year's growth, as the buds for leaf and inflorescence, if there is one, are laid down by the end of the summer when the plant starts to die down for winter dormancy. From one year to the next though, there is no predicting what sex it will be and after a really poor year, or after the effort of flowering,[7] it may return to being asexual like a juvenile and produce only foliage. Occasionally, a plant even takes a sabbatical and remains dormant through a whole growing season.

The advantage of this degree of flexibility is that the plants in a colony can adjust their expenditure to their income, only splashing out on expensive female flowers and fruits when the living is easy, yet still contributing with male flowers in leaner years and at least ticking over when things are really hard going. Thus the population remains relatively stable as individuals are able to decline rather than fall and pick up again when conditions improve. And in years when female flowers are few, or when little seed is set, the colony continues to maintain its numbers by producing offsets.

Not all arisaemas have the same pattern of sex change as *A. triphyllum*. Some start off as males and then add female flowers to the spadix as they increase in size. The green dragon (*A. dracontium*), a rather uncommon species of northeastern North America, does this.[8] Others go through male, male-and-female and female roles. Relatively few are like Jack-in-the-pulpit; Jack or Jill but never hermaphrodite.[9]

—— An alpine *Arisaema* ——

If there is one aroid which might qualify for the description of alpine, it is *A. flavum*. In one guide to Himalayan flowers it is listed as occurring at up to 4500 m (14,765 ft).[10] It has also been recorded at 4400 m (over 14,400 ft) in China[11] and at 4000 m (over 13,000 ft) in a gorge on Mount Everest.[12] Although it has none of the adaptations associated with true alpines – prostrate compact growth, for example – plants are often very small and able to grow in quite open stony places at altitudes above continuous forest.

It also goes to the other extreme and grows in hot dry regions. In fact, it probably has the widest range of any *Arisaema*, from Afghanistan right across the Himalayas to Bhutan and southwest China, down through

Pakistan into Oman, Saudi Arabia, the Yemens and Ethiopia. It is altogether a remarkable little plant and quite unlike any other in the genus – seldom reaching more than 50 cm, with diminutive bright yellow inflorescences measuring only 3.5–7 cm long and sometimes as little as 2 cm. But its small size and squat colourful spathes (one collector described them as 'sprinkled with gold dust')[13] are not its only interesting features. In China its tubers are used to make pesticide and included in remedies for respiratory tract infections, tetanus and epilepsy.[14]

—— An aroid on the mountains of the moon ——

The Ruwenzori Mountains on the Ugandan-Zaïrese border between Lakes Edward and Albert are thought to be the Mountains of the Moon which Ptolemy described in his monumental 2nd century work on geography. Ten peaks rise over 4876 m and the vegetation is distinctly zoned. Below 2000 m there is grassland, followed by forest up to 2590 m and then bamboos to 3050 m. After this come the tree-heaths, succeeded by lobelias and senecios at 3810 m. Above 4420 m there are bare rocks where little but mosses, lichens and helichrysums can survive. The area is known for its high numbers of endemic species, among which is *Arisaema ruwenzoricum*, a lover of damp shady spots in the bamboo and tree-heath zones.[15] Sometimes it grows epiphytically in moss on the trees, a strange situation for an arisaema some 120 cm tall. It usually has two leaves with five to seven leaflets and a pseudostem mottled reddish-purple. The striped spathe is about 20 cm long and the spathe limb ends in a short tail, curving over the opening to the tube where the cylindrical appendix is just visible.

OPPOSITE]
Lysichiton camtschatcensis (coastal northeast Asia) The eastern skunk cabbage, slightly smaller, and later flowering than its western cousin. The two species are decisively separated by the Bering Strait.

OVERLEAF]
ABOVE LEFT *Cryptocoryne spiralis* (southern India, Bengal and Bangladesh) Cut-away section of the tube, showing clusters of female flowers at the base and male flowers at the top, with a long thin sterile portion in between. The entrance to the tube is controlled by the valve-like flap.
ABOVE RIGHT *Anubias giganteus* (tropical West Africa) The largest species of this amphibious West African genus.
BELOW *Lysichiton americanus* (western North America) 'One of the great joys of spring in far northern wetlands'; beneath leafless alders the bright yellow spathes of skunk cabbages emerge from the mud. This colony is naturalized in wet woodland in southern England.

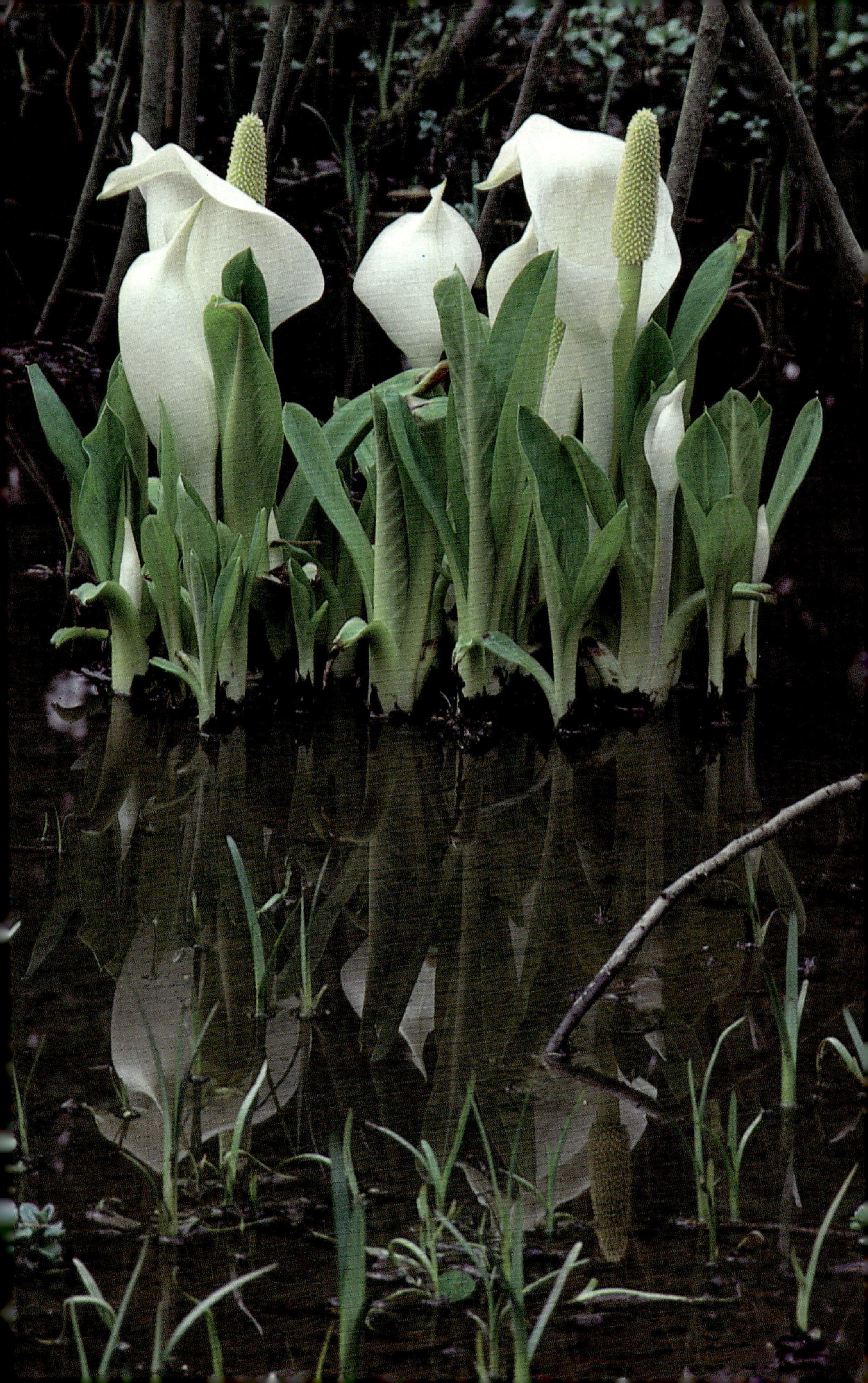

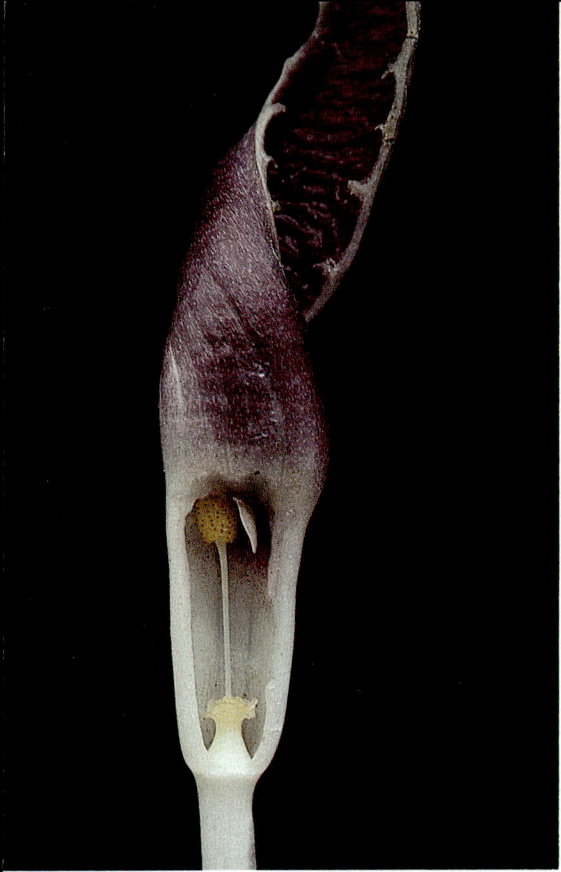
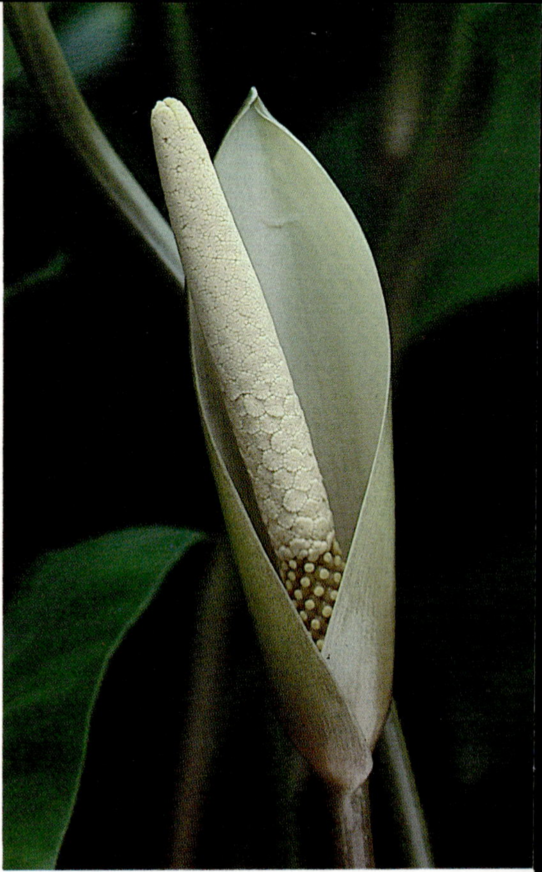
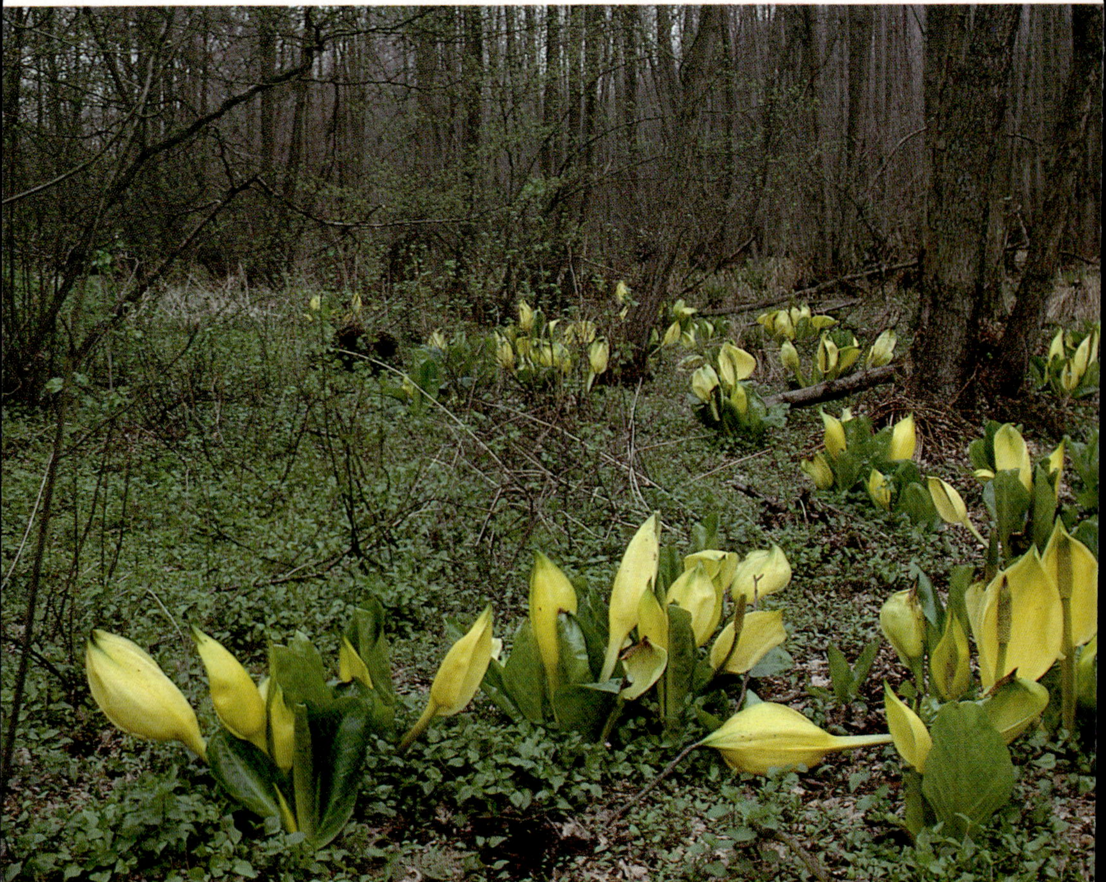

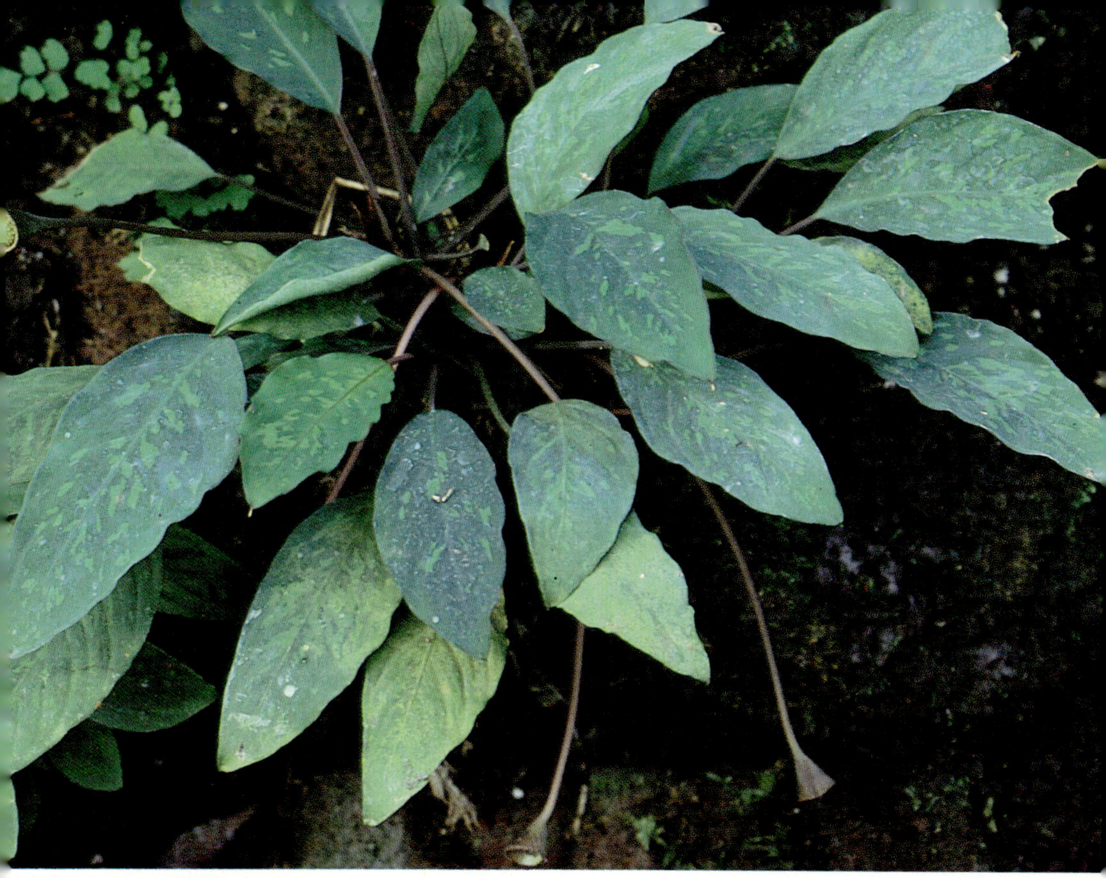

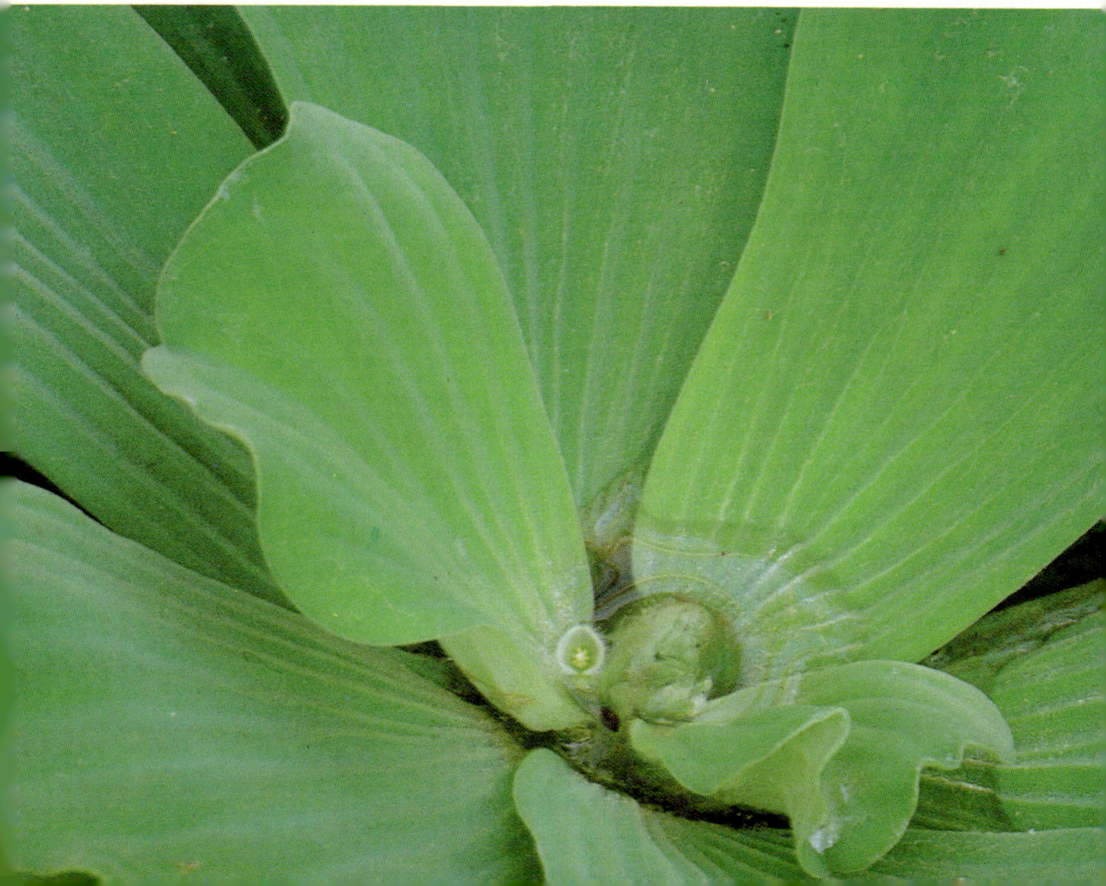

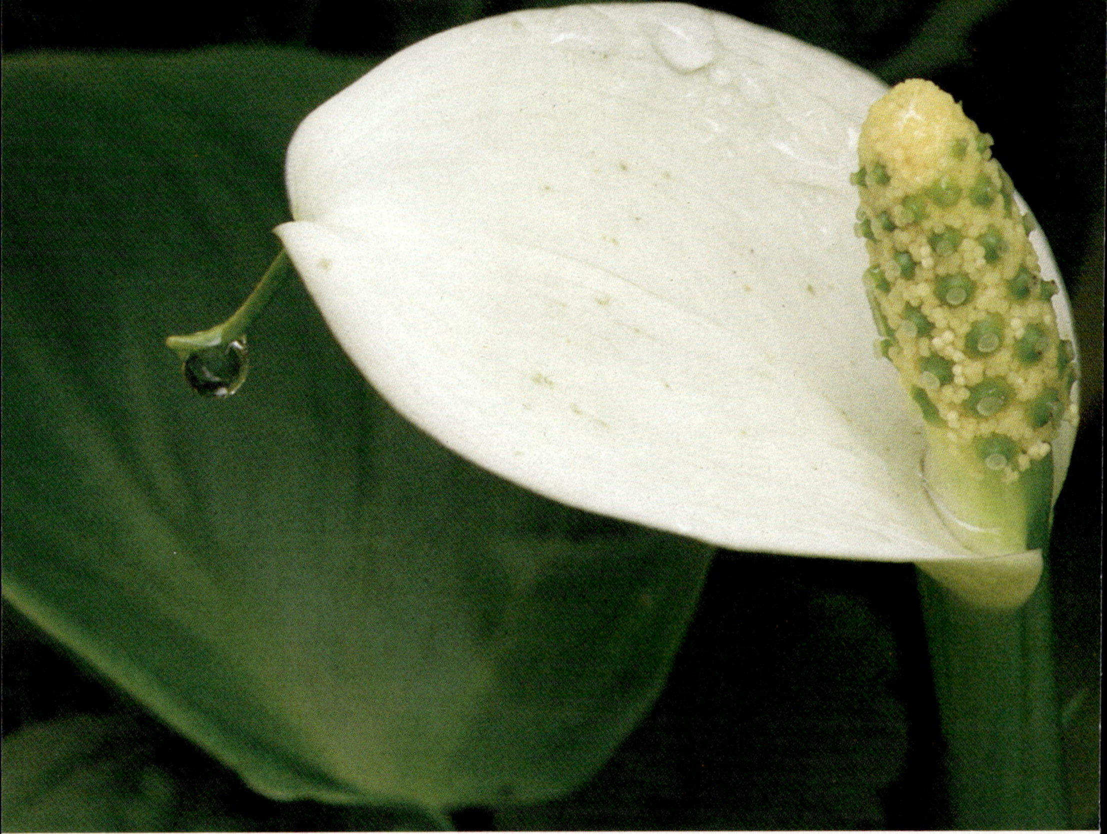
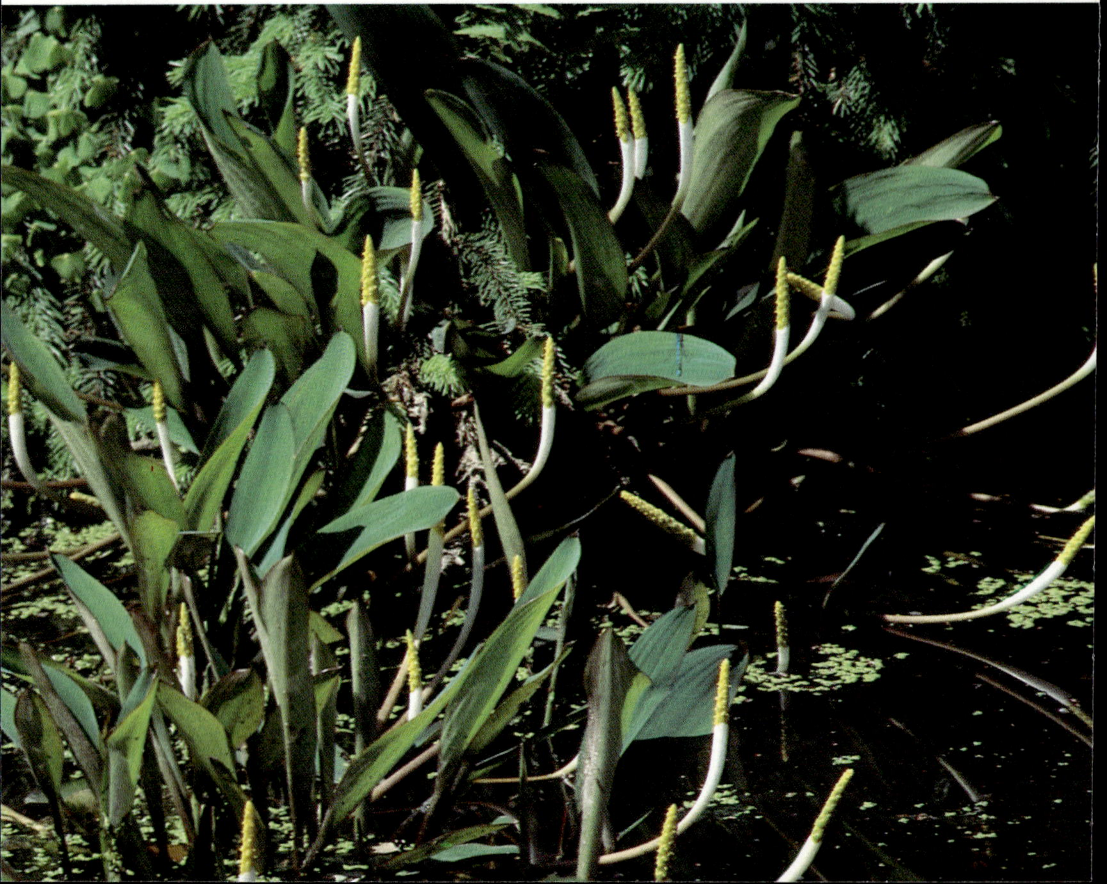

The fairest of them all

Several arisaemas are very popular in cultivation and make excellent subjects for the woodland garden. A group of plants in flower has a slightly sinister and inscrutable oriental dignity which holds observers spellbound and it is not surprising they are sometimes called cobra lilies. Yet one or two have, if anything, an air of innocence, being unhooded, open and brighter in colour. The loveliest is *A. candidissimum*, which has a spathe prettily striped in pale pink and white, and is delicately fragrant. It is hardy in many temperate areas because it starts into growth in very late spring (usually late May or even June) so that the inflorescences and broad trifoliate leaves are rarely damaged by frost. *A. candidissimum* is native to China. George Forrest, who introduced it into cultivation in 1914, found it at between 2400 m and 2740 m on dry rocky slopes and slate ledges in Yunnan.[16] Another collector recorded a pure white form at 3350 m.[17]

Less hardy and more handsome than pretty is *A. sikokianum*. The tube is blackish-purple outside with a contrasting pure white interior and the striped spathe limb, elegantly curved and pointed, stands up straight behind a dazzling white pestle-shaped spadix. The two leaves may have three, four or five leaflets, though not necessarily the same number on each leaf. It is a variable species found in eastern China and southern Japan and was named after the island of Shikoku.

The hitch-hikers

In contrast to *Arisaema*, the genera *Gonatanthus* and *Remusatia* are very small, with only three species between them. They are found at about

OVERLEAF]
ABOVE *Piptospatha ridleyi* (Malay Peninsula) A rheophytic species which grows in fast-flowing streams and waterfalls.
BELOW *Pistia stratiotes* (pantropical) 'A deceptively delicate drifter' – the only floating aquatic in the family and its most pernicious weed. Both leaves and inflorescence are covered in fine water-repellent hairs.

OPPOSITE]
ABOVE *Calla palustris* (north temperate) The bisexual inflorescence of the hardy bog or marsh calla often has a cluster of male flowers at the apex. This monotypic genus is not closely related to any other and according to some authors, forms a subfamily on its own.
BELOW *Orontium aquaticum* (eastern North America) Though apparently spatheless, this aquatic produces masses of showy inflorescences in spring. Its common name is golden club, which describes the yellow spadix.

1000–2500 m in subtropical forests, sometimes on mossy rocks and occasionally as epiphytes in trees. *Gonatanthus pumilus* is distributed from Himachal Pradesh throughout the Himalayas to southwest China; *Remusatia hookerana* from Himachal Pradesh to Sikkim; and *R. vivipara* from the Himalayas into Yunnan (southwest China), throughout most of Southeast Asia to northern Australia and in Oman and many parts of Africa, including Ethiopia, Zambia and Madagascar.

In their vegetative state they are difficult to tell apart. Indeed, they are so alike in many respects that it has been suggested that they should be combined as a single genus.[18] Both have heart-shaped peltate leaves, though those of *Gonatanthus* are often attractively blotched with purple. *Gonatanthus* flowers later than *Remusatia* – from June to August – when the leaves are fully developed after dormancy, whereas flowering in *Remusatia* takes place before or with the new foliage. In *Gonatanthus* the spathe is golden yellow and about 20 cm long, within which is a purple spadix. The green tube is inflated and persists to protect the yellow berries. *Remusatia* inflorescences are fragrant, the spathe pale yellow and the spadix creamy white. In *R. vivipara* they are completely reflexed after opening. They are followed by bright red fruits.

Both genera are tuberous and produce a number of peculiar stalks which are covered in scale leaves and clusters of hooked bulbils. This seems to be a most successful method of propagation, particularly in *R. vivipara*, which has an exceptionally wide distribution. It has hitch-hiked throughout the

Remusatia vivipara
a] ($\times \frac{2}{3}$): A solitary inflorescence is produced after dormancy.
In many places flowering is rare or unknown.
b] ($\times \frac{1}{4}$): The leaves are usually solitary (sometimes 2–3) and develop after the inflorescence.
c] ($\times 2$): Hooked bulbils are borne on aerial shoots. They catch on fur and feathers, enabling this species to travel afar. It is now widespread throughout Asia, Africa and Australia.

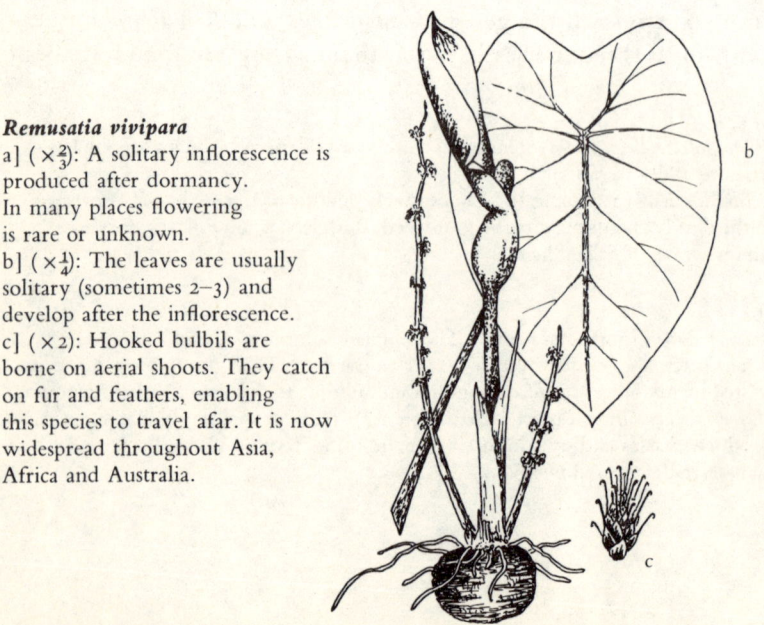

Himalayas, Southeast Asia and into Australia and Africa by these means and has even been found on Christmas Island.[19] The fact that it crosses land and sea with equal ease and is also found growing in trees must mean that burrs catch on birds well as mammals. It would appear that seed plays a lesser part, as plants in Java are reported to have sterile pollen,[20] and African populations are thought to flower rarely[21] or never at all.[22]

A northern *Amorphophallus*

An *Amorphophallus* is perhaps not quite what one would expect to find in the Himalayas but there is a species found at altitudes up to 1800 m in mountain forests from Sikkim to central Nepal. Until recently it was listed as *Thomsonia napalensis*. However, the genus *Thomsonia* has now been abandoned on the grounds that its rough-textured appendix is not in fact very different from several others in *Amorphophallus*.[23]

A. napalensis produces a nasty smelling greenish-yellow inflorescence before the new leaf emerges in the spring. The boat-shaped spathe is up to 45 cm long and the stout finely warted spadix may reach 25 cm.

Typhonium

The genus *Typhonium* is thought to have been named after Typhon, a monster and one of the whirlwinds in Greek mythology that is portrayed as having long erect pointed ears. *Typhonium* spathes, if not exactly ear-like, are mostly reddish with elongated tips. The leaves are commonly hastate, often with the basal lobes as large as the main part of the blade so that they are in effect lobed. It is a tuberous genus with 25–30 species native to Southeast Asia and Australia but several species are now naturalized in places as far away as Madagascar, Zanzibar, Brazil and Trinidad – to which they probably gained access as a result of being cultivated for medicinal uses. Typhoniums vary in size from robust species such as *T. trilobatum*, whose leaves reach over 20 cm across on stalks some 45 cm long, to some of the smallest aroids of all: *T. acetosella* from Thailand has 1 cm leaves and spathes just 4–5 cm in length, most of which consists of a long drawn-out tip. Although much of the genus is tropical, several species may be described as woodlanders.

Five species are endemic to Australia and range from the tropical north to the mountain ranges of New South Wales. *T. brownii* has trilobed hastate leaves and dark purple spathes. Plants reach about 35 cm and are locally common in wet sclerophyll forest. One species, *T. eliosorum*, may very well be deleted from the Australian flora soon. Again, the leaves are trilobed and the spathe is dark purple inside and comes to a narrow point. Restricted to a narrow range in New South Wales, this aroid is now threatened with extinction, its forest habitat having succumbed to clearance and urban development.[24] Its name means 'dormouse's tail'.

Two most interesting hardy typhoniums are *T. giraldii* (sometimes called *T. giganteum*) and *T. diversifolium*. The former is widespread in China, producing deep maroon inflorescences and bright purple berries. In Chinese herbal medicine it is used to relieve pain and spasms.[25] *T. diversifolium*, as its name implies, has very variable leaves which may be anything from heart-shaped, sagittate or hastate to lobed with five to seven segments. The spathe is typically reddish-purple. It is found in the Himalayas between 2500 m and 4300 m and so, like *Arisaema flavum*, is a near-alpine. The ability to grow at high altitudes is also seen in *T. alpinum* which is recorded as reaching 4000 m in China.[26]

The little green dragon – *Pinellia*

Pinellia is an eastern Asian woodland genus with six or seven species. Several have uses in traditional Chinese medicine which are described in chapter 10. The best known species in cultivation is *P. ternata*. The inflorescence is like a scaled-down version of the North American green dragon, *Arisaema dracontium*, with slender green spathes closely enfolding long green spadices which stick up like antennae above the trilobed leaves to emit alluring odours to passing insects. Another curious feature is that it produces a bulbil at the base of almost every stalk. It is theoretically capable of nearly doubling its population every year and in cultivation can spread rapidly and verge on becoming a weed.[27] In the wild, it has been found in a variety of locations, from shady thickets on a lava bed at 1500 m[28] and under pines beneath the Great Wall of China.[29] Apparently it seldom goes above 2500 m.[30]

The most unusual *Pinellia* is *P. cordata*. Although for a small genus, *Pinellia* exhibits great variety in leaf shape, from cordate and hastate to trilobed and pedate, the only variegated species is *P. cordata* from China. Its heart-shaped leaves, held close to the ground on dark purplish-brown stalks, are pencilled with silvery white around the veins and are purple underneath. Each has a horn-shaped bulbil at the junction of blade and stalk. The pale green inflorescences are sweetly scented.

The aroid that thinks it's a mushroom

A very odd little aroid, *Arisarum proboscideum*, has an appendix which mimics the appearance of a fungus. A number of arisaemas are pollinated by fungus gnats and invite their visitors by giving out mushroomy smells but none is known to go to this extreme. Commonly known as the mouse plant, it is popular in cultivation on account of its spathes which have brown tails about 17 cm long that protrude above the dense mass of glossy sagittate leaves, looking as if a family of mice has just dived for cover. It is found wild in woods in Italy and Spain but not in Albania as stated in Cecil Prime's *Lords and Ladies*; a confusion which probably arose from a collection in the herbarium at Kew. Made in 1867 and labelled *In montibus Albanis*, it was from the Alban hills southeast of Rome.

The inflorescences are on pale stalks about 15 cm tall and are held slightly inclined towards the ground. The lower portion of the spathe forms a tube 5–7 cm long with a slightly velvety brown hood and a white base. The only opening is a small hole beneath the point where the hood narrows into the elongated curve of the tail. To appreciate this inflorescence fully it is necessary to cut away one side. Inside is the crooked spadix with a few female flowers at the base and a larger number of male ones above. Then, fitting exactly into the hood of the spathe, there is the appendix which is spongy and white – a perfect replica of the underside of a fungus. Its faint mushroomy scent attracts female fungus gnats. Once inside the inflorescence they are stimulated to lay their eggs by the appearance and humidity of the imitation fungus.[31] They deposit them in the cavities of the spongy tissues and the larvae subsequently hatch, only to starve on the fake food. After egg-laying, the gnats enter the spathe chamber by either falling into it or being guided downwards by the bright refractive base. As they struggle to escape they drop any pollen they may be carrying from a previously visited inflorescence and pick up a new load before they leave.

The flowering of this cunning little aroid is timed to coincide with the first springtime generation of gnats which are compelled to breed when few real fungi can be found, so that in their desperate search for somewhere to lay their eggs they are readily taken in by its enticing guise. Such behaviour has earned the inflorescences the description of 'parasites on the ecosystem'.[32] This strange case of mimicry was first described in 1891 by the Italian botanist Arcangeli, who described *A. proboscideum* as 'the most peculiar herb in Europe'.[33]

Cuckoo pint

There are upwards of 100 common English names for *Arum maculatum*, a species found in woods and hedgerows in northern Europe. The earliest is probably cuckoo pint, a name which is rather obscure in modern English but in Anglo-Saxon was quite explicit: *cucu* meaning 'lively' and *pintle* being one of several words for a penis. A number of other names also describe the phallic appearance of the spadix, such as dog's dibble. There are different ways of looking at it though and some names have been suggested by the spathe itself (priest's hood), the position of the spadix within the spathe (parson-in-the-pulpit, babe-in-the-cradle), the leaves which often look as if they are spotted with blood (gethsemane, bloody-man's-finger), its habitat (Moll-of-the-woods) and pollination mechanism (flytrap). Some names are less easy to explain and in some cases may have arisen from mixing up elements of different names long after their original meaning had been forgotten: hence the unfortunate combination in priest's pintle.

The spadix is usually maroon but in some plants is pale yellow. This gave rise to the name lords-and-ladies, as English noblemen once wore rich burgundy and their ladies paler colours such as yellow.[34] Lords-and-ladies and cuckoo pint are the most widespread names in use today. Thomas

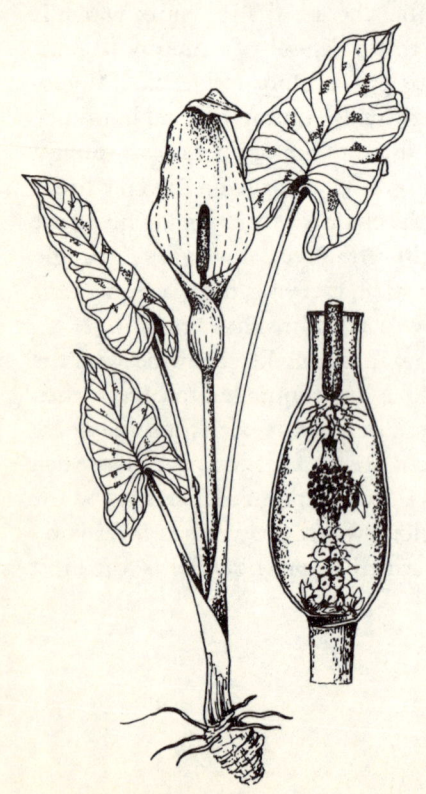

Far left: ($\times \frac{1}{4}$) ***Arum maculatum:*** this common wayside and woodland species of northern Europe is the only native aroid in several countries. In the British Isles its distinctive and curious inflorescences have given rise to more than 100 common names, more than for any other wild flower of the region. They include cuckoo pint, lords-and-ladies, starchwort, adder's tongue, babe-in-the-cradle and the delightful (if inexplicable) Kitty-come-down-the-lane-jump-up-and-kiss-me.

Left: ($\times 1$) ***View inside the spathe tube of A. maculatum:*** tiny flies enter the tube in response to the faecal smell produced by the appendix when the female flowers (at the base of the spadix) are receptive. They are unable to escape until the next stage, in which the male flowers release pollen on to them and the hair-like sterile flowers have withered to permit exit.

Hardy used both in his novels. In *Far from the Madding Crowd* he refers to 'the old cuckoo pint – like an apoplectic saint in a niche of malachite', and in *Tess* he describes the heroine 'gathering the buds called lords and ladies from the bank'.

Arum maculatum favours damp shady areas in deciduous woodland, hedgerows and thickets on slightly alkaline soils and is especially common in beech woods. It rarely grows much above 300 m, though is occasionally found at higher altitudes in warmer parts of the continent. It has the most northern distribution of any *Arum* species but is absent from the extreme north and from areas of the central mainland where summers are too hot and dry. In Britain it peters out around the Firth of Forth and plants this far north tend to be small and rarely fruit.[35] It is dormant from late summer to midwinter and sends up fresh green leaves very early in the year when little else is in growth. The conspicuously succulent foliage escapes being eaten through its potent mixture of oxalic and hydrocyanic acids, calcium oxalate crystals and saponins. Being in full growth when the trees and surrounding vegetation are still bare ensures high light levels before the sun is strong enough to scorch the foliage. By mid-April, when flowering begins, the necessary shade is provided as the canopy leafs out. The inflorescences are pollinated by midges (mostly *Psychoda phalaenoides*) which breed in dung and are lured by the faecal scent wafted from the warm spadix. The pollination mechanism is similar to that described in *A. nigrum* (p. 46), the midges being imprisoned in the spathe tube and prevented from leaving by the overhanging glassy smooth walls.[36]

When the fruits begin to form in early June, the woodland canopy is at its heaviest and the green berries are well camouflaged among the green shades. The leaves die down in July after a working life of little over four months and the fruits ripen to brilliant scarlet in August and September, showing up well against the fading, thinning vegetation of late summer. Blackbirds (*Turdus merula*), thrushes (*T. viscivorus* and *T. philomelos*), pigeons (*Colomba palumbus*), and pheasants (*Phasianus colchicus*) eat the berries and are thought to be the main dispersers,[37] but no studies have been made of whether the seeds are undamaged after passing through the digestive tract. A fruiting spadix produces as many as 30 seeds. Viability is largely dependent upon remaining damp and therefore packets of *Arum* seed usually give disappointing results. They germinate in early spring but show no sign of life for another year, the first season's growth consisting of only a root and tiny underground tuber. In successive years the tuber inches its way deeper into the soil to protect itself from frost, for in spite of having frost-resistant foliage the tubers are probably vulnerable.

A. maculatum is a variable species and several subspecies and varieties have been named, some of which have a restricted distribution. The sagittate leaves may be plain green or spotted with dark purple to black; the spathes narrow or broad, pale green, tinged, spotted or bordered purple, pink or white; the spadices longer or slimmer than average and various shades of purple or yellow.[38] In many populations there is a mixture of forms and it is interesting to count the proportion of spotted/purple spadixed plants to unspotted/yellow-spadixed. (A spotted plant may have a yellow spadix but this is quite rare.)

In the more southern parts of its range, *A. maculatum* overlaps with its close relative, the Mediterranean *A. italicum*. This has two subspecies, the best known being subspecies *italicum* which is widely cultivated for its beautiful white-veined winter foliage. The other is subspecies *neglectum* and although it can occur in the same regions as the former (but seldom actually together), it goes much farther north. It has plain leaves but is otherwise very similar. Both have vertical tubers (rather than horizontal as in *A. maculatum*) and come into growth in October. The growing season is about nine months – over twice as long as that of *A. maculatum*. Other distinguishing features are the larger, later flowering plants, and consistently yellow spadices. Where the two species do overlap (as in a few areas of southern England) hybrids have been recorded, especially in late springs when the last inflorescences of *A. maculatum* and the first of *A. italicum* are open together. Those observed appear to be exactly half-way between the two, even to the extent of starting into growth in the middle of the winter instead of autumn or early spring.[39]

In several European countries *A. maculatum* and *A. italicum* are the only native aroids. On the whole, they are common plants made familiar by the strange appearance and descriptive names of their inflorescences, which are quite unlike anything produced by the surrounding vegetation. Anyone in these countries who develops a special interest in plants, and in these curious woodlanders in particular, is bound to agree with the 18th century herbalist John Hill, who wrote that 'there is not in all the round of Nature a genus so singular as this'. Such a feeling of wonder and delight (or possibly dismay!) can only be increased with the discovery that in other habitats and countries there are numerous other related genera – both larger and smaller and, in many instances, even stranger.

4
Aquatics and Amphibians
Species of wetlands and water

One of the most surprising things about flowering plants is that after ages of painstaking adaptations to life on land, many have returned to the water. Aroids show this tendency more than most and some genera, even large ones such as *Cryptocoryne*, are wholly aquatic or amphibious. In order to go back to a watery existence, they have had to go through complex processes of re-adaptation to aquatic life which are, in evolutionary terms, very far from going back. Re-adaptation is going forward, even though new developments may resemble past stages. Fashioned by water, parts become simplified but not primitive: the rippled ribbon-like leaves of *Jasarum steyermarkii* waft in the water like algal fronds though they have a complex vascular structure, and free-floating *Pistia stratiotes* has thrown all ballast overboard, becoming a minimalist among flowering plants, yet highly advanced for all its reductions.

—— A love-hate relationship ——

In many cases the genetic capability for certain features is lost over millions of years of evolution: aroids could never, for example, come up again with the rootlessness and free-swimming eggs and sperms which characterized their dim and distant algal ancestors. So those which began to encroach on wet places had to devise something completely new in the way of buoyancy and reproductive techniques. Fortunately though, some capacities were retained. Leaves, stalks, stems, and roots adapt quite readily to the demands of constant wetness. The real stumbling block for flowering plants taking to water is the jewel in their crown – flowers. In meeting the needs of animal and wind pollination, the precious nectar and pollen must be protected from inundation. Even strictly terrestrial plants which flower when rain is an ever present threat have had to devise means of protecting their sexual parts from a temporary soaking. But far worse is the danger of submersion which confronts aquatic species. For the most part, flowering plants are doomed to a love-hate relationship with water, needing it for growth but having to avoid it for pollination. The hooded drip-tipped spathes of arisaemas which shed oriental monsoon rains are one device; the hermetically sealed flasks of underwater cryptocorynes another: umbrellas and wetsuits keep the water at bay.

It is always said that the earliest forms of life moved up from the oceans on to dry land, but in fact very little of the land is dry. It is riddled with the sweet water that falls as rain and condenses as dew, trickling into crevices, pouring down slopes and settling into hollows. Nevertheless, earth and air prevail on land and flowering plants are the culmination of adaptations to a more-or-less dry regime. But those which came to live in or near wet places had to contend with water too. Inevitably they were inundated now and again. Surviving ordeals by water depended on how deftly they coped with the various levels of wetting. Many could only survive with their heads well above water but others could go under for quite a time, and some found little cause to surface again.

—— What it takes ——

Plants are entirely dependent upon water as no growth is possible unless the cells are turgid. Surprisingly, light is of less importance. *Arisaema triphyllum*, a tuberous woodland species, grew experimentally for four successive seasons in total darkness when supplied with sufficient water and nutrients.[1] The most severe stress for land plants is therefore lack of water. Yet for those colonizing wet places the opposite is the case: too much, either in depth (which cuts down light) or in speed of flow (which tears and uproots them). In addition, oxygen diffuses over a thousand times more slowly in water than in air and is in especially short supply in stagnant water and waterlogged substrates. And without oxygenation roots rot and the plant drowns. On top of this, water is 800 times as dense as air, which puts enormous pressure on foliage and reduces the amount of photosynthesis that can take place.

To get round these problems, the rigid woody fibres which held them upright on land have largely been replaced with gas-filled cavities (aerenchyma) so that they attain greater flexibility, buoyancy, and improved oxygenation. Changes in the shape and thickness of leaves also help. Fish-shaped foliage offers less resistance to currents and varying leaf shapes on the same plant enable it to exploit different water levels – velocity being greater at the surface than in the depths. Photosynthesis in gloomy underwater worlds is made easier with a large surface area in relation to volume, so many aquatics have filmy leaves that are only one or a few layers of cells thick. These adaptations are clearly evident in *Cryptocoryne*.

Transpiration rates also need revising in aquatics. The average land plant has roots to search out water locked in the soil but it loses as much as 98 per cent of its water intake through evaporation from the leaves. A well-developed cuticle and pores to control water loss are therefore characteristic of terrestrials and generally the thicker the skin and deeper set the pores, the drier the conditions tolerated. In contrast, a plant entirely surrounded by water loses very little and can take it in over its whole surface. (If the leaves float or emerge then of course it will have to guard against water loss, though floating ones would only need protection on the upper surface.) The conduction of fluids from roots to leaves is therefore of little importance compared to the transport of oxygen. Roots are by no means redundant though. They are often tougher and more extensive in water plants for secure anchorage, as can be seen in *Jasarum steyermarkii*.

So far so good! However, one problem remains almost unsurmountable: flowering underwater. Very few genera of flowering plants have devised water-repellent and buoyant pollen so that their flowers can open in water. Most resort to sending up flower stalks well clear of the water or develop buds which only open when they breach the surface. In other words they are still dependent on wind or insects for fertilization and cannot, as yet, utilize the water for this purpose. Thus flowering for the majority of aquatics is a risky business and there is a tendency to avoid the issue. Vegetative reproduction often takes over and flowering only occurs when the water recedes and it is safe to do so.

The adaptations described apply mainly to the true aquatics which are dependent on water. Amphibians, which can take it or leave it – though not for very long – differ in their degrees of adaptation. Both aquatics and amphibians occupy a vast range of habitats, from marshes and water margins, shallow pools and meandering streams to torrential waterfalls, mighty rivers, brackish estuaries and swamps, and they vary just as dramatically in form. There are aroids which can be found in all these conditions and whether floating, submerged, fleshy, reedy, or even tree-like are finely tuned to the conditions.

—— Wetlands ——

Wetlands are biologically the most productive and important of all habitats. The plentiful supplies of water and nutrients together with an abundance of light (and in the tropics, of warmth, too) means that there are fewer limits to growth than anywhere else on earth. Most aquatics are rich feeders, able to utilize and neutralize potentially lethal levels of chemicals, especially nitrogen, phosphorus and heavy metals where run-off is from agricultural and industrial areas. They also provide such abundance of food and shelter that they have become the breeding grounds for a host of different animals, particularly fish and birds. In many cases these creatures spend the rest of their lives in other habitats which in turn are dependent on the wetland largesse.

The plants adapted to life in mud and water are valuable allies. Streams, ponds, rivers and lakes are the earth's gutters and drains; marshes and swamps are the sponges which soak up, filter and purify the surplus water and flushed-out sediments and wastes. Wetland vegetation slows the flow of water over soil, checking its scouring action and mopping up excess. Underwater root systems bind together sediment and debris and actually build up the level of the ground. (In fact, many wetland plants are pioneers which pave the way for their own demise, as annual layers of decomposed remains and daily transpiration leave them high and dry. Through them, water eventually gives way to marsh, and then successively to scrub and forest.) Slowly the water and dissolved substances they collect are released, to underground water-bearing strata after percolation, to the air by evaporation and to the rich diversity of plant and animal life through complex pathways of breakdown and assimilation. To many human concerns wetlands may seem like waste space, more usefully drained and filled in to grow crops and bear concrete but the truth is they are the land's only bulwark against dissolution and engulfment and its main space for waste.

⎯⎯ A deceptively delicate drifter ⎯⎯

When it comes to aquatic habitats, floating along on the surface would seem an easy option compared with holding fast against currents and floods. The fact that very few flowering plants manage to do this indicates that it is a very tricky situation and those that have mastered it show a high degree of specialization. Aerial parts must constantly resist inundation by waves and rain and lifting by wind. To achieve the delicate balance between sinking and blowing away the only feasible design is something akin to a cork with a rudder. *Pistia stratiotes* or water lettuce is the only floating aquatic aroid and is a superb example of what is needed for such a precarious existence.

In appearance *Pistia* bears no resemblance to any other aroid. Though variable, its leaves seldom reach more than 20 cm. They are wedge-shaped, pearly green, practically stalkless and occur in rosettes. If plants are well apart the leaves lie fairly flat on the surface but in crowded conditions may sit quite upright. The inner tissues have air-filled cavities and the outer surfaces are fluted and velvety with a dense covering of minute hairs. These features make it virtually unsinkable and as water repellent as a duck's back: if you push a plant under it bobs back instantly like a cork, shedding water in spherical drops. The rudder-cum-anchor is provided by a mass of feathery roots which dangle at some length from the base of the rosette. Flowering takes place dangerously close to water level. The minute hairy spathes are only about a centimetre long – the smallest in the family – and nestle in the axils. There is no spadix as such: just a single female flower and above it a few male flowers in a whorl which can be seen as a star at the opening of the spathe. Details of pollination have not been recorded but there is ample evidence of its reproductive success. Seeds are either washed into the water or picked up on birds' feet as they step across the rosettes in search of insects and aquatic prey. It also spreads with unnerving ease by stolons which produce plantlets at their tips.

—— The sudd ——

A single water lettuce plant is almost as delicate as a flower but *en masse* they play a formidable role in the ecology of tropical waterways. The earliest descriptions of a phenomenon known as sudd formation concerned the upper reaches of the White Nile in the Sudan. *Sudd* is an Arabic word

Pistia stratiotes
Right: (× $\frac{1}{3}$) Water lettuce produces a dense mass of roots and numerous stolons which develop plantlets at their tips.
Far right: (× 3) *Pistia* inflorescences are borne in the axils, very close to the water. They consist of a downy water-repellent spathe, a single female flower with numerous ovules and a whorl of 2–8 male flowers which are visible at the entrance to the spathe.

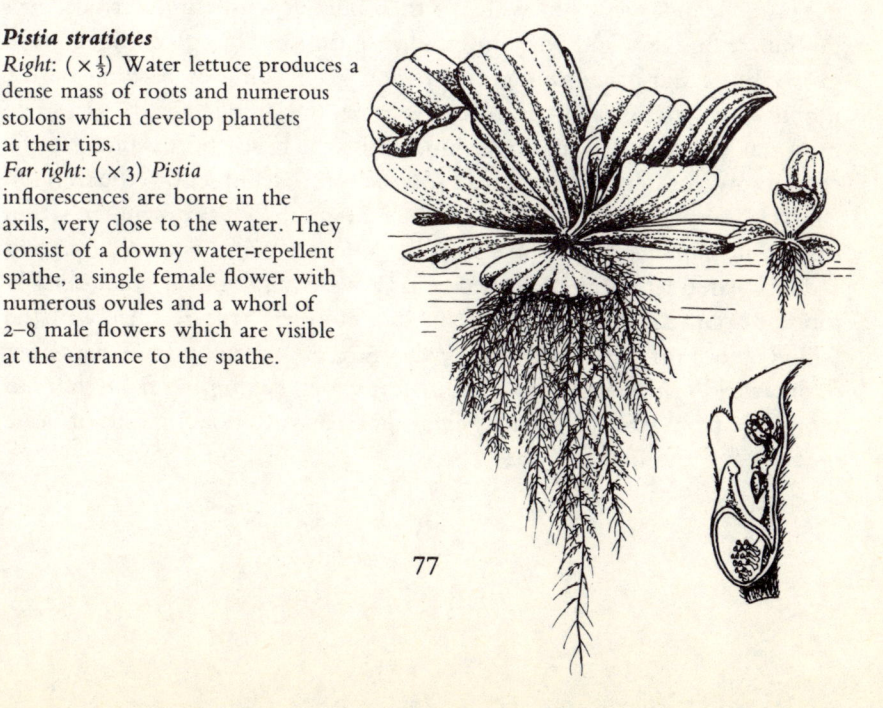

meaning obstruction and it refers to the great mats of floating vegetation which are characteristic of this region. Indeed, they are so common that the vast area of tributary rivers and swamps is generally known as The Sudd.

Pistia is not the only culprit in sudd formation. It is accompanied by even more fragile aquatics such as the tiny water fern *Azolla* and the carnivorous bladderwort *Utricularia*. Together they form thick layers around chunks of papyrus, broken reeds and other rotting vegetation. The main nurseries for *Pistia* are Lake Victoria and the complex of lagoons around it. From this network it is swept into the Nile by the strong winds and floods which occur during the rainy season.

Sudds are fearful obstacles to navigation. In 1878 Romolo Gessi, the Italian explorer and administrator in the Sudan, attempted to travel into western Ethiopia by boat. The voyage was a disaster. One part of the journey, which could be done in five hours, took over three months because of 20 major obstructions and he lost half his men from starvation.[2] Mary Kingsley's travels in West Africa fared better but she too dreaded the confrontation between *Pistia* and boats, in which the 'sweet green rose' with 'the manner of couch-grass' invariably won by mangling propellers with its dense roots and tangle of runners.[3]

── The tale of a rhino ──

The great swampy deltas of the Ganges and Brahmaputra rivers in India and Bangladesh are also home to *Pistia*. In the monsoon, mats of water lettuce join up with assorted debris forming sturdy rafts which can reach 75 cm thick and over 15 m wide. As they float downstream, various seeds germinate on them and with unfounded optimism young saplings start to grow. Birds and other animals come and go, feeding or basking on the mobile islets. Occasionally the cargo includes the stranded or the rare. The most unusual record is that of a rhinoceros which was borne along to the coastal town of Chittagong where it was captured and sent to the London Zoo.[4] Upon arrival it was found to be a new subspecies of the Sumatran rhino, *Dicerorhinus sumatrensis* subspecies *lasiotis*. Known as the Chittagong or hairy-eared rhino, this subspecies has a wide range throughout densely forested Southeast Asia. Unfortunately, its numbers have declined with the felling of the rain forest and it is now very rare. At one tonne it is the smallest of the world's rhinos and always lives near water, swimming from shore to shore, and from island to island,[5] even apparently, to floating rafts of *Pistia stratiotes* and the like.

A shocking end

One result of these dense mats of vegetation is that the water beneath receives very little light. In such dark turbid waters this has led to the evolution of some very unusual fish which navigate and hunt by sending out electrical discharges from special muscles in their sides or under their tails. Any object in the water causes disturbances in the flow pattern which are detected by receptor pores. Fish with this capability have evolved independently all over the tropics. In the Nile river system, the most notorious of all for sudds, there are over 100 species of the elephant fish (*Mormyriformes*), all with electric organs and well-developed brains and learning abilities.

Friend or foe?

The discovery of a new rhino and the evolution of electric fish are among the more bizarre consequences of floating aquatics such as *Pistia*. Other results are more mundane, some beneficial and some problematic. In its natural habitat *Pistia stratiotes* plays an important role in the nutrient cycle of tropical water systems. It mainly grows and reproduces in lakes and swamps, often in stagnant water which it helps purify, and is flushed out into streams and rivers during the rainy season. Sooner or later the mats of *Pistia* and other flotsam and jetsam become lodged: the debris decomposes, the viable plants take root and regions downstream are enriched.

The problems stem from the fact that human beings find water lettuce a very attractive plant and persistently cultivate it in ornamental pools, from where – in tropical and subtropical regions – it invariably finds its way into major waterways. This, together with its talent for exponential growth, has led to its pantropical distribution and a reputation approaching that of water hyacinth (*Eichhornia crassipes*) as a pernicious weed. A good example of its speedy transition from desirable ornamental to hateful weed took place in the 1960s in Australia. Apparently just three plants were set adrift on a four million gallon reservoir in Brisbane, Queensland. In no time at all the whole surface was covered and in spite of its attractions, the authorities were obliged to formulate plans for its extermination.[6] They were not alone: *Pistia stratiotes* is now regarded as a serious weed in many parts of the world, especially Australia, Africa, Asia, and the Caribbean. Many other aroids have escaped from cultivation, but only *Pistia* has become troublesome.

Of course, a weed is nothing more than a plant in the wrong place and is not inherently useless or harmful. Water lettuce has been indicted for reducing light penetration, oxygen levels and fish life. It is also said to increase water loss through transpiration and provide breeding grounds for mosquitoes. In its defence, it decontaminates stagnant and polluted water, shifts nutrients downstream, provides a spawning medium for fish and a refuge for fry and makes a feeding and resting platform for wetland birds, reptiles and amphibia (and the odd aquatic mammal). Its preference for nutrient-rich, slow-moving or stationary water suggests that it has potential in sewage treatment, and its high potash content makes a valuable fertilizer.[7] In the Philippines it is fed to pigs, and in many countries it has a wide variety of medicinal applications (which are described in more detail in chapter 10), the strangest being an East African cure for insanity in which the roots are bound around the head of the afflicted and an infusion of the leaves summarily poured over.[8]

Pistia and the duckweeds

As an aroid, *Pistia* is out on a limb in more ways than one. Although its inflorescence shows the hallmarks of *Araceae* the leaves and floating aquatic lifestyle are unique in the family. The fact that it shows some affinities to a different family altogether – that of the duckweeds (*Lemnaceae*) – is a long-standing controversy in botanical circles. (*Araceae* and *Lemnaceae* together make up the order *Arales*, so the relationship is acknowledged if not clear-cut.) Duckweeds are one of the oddest of all families of flowering plants, consisting solely of tiny aquatics, many of which are only noticeable as a green film on the surface of water. Common duckweed (*Lemna minor*) has just a single root and its stem and leaves are reduced to a flake of vegetation which is a mere 2–4 mm across. Yet even these lilliputian dimensions dwarf the smallest of all duckweeds, and of all flowering plants, the rootless *Wolffia*. An individual plant is scarcely visible to the naked eye, a dozen could fit on one 'leaf' of common duckweed and countless millions are nothing more than surface scum.

By comparison, *Pistia* is a giant and it is hard to see what they could have in common, other than floating on water. However, embryological studies have shown that *Spirodela* (a genus of *Lemnaceae*) is intermediate between *Lemna* and the aroids.[9] One theory is that the duckweeds evolved as a separate genus from *Pistia* seedlings which became sexually mature while remaining at the juvenile stage of growth.[10] This process, termed paedomorphosis or neoteny, crops up elsewhere in evolutionary theory and has been put forward as the possible origin of herbaceous plants from trees or even of human beings from young ape-like primates. Should this theory eventually be abandoned as explaining the relationship between *Pistia* and the duckweeds, a close contender is that they had a common ancestor in the not-too-distant past. Whatever the explanation, there is no doubt that the resemblance between them is strong, so much so that it has been suggested that *Lemnaceae* might be included in *Araceae*.[11] In which case, the family would include both the smallest known inflorescences (*Wolffia globosa* and *W. microscopica*), and the largest (*Amorphophallus titanum*).

—— A tree-like aquatic ——

An aroid completely different in appearance from *Pistia* also plays a dynamic role in tropical waterways, though it is confined to tropical America. *Montrichardia arborescens* grows in dense colonies of tall slender woody stems which are clad in vicious spines, especially towards the base, and at intervals sprout sagittate leaves. The inflorescences are white, 10–15 cm long, intensely fragrant and pollinated by bees. Infructescences the size of a child's head bear about 80 large (4 cm) yellow fruits, each with a single seed as large as a thumb nail. The fruits are edible when ripe[12] and both they and the leaves provide food for the hoatzin (*Opistocoma cristatus*), a primitive bird with hooked claws on its wings.[13] Very closely related is *M. linifera* with cordate blades and thicker, smoother stems. It is suspected that where their ranges overlap they hybridize, as some plants have indeterminate features.[14]

The Essequibo, which drains over half of Guyana, is one river where *Montrichardia* occurs abundantly. Locally the plant is known as 'mocca-mocca'. An early description tells how

> growing in the water, this monster arum develops great club-like stems, which come as close to each other as they can pack, and rise like rows of palisades to the height of 12 feet or more above the surface. As if this were not sufficient encroachment on the open space, the floating island grass (*Panicum elephantipes*) anchors itself to the mocca-mocca or bushes, and extends just as far across as the rapid current will allow. In dry weather, the extensions from either side meet in the centre, and close the passage-way for a time – only, however, to be torn away in great masses as the floods come. At such times great patches, 50 feet or more in

diameter, are seen floating downstream, sometimes carrying with them monster camoudies (*Boa murina*) or other snakes.[15]

These floating islets either get stuck on projections or are carried down to the estuary where they are repeatedly thrown back on the shore by the tides, forming huge rotting mounds which build up the coastline. Where they lodge in the river permanent islands are formed as the current is slowed and diverted around them. Once stationary again, the *Montrichardia* roots into the silt and debris, and the whole process begins all over again. 'The great rivers of Guiana all contain islands of different sizes, some as many as 10 miles long, and it may be confidently stated that all have been built up in this way, by means of the mocca-mocca.'[16]

The stems of *Montrichardia* are pliable so that they are bent by the current but not broken. If the force of the water is strong enough, whole plants – roots and all – are torn away, which makes for effective vegetative propagation. Its spread by seeds is no less efficient. They are large, buoyant, abundantly produced and capable of high-speed germination. The deltas of the major South American rivers are tidal but freshwater for some distance. In this zone *Montrichardia* is known to germinate and take root within the ebb and flow of a tide.

Colonies of *Montrichardia* can be the start of a succession of plant communities. When they are well established, mangrove (*Rhizophora racemosa*) moves in, followed by trees such as palms and cecropias. Each new invader suppresses the growth of its predecessor and finally a swamp forest is established with trees 30 m tall where first grew the pioneer shoots of *Montrichardia* seedlings.[17]

Montrichardia is found not only in wildernesses of brown water and green forest. It invades the urban jungle with equal vigour. In the city of Manaus (the capital of Amazonas, Brazil) it progresses along sewage-enriched ditches as an impenetrable 7 m high thicket.[18] If cut down, within three months it is as tall again. In some habitats, however, *Montrichardia* is less intimidating. It has been found only knee-high, but still flowering, in wet savannahs in French Guiana.[19]

—— *Lasia* ——

Another thicket-forming aquatic is *Lasia spinosa*. It belongs to a Southeast Asian genus of possibly two species (the other being *L. concinna*, known only from a single plant in cultivation at Bogor Botanic Gardens, Java, which may turn out to be a hybrid with *Cyrtosperma merkusii*).[20] The whole plant bristles with thorns and forms daunting barriers in swamps and along waterways. Even the fruits are spiny, a feature uncharacteristic of aroids on the whole, though seen also in *Pycnospatha*. Only the inflorescence and upper surface of the leaves are smooth. The slender spathe is dark maroon with a crimson lining and is drawn out into a long erect point. The foliage is variable: usually divided but plants with entire sagittate leaves are not uncommon. This tough aroid has upright or decumbent stems and colonizes by stolons,[21] as well as by pieces broken off in floods, and by its spiny quadrangular fruits.

—— *Urospatha* and *Dracontioides* ——

Two of the most striking wetland genera are *Urospatha*, with a dozen or so species, and *Dracontioides* with just one. *Urospatha* is outstanding for its mottled elongated spathes which characteristically end in a spiral twist. The leaves are strongly sagittate and held point uppermost. *U. sagittifolia* has a widespread distribution in northern South America, where it occurs in large stands on river banks and produces spectacular corkscrew-tipped inflorescences and triangular leaves with large flaring basal lobes.

The glossy sagittate leaves of *Dracontioides desciscens*, a native of eastern Brazil, are borne on spongy stalks 2 m tall arising from a thick vertical rhizome. The foliage is quite distinctive, being pierced (sporadically and sparsely) by narrow oval holes and it bears bulbils in the axils. The inflorescences are hooded like those of *Dracontium*. Schott regarded this species as a *Urospatha* but Engler thought it sufficiently different from *Urospatha* and the closely related *Dracontium* to have a genus of its own.

—— *Cyrtosperma* ——

Cyrtosperma, with 11 species, belongs to the same subtribe as *Lasia* and likewise is armed with prickles. Its centre of distribution is New Guinea, which gives it the distinction of being the only sizeable genus in the family whose origins are east of Wallace's Line, the hypothetical boundary which separates the characteristic Asiatic flora and fauna from that of Australasia.[22] Though long regarded as pantropical, a recent study indicates that the African and South American species have distinctive characteristics which warrant segregating them into separate genera.[23]

The problematic African species is *C. senegalense*, a dominant plant of West African freshwater swamps. Its long stalks have the usual prickles but they are quite different from others in *Cyrtosperma* (and in the rest of *Araceae*), being in uniform rows down the stalk. This species was originally known as *Lasiomorpha senegalensis*, then moved into *Cyrtosperma* and now back to *Lasiomorpha*.[24]

The neotropical cyrtospermas in question proved to be a mixed bunch: *C. spruceanum* has been put into *Dracontium* and *C. wurdackii* into *Urospatha*. *C. americanum*, a rare species from the Guianas, was only accommodated by the creation of a new genus: *Anaphyllopsis*.[25] Superficially it resembles *Anaphyllum*, a genus of two marsh species native to India which also have divided leaves (with some leaflets pinnate) and twisted spathes.

In addition, there is now evidence that a number of *Cyrtosperma* species, such as Oceanic taxonomic groups described as *C. chamissonis* and Malaysian *C. lasioides*, are all variants of the one species, *C. merkusii*.[26] It is widely grown as a crop in Oceania, Micronesia alone having 64 cultivars.[27] Commonly known as swamp taro, it can reach 4 m and produces enormous starchy tubers, even in brackish or stagnant conditions. (More about this food plant in chapter 9.)

The commonest *Cyrtosperma* in ornamental cultivation is *C. johnstonii* from the Solomon Islands. It is particularly attractive as a young plant, having prickly stalks marked with chocolate brown, broad sagittate leaves with pink midribs and colourful pink, green and brown leaf undersides. It is not, however, common in the wild, and only one collection (of seedlings) has ever been made.[28]

Submerged aquatics

There is more difference between the eastern and western hemispheres when it comes to submerged aquatic aroids than with any other category. The neotropics (South America, the West Indies and tropical North America) have only one species and the Asian tropics over 70, though most belong to one genus: *Cryptocoryne*. Why tropical America should be almost devoid of them when they abound in rivers of Southeast Asia is a mystery.

Jasarum

The lone South American submerged aquatic, *Jasarum steyermarkii*, is the only species in the genus. It is closely related to *Caladium* and is thought to have evolved from those which inhabit seasonal swamps. The leaves are thin and linear with a prominent midrib and numerous deep-set lateral veins which leave it almost at right angles. The blades reach up to 30 cm

Jasarum steyermarkii ($\times \frac{1}{5}$): the only submerged aquatic aroid in the New World: a monotypic genus described in 1977.

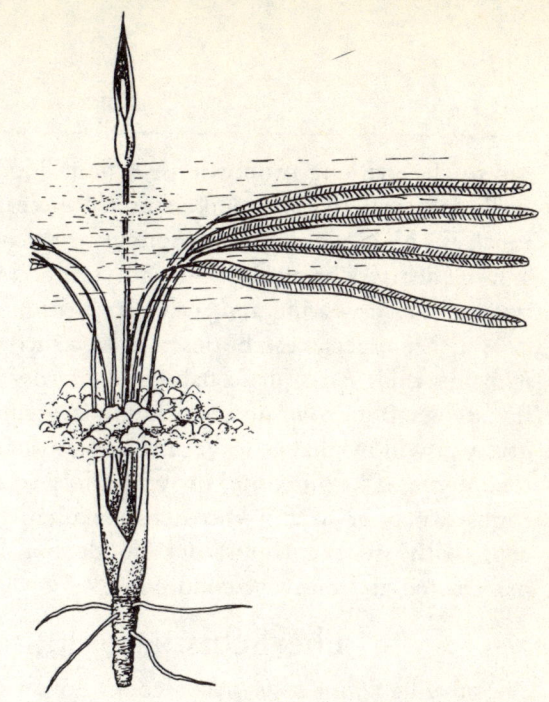

long and have stalks at least as long again, though much of their length is buried in the river bed. The thick vertical stem and stout spongy white roots go so deep that plants are very difficult to extricate. Flowering occurs in well-lit positions and the inflorescences have a decidedly terrestrial look about them, with elegant pointed spathes about 15 cm long which open above the surface. Fruits develop beneath the water and may possibly form colonies by germinating where they sink. (Unlike most aquatics, it does not appear to spread vegetatively, though plants are often very close together.)

Jasarum grows in acidic blackwater in two separate river systems in Venezuela and neighbouring Guyana. Though first discovered in 1960, this species long remained an enigma and was not described until 1977. As the only known submerged aquatic aroid in South America, it is also one of the most interesting finds of recent years.[29]

—— Rheophytes ——

Rheophytes are superaquatics: plants which thrive in swift currents of 1–2 m per second, and flash floods several times that speed. The word means 'lovers of the current' and theirs is one of the most demanding habitats on earth. They survive such force of water by offering on the one hand a minimum of resistance to the flow with their narrow flexible – literally 'streamlined' – leaves and on the other hand a maximum of resistance with a large and tenacious root system. Tropical rivers in spate can reach terrifying proportions. A flash flood rumbles downstream as a wall of water with a sound like thunder, increasing the level abruptly by as much as 5 m. Not only can rheophytes withstand such a battering of water and the debris it brings but they also respond rapidly to low water levels with speedy flowering and seed germination.

Whether a species can be described as a rheophyte depends on observing it in the wild. An aquatic habit and narrow leaves are not sufficient in themselves. If it grows nowhere but in fast-running streams and rivers, has grassy or willow-shaped leaves and a deep root system, chances are that it is rheophytic. If it sometimes grows in slow or stationary waters or on the banks, it may be what is known as a facultative rheophyte; that is, able to cope with swift currents and floods but not dependent on highly oxygenated and buoyant conditions.[30]

—— The Schismatoglottid group ——

The subtribe *Schismatoglottidinae* consists of seven predominantly Southeast Asian genera which include the bulk of rheophytic aroids. Largest by far is *Schismatoglottis* which, with about 120 species (mostly in Asia but a few in the neotropics), is one of the five largest genera in the family. It is by no means wholly aquatic, but several of its Bornean members are true rheophytes. These include *S. homalomenoidea* and *S. parviflora* which are peculiar in that, unlike the rest of the subtribe, they do not shed the upper part of the spathe after flowering.

The other genera in the subtribe are wholly rheophytic and likewise from Borneo. Not that there are many species between them – *Heteroaridarum* and *Bucephalandra* are monotypic, *Phymatarum* has just two species, *Hottarum* four, *Aridarum* seven, and *Piptospatha* nine – though there are undoubtedly more awaiting discovery as many of these have been found only recently. Though generally rare and rather undistinguished plants with small elliptic to lanceolate leaves, a few have potential in cultivation as specimens for waterfalls in tropical conditions. One of the prettiest is *Piptospatha ridleyi* with blue-green variegated leaves. It grows in streams in the Malay Peninsula. The smallest must be *Bucephalandra motleyana*. The 19th century Italian explorer Beccari was the first to describe it, naming it *Microcasia pygmaea*. He wrote how 'in one place a little streamlet formed a waterfall and I found the rocks wetted by its spray covered with a diminutive *Aracea*, hardly attaining a height of $\frac{3}{4}$ in.'[31] Its thin rhizomes send down a profusion of roots and at intervals shoots arise which produce about a dozen elliptic leaves. They can reach 8 cm but are often much smaller – sometimes as little as 7 mm in length, with inflorescences just as minute. *Bucephalandra* forms dense mats over wet rocks and stones in streams, sometimes above water level but still in constant spray. It can also withstand rapids and the turbulence of waterfalls. The name of the genus is from a Greek word meaning 'bull's head', referring to the clearly visible horn-shaped anthers.

Lagenandra and *Cryptocoryne*

From an aquarist's point of view, the two best-known aroids are *Cryptocoryne* and *Lagenandra* which are cultivated worldwide as foliage plants for tropical fish tanks and pools.[32] The latter has 13 species ranging from Sri Lanka to northeast India. Most are amphibious, rising above the surface in swamps and along riversides, but those commonly in cultivation, such as *L. thwaitesii*, tolerate long periods of submersion. Lagenandras are not unlike cryptocorynes but are rather larger, with longer lasting, stouter inflorescences, more female flowers and no tube between the blade and the chamber. The spathes are most commonly maroon and warty with a pronounced twist and a slit-like opening. The berries are peculiar in that they split open from the base when ripe. Lagenandras can easily be told apart from cryptocorynes when not flowering by examining an unfolding leaf: lagenandras have both margins rolled in towards the midrib (involute vernation), whereas those of cryptocorynes have the more usual scroll pattern (convolute vernation).[33]

Cryptocorynes are so popular as aquarium plants that they are grown on a large scale for this purpose. Their slender oval or ribbon-shaped leaves, often with a blistered texture or undulating edges, grow in rosettes at intervals along rhizomes. The inflorescences are some of the most fascinating in *Araceae* and it is unfortunate that they are rarely produced in cultivation. In the wild they usually appear only when the water level falls. If this is imitated in cultivation, there is a good chance flowering will take place, but in fish tanks this is often impracticable.

A *Cryptocoryne* inflorescence is air-filled and watertight as it develops under the surface and most of it remains underwater even after the blade of the spathe, which acts as a lid for the bud, has unfurled. In many species the tubular lower section varies its length according to the depth of the water so that the entrance to the spathe can be raised just above the surface. Inside and entirely enclosed by a valve-like flap is the spadix – the most delicate in the family – which can be viewed if the front half of the spathe is cut away with a razor blade (and the utmost care). It has a single whorl of female flowers at the base, then a long thin sterile zone at the top of which are the male flowers. The blade of the spathe is colourful and sometimes fringed or warted to give a larger surface area for the evaporation of the faintly unpleasant scent which is produced by olfactory bodies situated just above the female flowers. It ends in a point or a twist and in most species is held erect or bent backwards, though in *C. thwaitesii* it folds forward at a stiff right angle to the tube. The mouth of the tube is often delineated by a ring of tissue referred to as a collar. Some species have no actual collar but a conspicuous collar zone and a few, such as *C. lingua*, have neither.

The entrance to the tube is below the collar through a valve which opens once the female flowers are ripe. Tiny fruit flies (*Drosophilidae*) are attracted by the scent, crawl through the valve and then fall into the bottom of the tube. The top of the tube is often patterned with translucent blotches which constantly lure the flies upwards but the inner walls and sterile zone are glassy smooth, so they can only clamber about over the receptive female flowers in their efforts to find a way out. In doing so they deposit any pollen they may be carrying from a previously visited inflorescence. After 12 hours or so, the male flowers take their turn. Simultaneously the slippery lining breaks down, the valve withers and the insects escape past the male flowers where they are coated with tacky pollen. The fruits are unique in the family, being fused together and opening like stars when ripe. The wax-coated seeds have the same specific gravity as water and float, germinating within a few days at 28°C but dying if they dry out, making overland dispersal to other river systems unlikely.[34]

Flowering does not always result in fruit and in many species no fruits have been recorded. In fact, vegetative reproduction is more common in cryptocorynes than seeding. As a result, the 50 or more species have evolved very much as if they were marooned on different islands, even though many have overlapping or contiguous habitats. They are the Darwin's finches of *Araceae*: closely related and superficially resembling each other but highly idiosyncratic and precisely tuned to their particular patch.[35] Along a river, the species show zonation, just as seaweeds do, each with its own requirement for light level, substrate and quality of water.

Cryptocorynes may be divided into four ecological types: species of seasonally flooded areas which are reduced to dormant leafless rhizomes in the dry season (*C. nevillii* of eastern Sri Lanka, for example); species of rivers and streams, such as *C. wendtii*; tidal species (freshwater except for *C. ciliata* which inhabits brackish zones); and those which grow emersed, rising above the surface, in swampy areas of forest. Some species in the last category, such as the easy-to-grow *C. beckettii*, are at their most vigorous with their leaves above water and often diminish in size when submerged. *C. retrospiralis*, a species which grows both in and above water in sandy places along rivers in India, Bangladesh and Burma, is peculiar in having a resting form of filiform, reed-like leaves. It also produces a most distinctive inflorescence whose spathe limb is tightly twisted.

Whatever the habitat, cryptocorynes usually occur in sizeable colonies. Typically, a colony is derived from a single 'mother' plant which spreads by a creeping rhizome. Unique clones are established by the vegetative propagation of mutations which would probably be obscured if cross-fertilization took place. Continued isolation may result in the evolution of separate species from such clones. When a colony does flower, seed may not be set because all the inflorescences belong to the same self-sterile 'mega-plant' and may be separated from the nearest compatible colony by terrain which weak-flying pollinators cannot cross, or by a different flowering period. Hybridization is rare too, as species living close to each other often have different flowering periods, scents and pollinators.[36]

Not surprisingly, most *Cryptocoryne* species are variable and several have subspecies: often a sterile triploid form that can only reproduce vegetatively. Even the mode of vegetative propagation can vary according to the number of chromosomes: *C. walkeri* var. *walkeri* and *C. walkeri* var. *lutea* multiply by runners, but *C. walkeri* var. *legroi* does so by dormant buds on the stem.[37] It is no wonder then that cryptocorynes are confusing, even to the specialist and often prove difficult to grow. Most can only be identified with certainty if they flower, and unless you can identify the species, cultivation is bound to be hit and miss. They may survive the wrong conditions, but will never thrive and reveal their real beauty as aquarium plants unless the right substrate, pH, and light are provided.

—— Tide and timing ——

Plants growing in running water have a hard enough life. Reduced light and extreme pressure on submerged foliage, constant tugging at roots, and wind rush around emergent parts are exacting demands on plant engineering. Tidal zones, with their perpetual fluctuation between flooding and drying out, are tougher still. It is a habitat in which flowering plants are often conspicuously absent, but about a quarter of all *Cryptocoryne* species can tolerate the punishing routine of ebb and flow. *C. ciliata*, which of all 'crypts' is the least happy submerged for long periods, can even cope with brackish tidal conditions.

At 1 m tall, *C. ciliata* is the largest of the species. It also has a wider distribution, ranging from India to New Guinea, as its liking for the rich mud of estuaries and mangrove swamps enables it to spread along coastlines. This is achieved by viviparous seeds, a distinct advantage in the colonization of tidal areas. Soon after dropping from the plant, the seed coat splits to liberate an embryo with a tuft of miniature leaves and a rootlet, all ready to grow when it lodges in the mud. Cutting out the germination process combats being washed away (a little plant lodges more readily than even a large seed) and the problem of mud deposits which could easily overwhelm a seed as it started into growth. Unlike most cryptocorynes, *C. ciliata* grows in full sun, flowers frequently and seeds plentifully (although there is a sterile triploid form which spreads by short brittle runners).[38] The attractive inflorescence is most unusual as the dark red blade of the spathe has a fringed margin.

In common with many cryptocorynes, *C. ciliata* has an extendable spathe tube to enable it to keep its head – or rather, its throat – above water

Cryptocoryne ciliata
a] ($\times \frac{1}{3}$) The largest species of the genus inhabits brackish tidal zones and produces colourful fringed inflorescences.
b] ($\times 1$) Cryptocorynes have capsular fruits, unlike most other genera in the family, which have berries.
c] ($\times 1$) When ripe the capsule opens into star-like segments.
d] ($\times 1$) The seeds of *C. ciliata* are viviparous and have a tuft of rudimentary leaves and the beginnings of a root so that growth can get underway rapidly in shifting tidal mud.

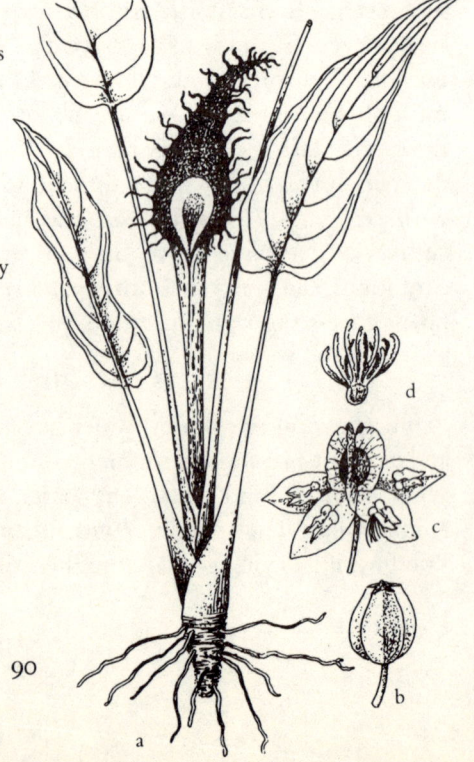

for flowering, regardless of the level. But some tidal species have unvaryingly short tubes which appear to invite disaster. In fact, their flowering is so precisely timed that they have no need to stretch their necks if they find themselves in deep water. The explanation is that in Southeast Asia there is only one high tide a day which reaches its highest midway between new moon and full moon. Consequently, straight after new moon and full moon, tides are at their lowest and the upper reaches of the tidal zone are barely touched by the ebb and flow. Somehow, species such as *C. ferruginea* and *C. pontederiifolia* can correlate flowering with these periods of specially low tides and achieve flower development, opening (anthesis) and pollination in under a week. Furthermore, they restrict flowering to the dry season, as even the lowest tide during heavy rain would not guarantee freedom from flooding.[39]

── *Aglaodorum* ──

Another aroid which has conquered tidal reaches is *Aglaodorum griffithii*. It is the only species in its genus and the differences between it and *Aglaonema* are small: one whorl of female flowers instead of several, a longer inflorescence stalk (peduncle), and green rather than red fruits. Marooned on inaccessible tidal mudflats in Malaya, Borneo and Sumatra, it is a little-known, let alone cultivated species and has seldom been collected or observed in the wild.[40] In common with other tidal aquatics, it has large seeds which germinate before leaving the plant.

── Skunk cabbages ──

We always associate tropical plants with luxuriant growth and bright colours, so the hardy semi-aquatic genus *Lysichiton* comes as rather a surprise. Its two species, *L. americanus* and *L. camtschatcensis*, produce lush leaves over a metre tall and dazzling inflorescences 35 cm long, yet grow as far north as the coasts of the Bering Sea. They are one of the great joys of spring in far northern wetlands and, as testimony, the *Flora of Alaska*[41] flaunts *L. americanus* on its cover.

Lysichitons are found in the rich mud of marshy woodland and alongside shallow water. Both species lose their huge leaves in the autumn and remain dormant through the winter. It is obvious from their appearance that they are closely related but in the wild they have no chance of meeting as their populations are decisively separated by the north Pacific Ocean.

L. americanus is the larger of the two. It is found in western North

America from Alaska to California, forming robust clumps above stout rhizomes. The mid-green, rather fleshy leaves are oval to oblong and are carried on broad flat stalks 30–40 cm long. They start into growth about the same time as the large bright yellow spathes which erupt through the mud in early spring. A colony in full flower beside sparkling water in the wintry bareness of deciduous woodland is one of the finest sights in all the family. The spathe arises directly from the rhizome, an unusual and possibly primitive feature. It begins as a narrow sheath enclosing the long stipe or flowerless portion of the spadix and expands into the concave showy upper part. The green bisexual spadix is completely covered with hard spiky flowers whose texture indents a pattern on the developing spathe. The smell, reminiscent of skunk and over-cooked cabbage, pervades the whole woodland on a sunny day. The odour comes from the spathe itself, not from the spadix which is more often the case. The spathe apex contains indole, a compound uncommon in higher plants, though several aroids are known to produce it. It is used as an aroma fixative in the perfumery, food, and pharmaceutical industries, though the commercial source is coal tar, not skunk cabbage![42]

The anthers produce masses of pollen which mainly attracts beetles. After flowering the spathe falls away, trailing in the mud and soon decaying. The one-seeded fruits are green when ripe in summer. The outer surface is rectangular and sharply pointed; the inner mealy and embedded in the feather-light tissue of the spadix.[43] The ripe seedhead is very brittle, disintegrating readily and falling from the stalk at a touch. The fact that plants are usually found in large colonies with many seedlings among them indicates that many seeds germinate where they fall.

The eastern skunk cabbage, *L. camtschatcensis* is almost identical in appearance to its relative apart from the obvious colour difference of the spathe, which is pure white. In cultivation it can also be seen that the young leaves have a more glaucous sheen and the inflorescences appear a little later. If the two species are grown together, occasionally a hybrid with cream inflorescences is produced, which unfortunately does not bear comparison with the brilliant white and yellow of its parents. In the wild *L. camtschatcensis* has an entirely different distribution from *L. americanus*, being found in the alder (*Alnus hirta*) marshes of coastal northeast Asia, from Kamchatka and the Kurile Islands in the Soviet Union, to Honshu in central Japan. Presumably the two species evolved when the great Asian and American land masses became separated by the Bering Straits.

The central heating plant

The third aroid to go under the name of skunk cabbage is *Symplocarpus foetidus*, though it is also called collard, meadow cabbage, polecat weed and swamp cabbage. Its leaves are similar to those of lysichitons but thinner, brighter green and smaller, reaching only 30 cm or so. Flowering takes place during winter and early spring according to local conditions. The inflorescences emerge before the leaves and are borne at ground level. Their streaky purplish-brown coloration camouflages them against the dappled woodland floor. The spathes are leathery and in bud have tough horn-like apexes designed to push through ice and snow. This feat is aided by the temperature of the spadix – a phenomenal 15–35°C above the surrounding atmosphere[44] – which is sufficient to melt frozen ground and protect the spathes themselves from freezing. Centrally heated, the inflorescences sit within dark circles of thawed mud when the surrounding earth is white with frost.

The rise in temperature also serves to dissipate foul-smelling compounds which attract pollinators (the details of which are described in chapter 2). Once emerged, the tough spathe horns curve inwards and the deeply concave inflorescences look not unlike large cowrie shells. This structure plays a vital role in the success of ground level winter flowering as the shape causes a vortex, spiralling air around the stubby spadix whatever the direction of flow. Thus the heat is retained and circulated, providing a subtropical micro-climate for flowering when the surroundings are in the grip of winter.[45]

In order to remain bogged down, both *Symplocarpus* and *Lysichiton* have of necessity large and deep root systems to anchor the broad and heavy foliage. They also have contractile roots for adjusting plant level after frost heaving or flooding.

Golden club

Very different in appearance from the skunk cabbages, but part of the same tribe (*Orontieae*) is the genus *Orontium* which has just one species: *O. aquaticum*. It is more truly aquatic than its relatives, growing in ponds, streams and shallow lakes throughout eastern North America and elsewhere in temperate zones as an ornamental water garden plant. The pointed oval leaves are blue-green with water repellent surfaces and are a perfect foil for the striped inflorescences and for any iridescent dragonflies that happen to settle on them.

No other aquatic aroid looks anything like this species when flowering, for it appears to have no spathe at all. Careful inspection reveals one well below the spadix but its tiny green blade withers early in the flowering process. In spite of being to all intents and purposes spatheless, the inflorescences are very showy and are produced abundantly in the spring. The long reddish flower stalk curves through the water and rises to display a broad white band below the bright yellow spadix, the part described by the common name of golden club. The flowers are bisexual and packed tightly to the tip of the spadix. Large starchless pollen is extruded by the anthers, which suggests that it may be pollinated by bees.[46] Round green berries, about 1 cm in diameter and each containing a single seed, develop underwater and sink down into the mud when ripe, germinating by late summer.

⸺ Calla ⸺

The one species of *Calla*, *C. palustris*, was once thought to be closely related to *Lysichiton*, *Symplocarpus* and *Orontium*, but recent work has shown significant differences. In fact, *Calla* appears to have no close relationship with any other genera and has now been assigned to a subfamily of its own: *Calloideae*.[47] So *C. palustris* stands alone: an attractive and very successful little aroid that creeps through wetlands all over the northern hemisphere.

It has shiny bright green leaves up to 10 cm across which are heart-shaped to kidney-shaped. They are held well clear of the water on stalks 15–20 cm long, arising at intervals from a flattish green rhizome. In the winter the leaves die off, leaving the rhizome, clad in the persistent brown sheaths of the leaf bases, to withstand the rigors of northern and subartic frosts under the protection of mud and water. Flowering takes place in summer when the new leaves are fully developed. The spathe is white, about 6 cm in length, and erect, backing the short thick spadix of green, mostly bisexual florets (a few at the top are usually male). After flowering closely packed dull red fruits develop, each containing several cigar-shaped red seeds. The spathe remains but turns reddish-green.

C. palustris is a very popular water garden plant: very hardy, easy to grow and rapidly forming neat colonies in shallow water. Its common names include bog arum and marsh calla, inevitably causing confusion with the larger and even better known semi-aquatic, *Zantedeschia aethiopica*, which goes under the name of arum or calla lily.

Zantedeschia

There are about half a dozen species of *Zantedeschia*, all native to southern Africa and all popular in cultivation since the earliest days of European settlement. The genus was formerly named *Calla* (by Linnaeus) and *Richardia*.

One of the most striking zantedeschias is *Z. elliottiana*, which has golden yellow spathes and sagittate leaves marked with translucent spots. The origin of this species is a mystery as it is unknown in the wild. It has been suggested that it might be a hybrid between *Z. pentlandii* from eastern Transvaal and *Z. rehmannii*,[48] which has a wider distribution in Transvaal, northern Natal and Swaziland. The former has unspotted yellowish-green leaves and is also yellow-spathed but with a dark purple blotch at the base. *Z. rehmannii*, on the other hand, has pink spathes (sometimes white or even dark maroon) and is less often found in marshes, preferring more shady conditions among rocks or at the edge of woodland. It is also distinctive in having lanceolate leaves.

The hardiest and best known is of course *Z. aethiopica* – one of the most widely cultivated of all aroids and of considerable economic importance as a cut-flower. Native to South Africa (from western Cape Province to northeast Transvaal) and Lesotho, it grows from tuberous rhizomes and reaches 60 cm or more, with lush foliage which remains evergreen in genial climates (unlike the other species which invariably have a dormant period).[49] Its handsome bright green sagittate leaves and exquisitely sculptured inflorescences with recurved pure white spathes are frequently depicted in art and design, and grace gardens and marshy areas in the wild in many different countries. In contrast, one of its common names is pig lily, which refers to the fact that in some places it is used as pig food.

Although *Z. aethiopica* is now naturalized worldwide in subtropical regions, in its native southern African wetlands it keeps company with a small long-legged climbing frog, *Hyperolius horstockii*, which can change its colour according to the background. Down in the mud or water it is dark brown, but the males scale arum lilies to serenade prospective mates, turning cream in order minimize heat loss and to remain inconspicuous against the pale spathes.[50] Perhaps they also take advantage of their camouflage to snap up insect visitors attracted by the faint scent and profuse stringy pollen.

Anubias

The genus *Anubias* is also African. There are about 10 species, characteristically with creeping rhizomes, dense lanceolate to hastate foliage and a persistent spathe which is shorter than the spadix. They are found beside or in streams and rivers in the coastal forests of tropical West Africa.[51] Though seldom seen in cultivation they are neat plants with attractive leaves and inflorescences. *A. hastifolia* from Cameroon has a pinkish-brown spathe and shell pink male flowers. The commonest and most variable is *A. barteri*. The variety *nana* has oval leaves and does well as a submerged aquarium plant but becomes larger (to 20 cm or more) and flowers if grown emersed. The largest is *A. giganteus* with bold hastate, often almost trilobed, leaves and pink-flushed inflorescences with elegant pale green spathes and a fine mosaic of scalloped ivory male flowers.

Arrow arums

Eastern North America is home to the wetland genus *Peltandra*. The common name of arrow arum describes their sagittate or hastate leaves. There are three species, all with pointed hourglass inflorescences. In *P. virginica* the spathe is a delicate pale green, fading to almost white along the undulating margin and opening slightly to reveal the cylindrical white spadix.

OPPOSITE]
Arisarum proboscideum (Italy and Spain) A cross-section of the long-tailed inflorescence, showing the spongy white appendix which mimics the underside of a fungus to attract fungus gnat pollinators.

OVERLEAF]
ABOVE LEFT *Sauromatum venosum* (northern Asia to West Africa) The voodoo lily or monarch of the east: a tuberous species which produces its foul-smelling ground-level inflorescence at the end of winter dormancy, before the roots and graceful pedate leaf develop. Cream-coloured male flowers are visible at the base of the sterile appendix which reaches 30 cm or more in length.
ABOVE RIGHT *Helicodiceros muscivorus* (Corsica, Sardinia and the Balearics) The hairy arum: mimicking the tail end of a dead animal in more ways than one.
BELOW *Helicodiceros muscivorus* (Corsica, Sardinia and the Balearics) The elegant profile of the hairy or dead horse arum, revealing the spathe blade at 90° to the tube and a hairless, boldly patterned outer surface.

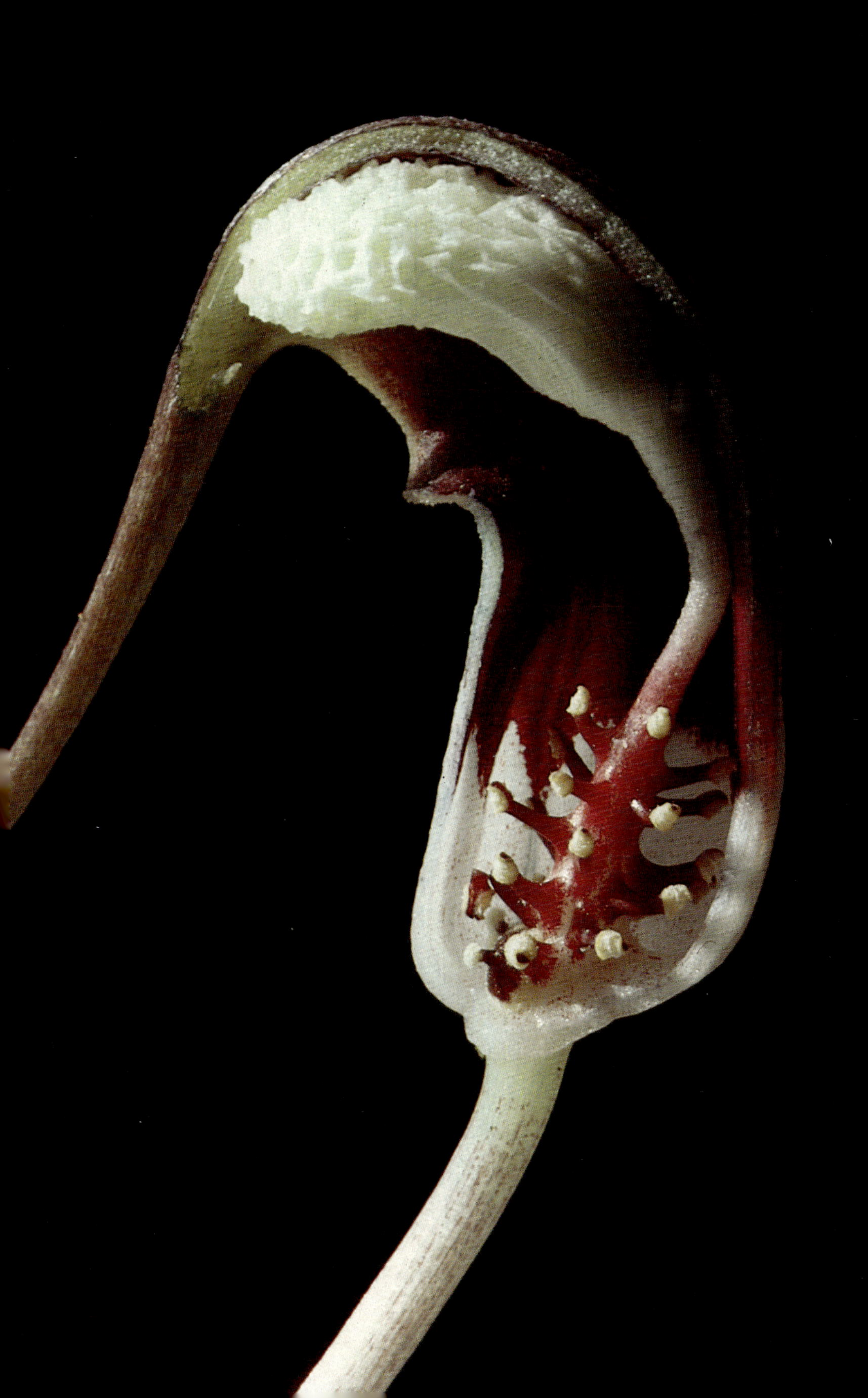

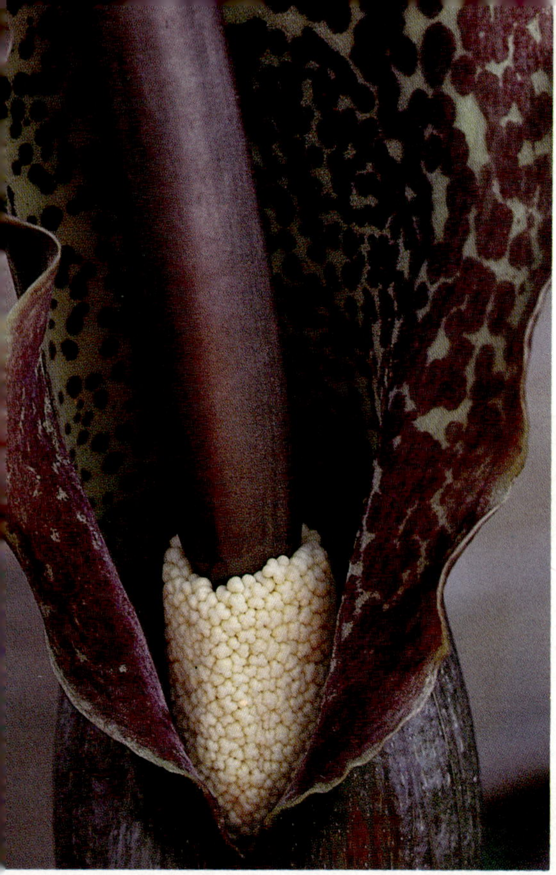
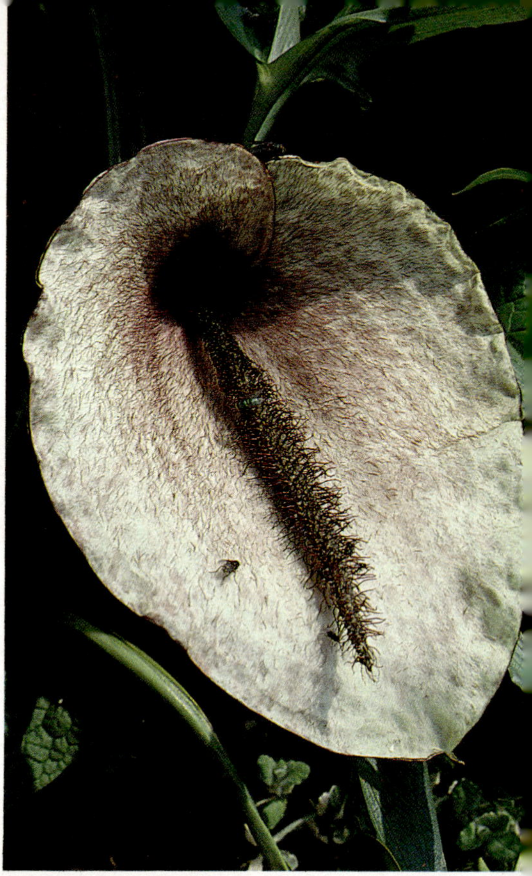
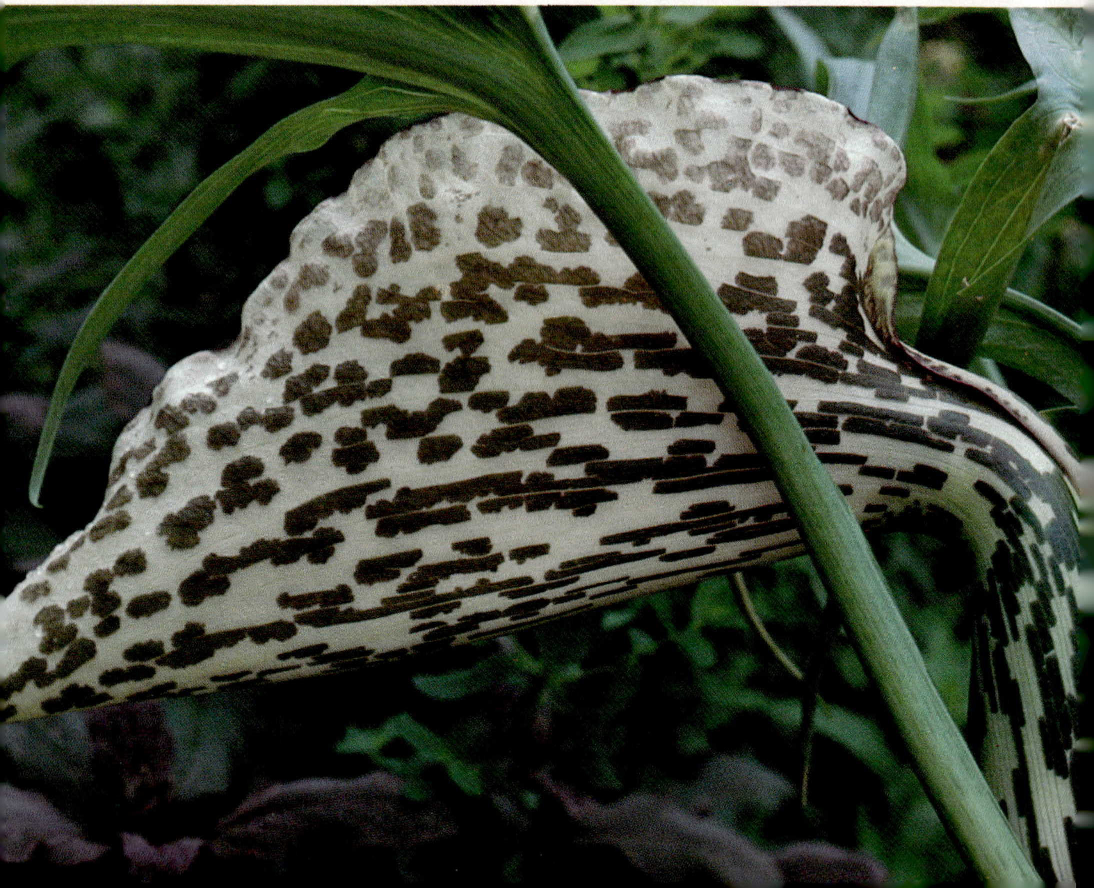

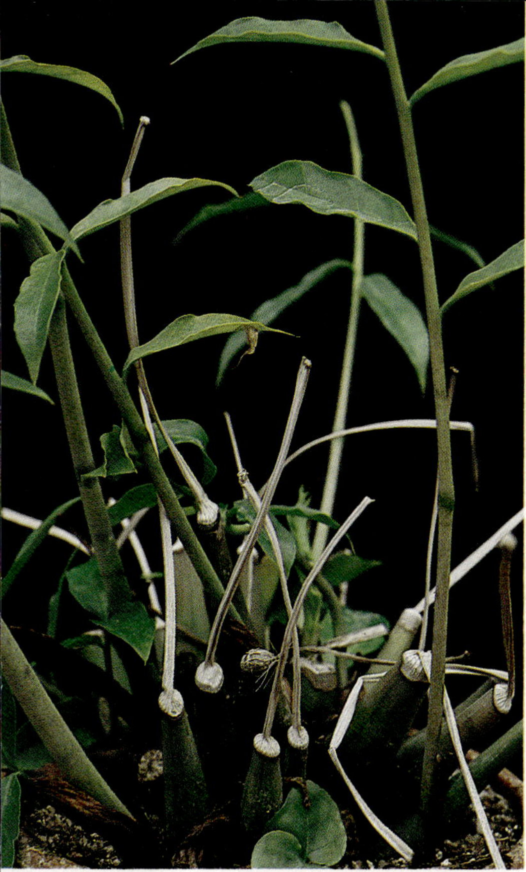
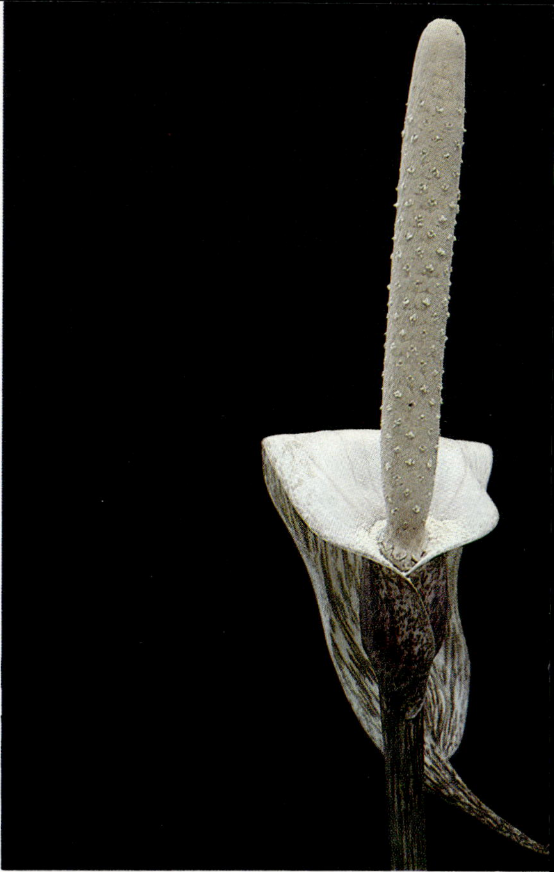
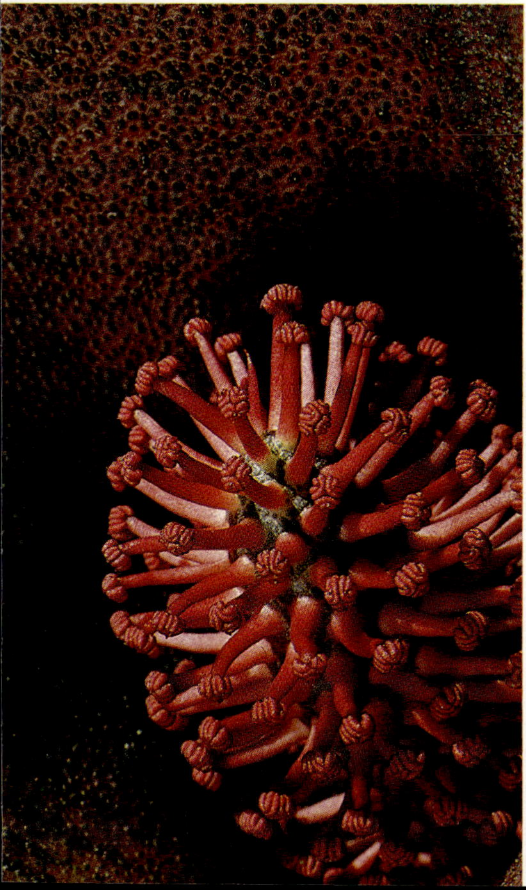
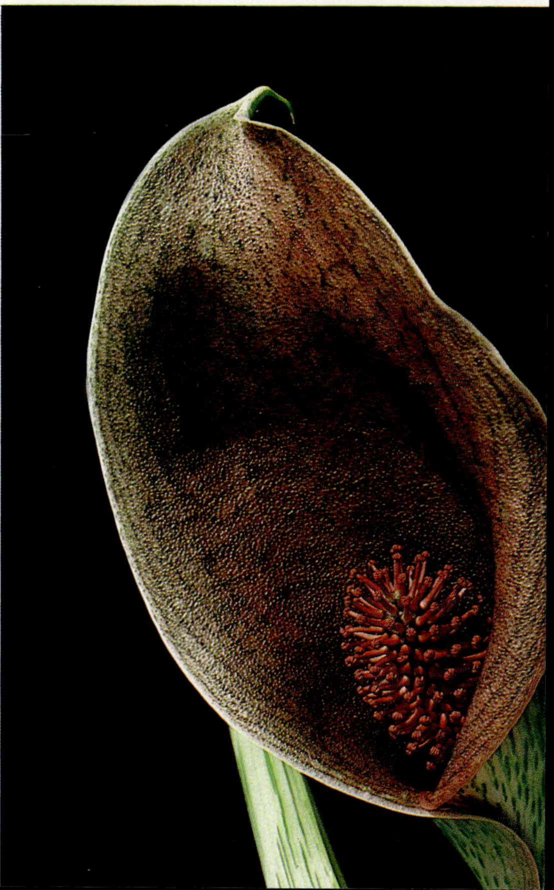

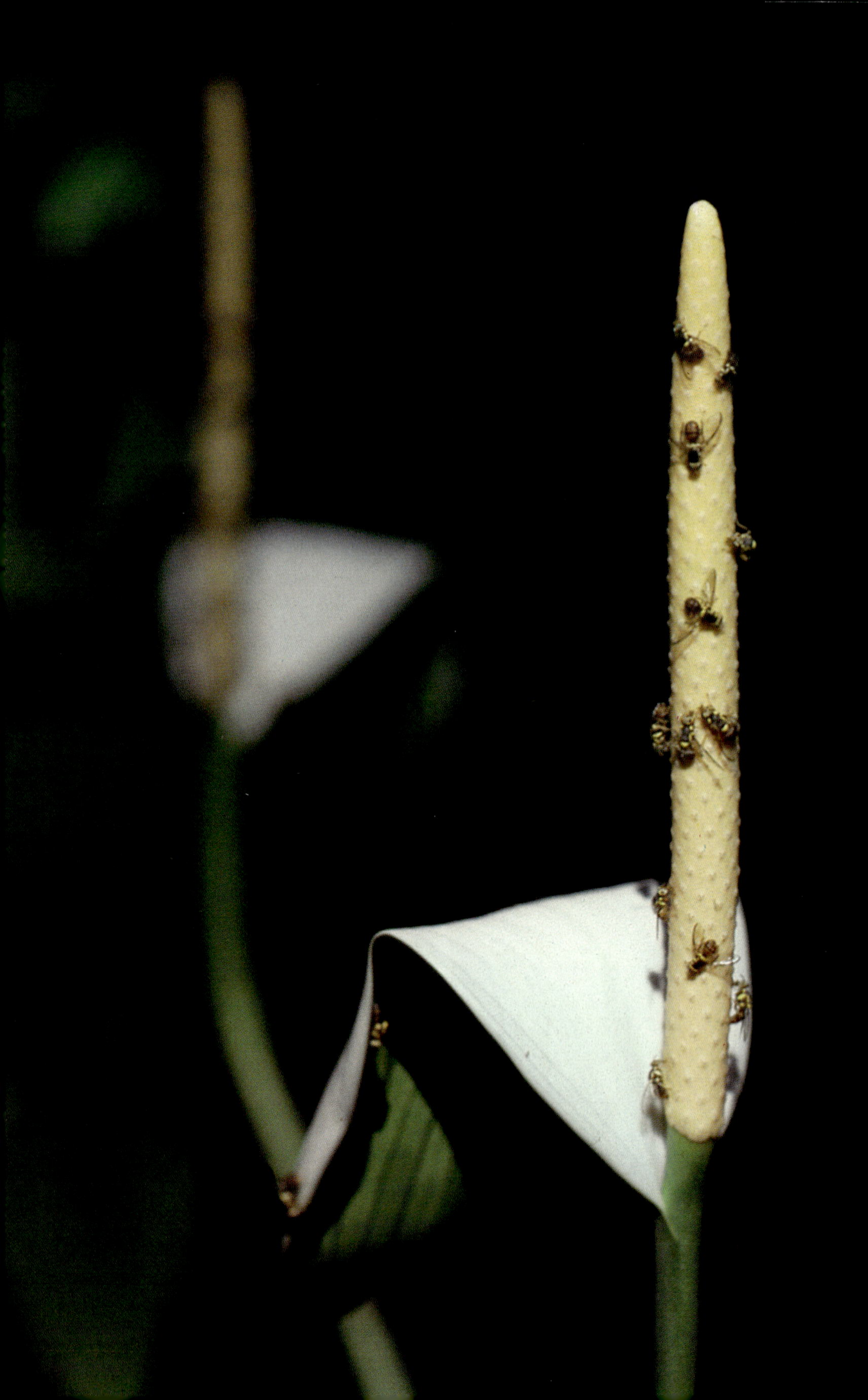

The giant arrow arum of Madagascar

On the other side of the world grows an aroid which in many ways resembles American *Peltandra*. *Typhonodorum lindleyanum* is a semi-aquatic species native to Madagascar and the surrounding islands. It usually occurs in large stands in swamps and spreads by viviparous seeds which are about 3 cm long. Apart from its size which, at 4 m or so, greatly exceeds that of the knee-high peltandras, it has the same arrow-shaped leaves and constricted spathe and similar flowers, pollen and seeds. Only the sterile stamens (staminodes) differ to any degree; those within the female zone in *Peltandra* are fused together, whereas in *Typhonodorum* they are free.

Discovering the relationship between these far-flung genera demonstrates one of the most interesting sides to taxonomy. Without fossil evidence there can only be speculations as to how it came about, but one scenario is that they had the same ancestor in Africa before the continents drifted apart. When the land mass was still continuous (that is, before the Eocene epoch, some 50 million years ago), populations spread into what is now Asia and Europe and then across to North America. Those that remained on the mainland of Africa later became extinct during the ages of drought which afflicted the new continent, leaving survivors in North America and on islands east of Africa. These separate populations continued to evolve: those in North America becoming *Peltandra* and the Madagascan dynasty forming the monotypic *Typhonodorum*.[52]

OVERLEAF]
ABOVE LEFT *Zamioculcas zamiifolia* (Kenya to northeastern South Africa) A bizarre monotypic genus with pinnate leaves. In the dry season the leaflets fall and the stalks wither, leaving swollen bases to tide the plant over to the next rains.
ABOVE RIGHT *Gonatopus boivinii* (Central and southeastern Africa) The small tuberous African genus *Gonatopus* has unisexual flowers. In this species the male portion of the spadix has a smooth outer layer of cream tepals, four to each flower. Here the anthers are extruding pollen.
BELOW LEFT *Synandrospadix vermitoxicus* (Argentinian and Bolivian Andes) The male flowers are unusually long.
BELOW RIGHT *Synandrospadix vermitoxicus* (Argentinian and Bolivian Andes) The deeply hooded spathe has a smooth pale green outer surface marked with short dark green lines and a leathery inner surface studded with warts in the typical carrion flower colours, maroon and brownish green.

OPPOSITE]
Spathiphyllum cannifolium (northern South America and Trinidad) The fragrant waxy inflorescences of spathiphyllums are generally pollinated by bees but in cultivation (here in Oahu, Hawaii) attract large numbers of other insects.

The oddest aroid of all

However puzzling the relationship between *Peltandra* and *Typhonodorum*, it has caused nowhere near as many headaches to botanists as the genus *Acorus*. There is only one other aroid genus that resembles it in appearance and that is the odd monotypic *Gymnostachys* from Australia. Both have reed-like leaves which are quite untypical of aroids but the resemblance is superficial and a closer comparison reveals significant differences. In floral structure they are both primitive: *Gymnostachys* possibly because of its geographical isolation and *Acorus* on account of its predominantly vegetative reproduction, but again with few characters in common.

Various details of the flowers, pollen and fruit of *Acorus* have few or no parallels in the rest of *Araceae*. Indeed, it shares some features with another family altogether, having the same flat sword-shaped leaves, volatile oil and parasitic fungi as the bur-reeds (*Sparganiaceae*). As a result, some botanists feel that *Acorus* would be better off in its own family or, failing this extreme, tagged on to another family of monocots altogether.[53] The more conservative give it the benefit of the doubt and leave it as an aroid.

For more reasons than taxonomic, *Acorus* is a remarkable genus. It is generally accepted that there are two species: the smaller being the variable *A. gramineus* which is often sold as a house or garden plant in its variegated form. The leaves are grass-like and may reach 50 cm but are often much smaller. The dwarf non-flowering form, which is only about 10 cm tall, was named 'Pusillus' by Engler. *A. gramineus* is found wild in India, China, Hong Kong and northern Philippines, where it grows on wet rocks in streams and damp ravines. The variety *macrospadiceus* from southern China, which has also been called *A. tatarinowii* and *A. macrospadiceus*,[54] has a strong scent of anise.

A. calamus or sweet flag is a much larger plant altogether. It has a stout creeping rhizome from which tufts of flat parallel leaves arise at close intervals. A non-flowering plant can be distinguished from most reeds and reedmaces by the prominent midrib which can be felt as a ridge down the centre of each leaf. In addition, though the leaves are linear and quite stiff, the blade often has a slight puckering here and there.

Both leaves and rhizome are fragrant when bruised and contain an essential oil which has medicinal uses (these are described more fully in chapter 10). In the days before carpets when floors were covered with resilient leaves, sweet flag was also in great demand as a strewing herb, for it combined toughness with scent and insecticidal properties. Plants have therefore been cultivated and traded from earliest times, so it is difficult to be sure of its origin. The two most likely regions are central Asia and northern North America; quite possibly it occurred wild in both. It probably reached Europe from Mongolia and Siberia. Apparently by the 13th century it had spread to Poland and in 1557 was brought into Hungary from Constantinople (now Istanbul). It progressed to western Europe through the Viennese botanist Clusius who obtained plants from Asia Minor and distributed them to fellow botanists in France, Germany and Belgium. The English herbalist Gerard recorded receiving it from Lyons.[55]

For an aroid, sweet flag is decidedly plain in appearance (though the variegated form is popular in horticulture). Its leaves look like those of countless other waterside plants and the inflorescences which appear incognito in summer do nothing to enliven them. The spathe is virtually indistinguishable from a leaf and the cylindrical spadix, with its close-set greenish bisexual florets, juts out from the ridged flower stalk at about 45°. Flowering is hard to spot but not uncommon, though fruiting is virtually unknown in Europe as the population consists mainly of sterile triploids. Distribution is mainly by pieces of rhizome which are broken off during floods and erosion of river banks. This way it has colonized wetlands throughout the world and now has the widest distribution of any aroid, from just short of the Arctic Circle in Scandinavia to Celebes and New Guinea, south of the equator.

In spite of its unprepossessing appearance, *A. calamus* is historically one of the most widely revered of all aroids for its tangerine-like scent and therapeutic uses. It may not fill a present-day collector's heart with poetic sentiments, but Walt Whitman praised it at length in his poem 'Calamus', calling it 'scented herbage of my breast' and describing how he waded into a pond and gathered the 'pink-tinged roots' to give as tokens of friendship. And no other aroid has such inviting common names: sweet cane, sweet flag, sweet rush, sweet sedge, sweet root. . . .

Tree-like *Montrichardia*, floating *Pistia*, tidal-flowering *Cryptocoryne*, and ice-breaking *Symplocarpus* – to such extremes have aroids gone in pursuit of niches where the necessities of life are available and the competition reduced. They have stopped short only at the open sea and the tundra. Relatively few have ventured far from the tropics but, of those that have, the aquatics and amphibians have gone furthest north: *Lysichiton*, *Calla* and *Acorus* are advancing on the Artic Circle.

5
A Place in the Sun

Species of arid and seasonally dry regions

'Truly it may be said that aroids are not for the faint-hearted, and yet what rich rewards await the curious.' This was written by Simon Mayo[1] as he lamented the fact that his colleagues at Kew frequently regard flowering specimens on his desk with the greatest suspicion, recoiling as if they were something venomous and likely to attack. It must be admitted that their reluctance to approach is sometimes justified, as some of the more unsavoury-looking inflorescences may well assail their nostrils with odours to match.

—— Rotters and stinkers ——

Some of the aroids most likely to arouse such distaste are those belonging to the subtribe *Arinae*, a group of genera in the subfamily *Aroideae* which have in common the production of inflorescences which usually – though not always – are reptilian or carrion-like in appearance, smell quite disgusting and live in places with long dry seasons which they survive by 'resting' within their underground corms. But as Dr Mayo so rightly says, if you can steel yourself to observe these plants closely and avoid inhaling too deeply as you examine their superficially repellent inflorescences you may well discover things about them which will cause you to hold your breath for other reasons, for their floral details and lifecycles are some of the most fascinating in the family.

—— The dead horse arum ——

Take the dead horse arum (*Helicodiceros muscivorus*) for example. E. A. Bowles wrote that this is 'the most fiendish plant I know, the sort of thing Beelzebub might pluck to make a bouquet for his mother-in-law – a mingling of unwholesome greens, purples, and pallid pinks, the livery of putrescence, in fact . . . It only exhales its stench for a few hours after opening, and during that time it is better to stand far off, and look at it through the telescope'. His description is accurate, though it still falls short of conveying the sheer horror of the inflorescence. The smell may well resemble that of a dead horse but it is another of its common names, the hairy arum, which reveals the most unpleasant aspect of all: that the inside of the spathe is covered with coarse hairs and even the spadix is shaggy. If such an inflorescence were small, scarcely opening and discreetly concealed by the foliage, it would not be so bad, but the fact is, it is large (over 35 cm long and almost as broad), gapes widely and is held clear of the leaves at right angles to the stalk so that when you come upon it you are confronted with a full frontal view which is probably one of the most shocking sights in the plant kingdom.

It lives only on a few islands in the Mediterranean – Corsica, Sardinia, and the Balearics – typically in the shelter of rocks along the coast where seabirds breed. The details of this plant's lifestyle do little to enhance its image. It flowers at the height of the nesting season in spring, having produced leaves throughout the relatively cool damp winter. Seabird colonies are not noted for their hygiene and the ample waste matter and accompanying stench are an ideal breeding ground for carrion beetles and blowflies, whose larvae perform the useful service of consuming the rotting debris. And this flyblown site also provides the hairy arum with a ready supply of pollinators,[2] though whether flies or beetles is uncertain.

For pollination to take place the plant has to compete for the attentions of the insects against such attractions as rotten eggs, dead chicks and undigested fish, as they search frenziedly for places to lay their eggs. It does this not only by the putrid smell but also by resembling a corpse in great detail. The best place for the insects to deposit eggs is in or around a wound or an orifice which provides humidity, protection and easy access to decaying flesh for its offspring. As you watch them home in on the foul-smelling inflorescence and make their way against the direction of the hairs into the dark hole where the spadix emerges from the spathe chamber, the inescapable conclusion is that the plant has devised a resemblance to the hide of an animal, complete with tail and anus.

A fate worse than death

The insects are completely taken in by the inflorescence. To gain entry to the foetid chamber they have to push past a zone of upward-pointing filaments. They then fall through a ring of horizontal filaments which separates the unripe male flowers from the ripe female flowers at the base. Once inside, the slippery chamber walls prevent them from climbing out.

If they are carrying pollen from another inflorescence, it falls on to the female flowers as they struggle to escape. Many lay their eggs on the floor of the chamber during their imprisonment and some in desperation deposit them all round the base of the 'tail', though the maggots will starve to death when they hatch. It is not a pretty sight, though proof of the plant's success at deception. The insects are held two or three days and are fed by nectar from the female flowers, but inevitably in the confined and overcrowded space, a number die, adding to the macabre spectacle.

When the female flowers are no longer receptive, the male flowers start to release pollen. This triggers changes in the structure of the chamber which enable the inmates to clamber past the filaments, picking up pollen as they crawl over the male flowers to freedom or, more likely, to another irresistible inflorescence. The pollinators show no ability to learn from experience and the plant demonstrates complete mastery in manipulating their behaviour for its own ends. And it may be no coincidence that it not only mimics carrion but flowers in the spring when most animals are breeding and casualties and predation – and insect populations – are highest.

—— Dragon arums ——

Less grotesque is *Dracunculus*, a genus which some authorities consider close to if not indistinguishable from *Helicodiceros*.[3] The smaller of the two is *D. canariensis*, endemic to Madeira and the Canary Islands. This isolated species has a creamy white spathe and spadix, unspotted leafstalks and leaves with 7–11 leaflets.

Larger in all its parts is *D. vulgaris*, the dragon arum, an imposing plant frequently cultivated for its metre-tall clumps of mottled stalks and white-streaked divided leaves with 11–15 segments. The dramatic inflorescence greatly exceeds those of other aroids in the European flora, measuring 30–60 cm on average and occasionally reaching 1 m. It leans back slightly above the foliage and has a dark velvety crimson spathe and a stout, shiny, almost black spadix. The appendix gives off what has been described as a 'terrifying odour'[4] of rotting flesh which rallies all the carrion- and dung-loving insects in the neighbourhood. Two plants in cultivation in Seattle have been recorded as accumulating no less than 298 beetles of 15 different species.[5] Though a splendid plant for the garden – and certainly a conversation piece – it needs careful positioning and is not recommended for placing under windows or beside garden seats, as the stench it produces when flowering in early summer is sufficient to disrupt one's social life (even more so if you take the advice of the 1st century Greek physician Dioscorides who recommended that 'being drunk with wine, it stirs up the vehement desires to coniunction').[6]

The dragon arum is widely distributed in the Mediterranean region from Corsica to southern Turkey and some interesting variations have been recorded. A specimen from the Canakkale area of western Turkey was found with both spathe and spadix covered in dense short white hairs.[7] Many plants in Crete have pronounced white markings on the foliage (var. *creticus*), and on one visit to the island I came across a colony of green-flowered plants. They were magnificent specimens, standing some 1.5 m high in the undergrowth beneath carob trees. Some had pale green spathes with black spadices and in others the spadix was yellow. A few were intermediate, with spathes mottled green and maroon. Perhaps this is the 'white-flowered form' mentioned in *Flowers of Greece and the Aegean* by A. Huxley and W. Taylor.[8]

Climate and habitats

The typical habitats for *Helicodiceros*, *Dracunculus* and several other genera of the subtribe *Arinae* are thickets, scrub, grassy waste ground and rocky or stony places. These kinds of areas are particularly common around the Mediterranean, which is where many of the species occur. Most are easily found along roadsides and at the edges of footpaths and fields. I came across a deserted village in a remote part of Crete where groups of dragon arums stood with their backs to the ruins, as if, like triffids, they had taken over after overpowering the inhabitants. Rubbish dumps are another favourite spot: damp and foetid, with a rich supply of nutrients and insects, they provide luxury accommodation for the discriminating, if unrefined, tastes of these plants.

The Mediterranean is a mere fragment of the much larger Tethys sea which once extended from the western limits right across to India. After the continents collided, buckling the earth's crust into the great mountain ranges of Europe and Asia, the sea diminished and the climate became drier. This effectively cut off species which had reached Eurasia and northern Africa from their tropical relatives, and further migrations from tropical regions to the north were prevented by the encroaching deserts of Africa and the Middle East.[9] From then on they were on their own, struggling for survival in an increasingly hostile environment.

The changes in climate and habitat around the Mediterranean became more extreme with human impact as the almost complete deforestation of southern Europe took place. What we see today is land bordering on aridity, with hot dry summers and relatively warm winters when most of the rainfall occurs, and a degraded vegetation of tough, often spiny, shrubs and a resilient herbaceous layer. There must once have been a far larger population of aroids but species shrank into extinction as the conditions worsened. All that remains is some half dozen genera with perhaps no more than 30 or so species: the survivors who were able to exploit the few niches available to a family largely dependent on a plentiful supply of moisture and in doing so reached new evolutionary heights.

Adaptations

For aroids, keeping water loss to the minimum in hot dry regions means growing mainly under trees and bushes, at the bases of rocks and walls, or near springs, watercourses or the coast – all of which provide damp microclimates. On the whole they are unable to produce the reduced leaves and thickened cuticle of true xerophytes (plants structurally adapted to a limited water supply – such as succulents). They therefore hold on to their place in the sun by avoiding the worst of its desiccation, coping with drought by dormancy in starchy tubers and returning to growth with the onset of cooler wetter weather.

For the most part, they have comparatively lush foliage – an obvious target for browsing animals but one they get away with by possessing the ultimate deterrent to herbivores: chemical toxins together with needles of calcium oxalate which pierce the delicate mucous membranes. Goats and camels have hard palates for munching thorny vegetation yet they have no defence against this chemistry, which devastates the unprotected lining of their throats. Thus many of these aroids grow conspicuously green in a landscape long impoverished by overgrazing.

The genus *Arum*

When it comes to flowering, most dry land aroids rely on flies and beetles for pollination, attracting them by superb imitations of dung and putrescence. Many display liver colours and manufacture a range of powerful odours, heated up for greater effect, which they waft into the surrounding countryside from the tips of pointed spathes and spadices. Just how specific these mechanisms are is still being discovered. The Middle Eastern *Arum conophalloides* lures blood-sucking midges or gnats (*Ceratopo-*

gonidae) and the ferocious black flies (*Simuliidae*) which swarm round warm-blooded creatures in their millions. One spathe can collect as many as 600 of these tiny monsters.[10] *A. palaestinum*, the black arum or Solomon's lily, has two ecotypes: one growing on calcareous rocky ground in Judaea and Samaria which has inflorescences smelling of fermenting fruit and is pollinated by fruit flies (*Drosophila*); the other on alluvial soil and basalt around Mount Hermon in southern Syria, which has a foetid odour and attracts dung and carrion beetles and flies.[11]

The genus *Arum* demonstrates how even in an arid landscape there are suitably damp and shady spots. Niches such as cemeteries, areas around fountains and wells, olive and citrus groves, undergrowth and thickets of plants such as shrubby jasmine (*Jasminum fruticans*) and Christ's thorn (*Paliurus spina-christi*), and higher altitudes in mountains provide more moisture and shelter than in the open. *Arum hygrophilum*, a species found in Cyprus and countries at the eastern end of the Mediterranean, grows near watercourses and is found up in mountains where melting snow provides the necessary moisture and humidity. It has a green, purple-flushed spathe and dark purple spadix. The vile-smelling black-spotted spathes of *A. dioscoridis* appear at ground level in April and May in damp places on islands to the east of the Aegean and from Turkey and Iraq to Israel. I found it in profusion in dank undergrowth and rubbish heaps at the foot of the massive walls of the old city of Rhodes. Presumably the species was named after Dioscorides, who, in his influential work *De Materia Medica*, stated that the root, applied with bullock's dung, was an effective treatment for gout.

Coastal areas have a higher humidity than inland and several arums favour these: *A. cyrenaicum*, the only arum endemic to Africa, grows near the Mediterranean in Libya and *A. nigrum*, which attracts dung beetles and flies, is found in stony, shrubby places along the Adriatic coast of Yugoslavia. *A. pictum* is restricted to the Balearic islands in the western Mediterranean. It is quite unlike other arums in that it flowers as the new leaves appear in the autumn. In spite of the disagreeable smell it is an attractive plant with velvety maroon spathes, blue-black spadices and shiny dark green leaves with a network of paler veins. There is a fine clump of it at Cambridge Botanic Gardens, so it would appear to be reasonably hardy.

A. creticum is unique in the genus for its fragrant pale yellow inflorescences. Its large bright green leaves develop in the autumn and the recurved inflorescences are borne above the foliage in early spring. It has a very limited distribution, growing in the shelter of rocks on Crete,

Karpathos and Samos, with a small population in eastern Turkey too. Occasionally plants at higher altitudes in the mountains are found with pale cream spathes and purplish spadices.

Arums produce spikes of glossy orange-red fruits which are dispersed by birds. They are poisonous in excess to human beings, as are other parts, but locally the plants are used for a variety of purposes. In some areas of Turkey, for example, the ripe fruits are considered a good treatment for haemorrhoids: two or three are swallowed with a glass of water each morning for a week. They and the powdered tubers are taken as an antidote to the venomous bites of snakes and scorpions. The young leaves are gathered as a vegetable; you can see bunches of them for sale in markets around Ankara. Making them safe for consumption is done by boiling them in two changes of water into which a heated iron rod is plunged several times. They are then strained and served with the dried pulp of plum or apricot, which takes away the slight burning sensation that even thorough cooking fails to remove completely. In the Antalya district of western Turkey, the tubers are dried and fed to goats and sheep to increase the milk supply and are also used to treat eczema. The fresh tubers – ideally dug from the ground before sunrise in May – are sliced and applied to rheumatic joints. Perhaps the most novel use is found in the town of Mugla. Here the women tint their fingertips with henna and then tie on arum leaves to set the dye.[12]

The genus *Arum* has about 20 species and quite a wide distribution in Europe and eastern Asia. The majority of species grow in countries with hot dry summers but a few, such as *A. maculatum* and *A. italicum*, grow further north and are more woodland plants with winter dormancy (see chapter 4).

—— Biarum ——

With nearer a dozen species and restricted to the Mediterranean region is the genus *Biarum*, which is found from Portugal and Spain to Italy, Yugoslavia, Greece and its islands, into Turkey (which has the most) and the Middle East. They are mostly small plants with flattened spherical corms and are found on stony hillsides, rocky ground or at the edge of fields. Without exception they have undivided oval, oblong or strap-shaped leaves. One species, *B. syriacum*, has very narrow grass-like foliage, which inspired its former name: *B. gramineum*. The leaves are only 3–4 mm wide: a possible adaptation to the sun-baked regions of Syria where it grows.

Though fairly hardy, biarums need a hot dry rest in summer to induce flowering, rather like crocuses – with which they often grow. The leaves develop in late summer or early autumn, usually preceded by flowering. The inflorescence is borne at ground level or even with the base slightly below, so that the actual flowers may be underground. Some, such as *B. eximium*, are rather like arums in shape, but the most widespread species, *B. tenuifolium*, has slender spathes about 30 cm long. The outer surface is greenish, the inner a dark burgundy and they lie coiled and twisted on the ground with the even longer black spadix held stiffly in the air to dissipate the foul odour. A group of plants in flower can be smelt a distance of 20 m away[13] and probably attracts dung and carrion insects from much further afield. This species is very variable and var. *abbreviatum* (a southern form) is smaller, with a short, wide, hooded spathe.

One of the most attractive biarums is *B. davisii*. In September it produces tubby little cream spathes with mauve freckles and thin pointed brown spadices which droop from beneath narrow hoods. They sit upright on the ground, only about 7 cm high, emerging from similarly coloured bracts. In contrast, *B. pyrami* from Asia Minor and Iran is one of the largest and darkest. Its curled-back, almost black spathes can reach 30 cm in length, with lower margins joined to form an almost spherical chamber around the exquisitely detailed flowers.

Biarums fruit at ground level, bearing round clusters of white or pale green berries, sometimes marked with purple. Their means of dispersal raises some interesting points. One theory is that in an inhospitable environment dispersal may not necessarily be in the interests of survival, as the chances of the seed landing somewhere congenial are few and far between.[14] The fact that the parent plant has found a suitable niche for growth and reproduction means that its seedlings would probably thrive there too and so the seeds are earmarked for local burial rather than dispersal, probably by ants or small rodents. The light colour of the fruits camouflages them against rocks and stones, so that birds fail to notice them.

—— *Eminium* ——

There are about half a dozen species of *Eminium*, and they are generally found further south and east than biarums, growing in more barren regions than any other aroids. Their homelands are Turkey, Iran, Iraq, Israel, Egypt, Syria, Afghanistan and the southernmost regions of the Soviet Union (Tadzhikistan, Turkestan, Kazakhstan and Uzbekistan). Unlike biarums, they are rarely seen in collections and are one of the least known genera of hot dry habitats.

In appearance, eminiums are somewhere between *Biarum* and *Helicodiceros*, often with inflorescences and fruits resembling those of the former and foliage more like the latter. They usually grow in quite open situations against a pale background of stones, the caked surface of dry sand or clay or limestone, against which the typical purplish-black coloration of the inner spathe limb stands out clearly.

E. lehmannii is a common species of rocky slopes, sandy wasteland and heavily grazed grassland in Iran. The spathe is greenish outside and velvety black inside; the spadix is black. A spherical cluster of white berries develops below ground, pushing up above the surface when mature and flushing violet in the light. This species has simple lanceolate or barely sagittate leaves, whereas those of *E. intortum*, which is found on cultivated land and dry stony hillsides in Turkey and Iraq, are strongly divided. Again, the inflorescence is almost black and its foetid odour has been likened to the 'smell of rotting dogs'.[15]

—— The desert aroids ——

A hot dry rest in summer is rather an understatement for what some aroids are able to endure. The Negev is a semidesert region of southern Israel bordering on Egypt and the Gulf of Aqaba, which at one period of the year

Eminium spiculatum ssp spiculatum ($\times \frac{1}{2}$)
Some *Eminium* species tolerate arid conditions verging on desert. In this subspecies the blackish leathery spathe is often pressed to the ground with the tube embedded in the sand.

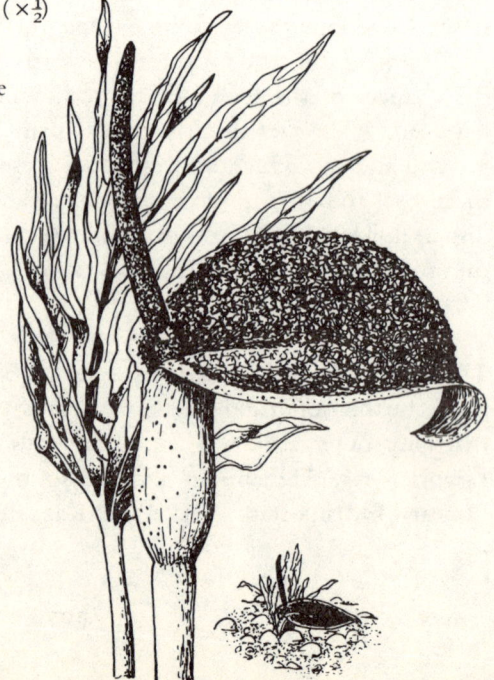

receives sufficient rain to support a covering of vegetation. The northwest is a low plain of scrub, gravel and sand dunes, much of the remainder consisting of higher rugged ground riven with wadis and sloping in the east to the Dead Sea. Thriving in regions such as these entails adapting to searing daytime temperatures, glaring light, little moisture or shelter and, in some areas, substrates with high mineral contents. In all, this is not the kind of habitat one would associate with aroids, but both *Biarum* and *Eminium* have representatives in these outposts.

B. olivieri has linear, often wavy leaves 2–10 mm wide and spathes up to 8 cm long which have drawn-out tips. It grows in the Negev and into Egypt, flowering in November and December and has been recorded with both foliage and inflorescences almost covered by sand.[16] It also appears to tolerate saline desert conditions.[17]

E. spiculatum has two subspecies. Both have complex leaves with lower lobes held erect which are produced before flowering takes place; broad black corrugated inner spathes and white fruits which are apparently not dispersed. The larger of the two is the subspecies *spiculatum*. It flowers from March to May, with the tube often embedded in the sand and the spathe pressed to the ground. At a distance it resembles dung. It is found in Iraq (where the tubers are sometimes dug up – often from great depths – boiled and eaten with sour milk),[18] and in Syria and Turkey. Smaller in all its parts is the subspecies *negevense* (the spathe, 5–11 cm long and 4–7 cm wide, is about half the size). It grows further south, in the Negev and the Sinai Peninsula, flowers earlier (in February and March) and has an erect spathe. Perhaps the most interesting difference, however, is that it flowers at night.[19]

—— *Ambrosina* ——

The smallest aroid genus in the Mediterranean region is *Ambrosina*. It has only one species, *A. bassii*, which is, in turn, the smallest aroid of the region. This curious plant has a widespread distribution: in Corsica, Sardinia and southernmost Italy (Sicily and Calabria); and on the African side of the Mediterranean in Algeria and Tunisia. In evolutionary terms it is generally regarded as the most advanced in the family. Its parts are much reduced: a few oval leaves 4–6 cm long which are rather like the first leaves of many aroid seedlings and an inflorescence scarcely 2 cm long on a crooked stalk, which has a spathe like a tiny brownish-green speckled egg ending in a recurved tail. Inside is but a single flask-shaped female flower and 8–10 males in a divided chamber. The seeds have appendages (elaiosomes) which

Ambrosina bassii
Left: (×1) A curious little Mediterranean species which in evolutionary terms may be the most advanced of all aroids.
Far left: (×2) The tiny brownish-green inflorescence measures only 2 cm or so. Inside it is divided into two chambers with a single female flower in one and 8–10 male flowers in the other.

contain nutritious substances that are eaten by ants, and are thus dispersed as the insects carry them away for food.[20]

A. bassii can be found on north-facing hillsides and in well-drained humus overlying limestone. It sometimes grows in association with *Arisarum vulgare* though each has its special niche, the latter favouring slightly damper, more nitrogen-rich spots. Often they nestle in crevices, *Ambrosina* above and *Arisarum vulgare* below.[21]

The friar's cowl

The advantage of small size in a barren landscape is demonstrated by *Arisarum vulgare*, popularly known as friar's cowl. It peeps out from rock crevices and huddles under ledges, which protect its fragile sagittate leaves and neat brownish-green and white striped inflorescences from drying sun and wind. The spathe is cylindrical and hooded with a downward pointing tip from which the brownish spadix protrudes. It is the farthest-flung of Mediterranean aroids, found all over the region and on the Canary Islands – and even as far out into the Atlantic as the Azores. As might be expected from this wide distribution, it is variable and some populations have spathes striped lime green or finely dotted pink. This species grows and flowers throughout the winter and by April the foliage is withering, ready for six or so months of dormancy.

The type of climate and habitat characteristic of the Mediterranean does of course occur elsewhere in the eastern hemisphere, in places as far apart as Australia and India. Members of the subtribe *Arinae* are found in these regions too, though for the most part they belong to genera which do not reach as far north as the Mediterranean.

——— The voodoo lily ———

By far the best known is *Sauromatum*, a genus with a range through Africa from highlands in the tropical west to those in Ethiopia, to the Yemen, the Himalayas and northern Burma. The commonest of the two species is *S. venosum* (formerly *S. guttatum*), which is frequently sold as a curiosity under the enticing names of monarch of the east and voodoo lily. The instructions on the packet containing the dormant corm are that it should be placed on a saucer where it will flower without soil and water. Indeed it will, but it is impossible to keep it upright on a saucer and it is altogether more impressive emerging from the earth, into which it can of course later send down new roots.

The inflorescence should only be approached by those of a robust

Sauromatum venosum ($\times \frac{1}{5}$)
The voodoo lily is often sold as a novelty to flower without soil or water. The strange appearance and evil smell of the inflorescence usually result in its being rapidly discarded, but those courageous enough to plant the tuber after flowering will be rewarded with an elegant pedate leaf on a finely patterned stalk.

constitution. One author wrote: 'A more extraordinary flower than this Asiatic arum cannot possibly be imagined. It is certainly the strangest of all bulbous plants . . . a real weed of Satan'.[22] I once grew three corms in a large plant pot. They were mature and flowering-sized, about 9 cm in diameter (though they can get almost twice this size). The buds began as hard grey horns and elongated at an almost visible rate (50 cm in nine days has been recorded)[23] into thin pokers with egg-timer shaped bases and short white lilac-spotted stalks. They were almost the same colour as dry soil, grey-brown but glossy. Then, on a sunny morning, they opened – the spathes splitting to reveal smooth purplish spadices like metal rods, ruler straight and even in width from top to bottom, with the cluster of cream male flowers exposed at the base. As the sun warmed, the spathes peeled back like strips of leopard skin, yellow with brownish-purple spots and a velvety sheen. With no green anywhere and nothing resembling what we usually expect of flowers, they looked quite demonic or like something from another planet; and I admit to having given several friends a very nasty shock.

These impressive inflorescences only last a day or less but in that time their output of malodorous compounds is almost without rival. It reaches its peak at midday and the spadix then is actually warm to the touch, having raised its own temperature by 5–10°C above that of the surrounding air to disperse the stench.[24] One spring, in order to appreciate this brief flowering fully, I kept a single inflorescence on my desk as I worked and I can vouch for the fact that the smell came in waves, soon inducing a feeling of nausea which rapidly curtailed my admiration.

No doubt many people, having witnessed the phenomenon on a saucer, are impelled (or repelled) into throwing the shrivelled corm away. If they planted it they would, for once, be in for a pleasant surprise. The leaves develop soon after flowering, the 7–11 leaflets neatly rolled and crossed over as they arise on shiny spotted stalks and unfolding gracefully when fully grown. A mature leaf reaches 60 cm or more and although only one is produced by each tuber, a group of several plants with their offsets makes an elegant display. Sometimes seed is set in knobs of dull purplish-black hard fruits, borne quite close to the ground.

Although *Sauromatum* has been recorded as growing in dry sunny situations – among rocks in full sun in the mountains of eastern Cameroon and in upland scrub and fields in Tanzania – it generally prefers shady places in humus-filled hollows at the base of boulders or under trees.

Stylochaeton

Stylochaeton is a wholly African genus. There are about 20 species and they grow on rocky hillsides, savannah, dry bush and woodland, in almost all countries south of the Sahara, from the Somali Republic and Ethiopia in the east to Sierra Leone in the west and South Africa at the southernmost tip. They can obviously survive arid conditions in regions where the annual wet season sometimes fails to materialize and which are subject to fires in the dry months. Characteristically they have hastate leaves, partly subterranean pale green spathes, white to grey subterranean fruits and rhizomes from which sprout thick fleshy roots. Species range in size from about 15 to 60 cm. Flowering is thought to be rare. One observer, who lived in Northern Rhodesia (now Zambia), reported that a plant of *S. puberulus* did not flower in six years. This could be explained by another report that it flowers underground.[25] One collector noted that the fruits are so completely buried that a pick axe was needed to unearth them.[26] Others (such as *S. crassispathus* and *S. lancifolius*) are known to flower when the rains come and they are still leafless after the dry season dormancy. The latter species is said to be pollinated by beetles.[27]

This genus has been little studied and yet has many interesting features, not least of which are the tuberous roots. (Elsewhere in the family it is the

Stylochaeton lancifolius ($\times \frac{1}{3}$)
The African genus *Stylochaeton* shows adaptations to harsh and sometimes prolonged dry seasons. A number of species have swollen fleshy roots and flower at the onset of the rainy season before the new leaves develop. In some the flowers – and later the fruits – are borne underground.

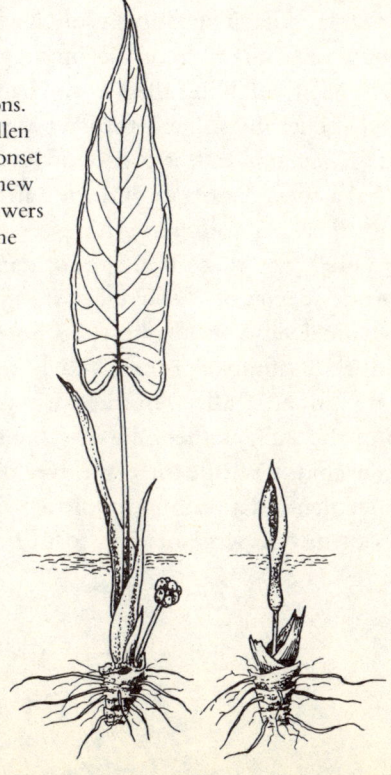

stem which becomes tuberous).) In *S. bogneri* and *S. natalensis* the roots are distinctly swollen and sausage-shaped and in a number of species – *S. euryphyllus*, for example – they are long, spreading and numerous. Several species, including *S. puberulus* of Malawi and Zambia, have been found growing on termite mounds,[28] but the significance of this is unknown. The latter is unusual in often having pubescent stalks, spathes and leaf undersides and, presumably, some interesting chemical constituents, being added to the local brew to produce powerful intoxication followed by severe depression.[29] Other are used medicinally. In the Upper Volta, the roots of *S. hypogaeus* are boiled, mashed and applied to boils,[30] and in Mozambique the roots of *S. natalensis* are crushed as a dressing for wounds.[31] It is also reported that gorillas and elephants eat the leaves of *S. zenkeri*.[32]

——— *Gonatopus* and *Zamioculcas* ———

Gonatopus and *Zamioculcas* are closely related bizarre African genera which together form a tribe of the subfamily *Pothoideae*. They are considered to be relicts from the ancient African flora that survived the extinctions caused by drastic changes of climate in the late Tertiary and Quaternary epochs.[33] Both are seasonally dormant and peculiar in many respects, not least that they are the only genera in the family with unisexual flowers which have tepals.

Gonatopus has five rhizomatous or tuberous species whose centre of diversity is Tanzania.[34] Each produces a solitary leaf, pinnately lobed, and a stalk with a central or basal pulvinus, a prominent swelling which performs the function of a joint, enabling the leaf to change position. In most cases the inflorescence (generally solitary too) appears just before the leaf and in *G. clavatus* and *G. marattioides* it is at ground level. The largest species is *G. boivinii* which has a wide distribution from Zaïre to Natal in South Africa. It has up to five tall cream-coloured inflorescences and a leaf which reaches 1.5 m and is divided into three branches of dark green elliptic leaflets.

There is just one species of *Zamioculcas*: *Z. zamiifolia*, which is found in eastern Africa from Kenya to northeastern South Africa. It grows in dry grassland and forest, often on rocks, and has several stout fleshy stalks bearing alternate pinnate leaflets. In the dry season the leaflets fall and the upper portion of the stalk withers, leaving the swollen bases – much like pseudobulbs in orchids – to tide the plant over to the next rains. As is quite common in succulent plants, but unknown elsewhere in *Araceae*, the leaflets can sprout into new plants and form tiny tubers at the base.

Theriophonum

Restricted to the Indian subcontinent and Sri Lanka is the small genus *Theriophonum* with under half a dozen species. They look like rather delicate arums, with more slender hastate leaves, tapered spadices, and spathes with a broader upper portion and smaller tube. They are in fact very closely related to *Typhonium* (see chapter 3), differing mainly in the arrangement of the ovules, though typhoniums are mostly much larger.

Theriophonums grow from small round corms in the shelter of scrub, trees and rocks. They are virtually never seen in collections. The most fragile of all is *T. minutum* with the most delicate narrowly hastate to trilobed foliage and diminutive 5 cm purplish-green inflorescences. It is the only species to extend into Sri Lanka. Two species have been named only recently: *T. sivaganganum* in 1968 (first published as *Pauella sivagangana* in 1966); and *T. fischeri* in 1981.[35] The latter is known only from a very small area at the southernmost tip of India.

High and dry in the new world

Across the Atlantic in the Americas there are also a number of aroids which have adapted to high light levels and drought. Some of these are not even tuberous and have devised other ways of combating a dry climate.

In Brazil, the state of Bahia has an arid interior which is subject to seven months or more of drought every year, with rain falling between November and April. The soil is thin and the vegetation a scrub in which orchids and bromeliads flourish side-by-side with cacti. It is not the kind of place one would think of finding evergreen aroids but this is exactly what happened when the 1977 Kew expedition visited the area.[36] *Anthurium affine* was quite common and one population had highly crinkled leaves with the two halves of the blade held close together: a modification which would reduce the amount of leaf surface exposed to the scorching sun. Growing nearby was the tree-like *Philodendron saxicolum* which resembles its close relative *P. bipinnatifidum* but has more rounded lobes and a bluish bloom to the leaf.

The most exciting discovery of all was a new species of *Philodendron* with an extraordinary way of eking out a living in drought-stricken country. *P. leal-costae*[37] is an erect or sprawling plant reaching about 1.5 m and it grows on granite outcrops in Bahia and other semi-arid areas of northeast Brazil. It has glossy compound leaves with around ten leaflets. As successive leaves age and drop off, the grey stem is left bare but for the yellowish-brown aerial roots produced at each node. These roots are the key to the plant's survival. They are up to 4 mm thick and sprout from the base of every leafstalk, growing quickly and with uncanny precision into nooks and crannies, especially the water-filled leaf bases of bromeliads even at some distance away. Some of the bright red berries with their pink seeds end up in these vases too – perhaps dropped by birds – as a number of young plants were seen actually growing out of the bromeliad reservoirs.

More conventional ways of surviving dry seasons are found in other South American genera. The technique of dormancy in underground tubers is practised by several small and rather strange genera. Two of them are little known: *Scaphispatha* and *Aphyllarum*, which both belong to the subtribe *Caladiinae*. *S. gracilis* is a rarity of savannahs and other open grassy areas in central and eastern Brazil. It flowers at the onset of rains before the arrow-shaped leaves are produced. First described in 1860 by Heinrich Schott from a single inflorescence, it was not seen by botanists again for over 100 years. In 1976 Josef Bogner collected flowering tubers and the next year the secret of their foliage was revealed when they came into growth in Munich Botanic Gardens.[38] A name meaning 'leafless arum' would have suited it in the circumstances, but instead was given to *Aphyllarum tuberosum*. This so-called 'leafless tuberous arum' lives in the Mato Grosso region of central Brazil. It does of course have leaves but, as is common with tuberous aroids, they emerge slightly after the inflorescence. Dormancy during the dry season on the grasslands where it grows has an added urgency as they are often subject to burning.

Far stranger in appearance and considerably larger is *Synandrospadix vermitoxicus*. The offputting name refers to its ability to kill pests and no doubt much else. It has white tubers weighing as much as 2 kg and grows on dry hillsides, along roadsides and in thorny scrub in the northeastern Andes of Argentina and Bolivia. Locally it is known as *'cano brabo'*, ferocious cane, which refers again to its highly poisonous properties.

The plant looks nothing like a cane though. Its three or four broadly ovate leaves undulate at the edges and are a matt medium green. The fleshy stalks, 30 cm or more long, are perhaps the most attractive feature: a delicate pale green marked with perfectly straight fine short lines in the same shade as the leaves. They are characteristically held at about 45° so that with only a few leaves the plant covers over 1 sq m. The inflorescences appear in spring along with the new leaves and are marked on the outside in the same way as the stalks. The inner surface of the deeply hooded spathe is leathery and covered with maroon and brownish-green warts and streaks. The spathe measures about 14 cm and overlaps slightly at the base, concealing the point at which it is fused to the female zone of the spadix. The visible part of the spadix is egg-shaped and covered with long spiky male florets in a rich crimson. The smell is unpleasant but not pervasive.

Closely related to the monotypic *Synandrospadix* is another strange and poorly known South American genus: *Taccarum*. It has five species, all tuberous and inhabiting seasonally dry areas. In *T. weddellianum* the spadix is most extraordinary, looking at a distance rather like an anaemic etiolated hyacinth and the spathe appears to be rather an afterthought, crumpled at the base of the spadix and barely covering the female florets. A spicy aroma issues from the 50 cm spadix,[39] but it is not known what pollinates it. After flowering a single large and complex leaf, delicately cut into numerous small leaflets, arises on a green and white streaked stalk some 1.2 m tall.

Far better known are the tuberous caladiums, whose hybrid descendants are now some of the most colourful foliage houseplants with every possible permutation of pink, red, green, and white variegation. Their origin is the species *C. bicolor*, a far more subdued plant but so variable that it has had no less than 40 different names assigned to it. Widely distributed from Panama

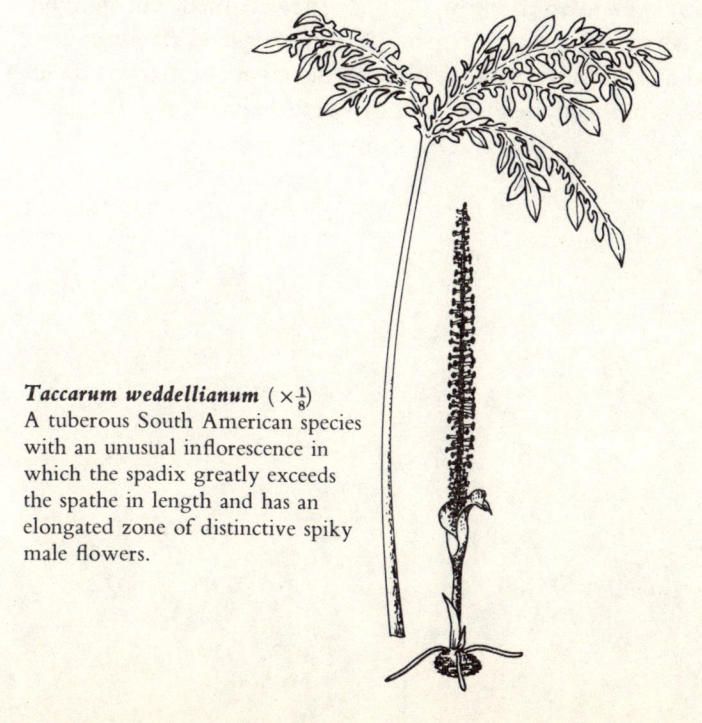

Taccarum weddellianum ($\times \frac{1}{8}$)
A tuberous South American species with an unusual inflorescence in which the spadix greatly exceeds the spathe in length and has an elongated zone of distinctive spiky male flowers.

to Bolivia and eastwards to the coast of Brazil, its peltate heart-shaped leaves may be sparsely or heavily spotted, blotched or streaked with one or more colours. Sometimes crossed with the far less common Guianan species, *C. schomburgkii*, which has smaller narrower non-peltate leaves with ruffled margins and similarly colourful variegation, it has produced countless variants and hybrids since its introduction to cultivation in the 18th century.

The tuberous caladiums grow in open grassy places, often on sandy soil, and die down for the dry season. The smallest species, *C. humboldtii* from Brazil and Venezuela, which has dazzling white-flecked emerald green leaves, has never been known to flower and propagates solely by tubers. *C. ternatum* is dormant for six or seven months of the year.[40] The habit is firmly established and even if plants are kept warm and damp, they still become dormant, so there is no way the beautiful angel's wings in cultivation can be kept growing all the year round.

Like the far more weird and wonderful genera already described, caladiums are prepared for the worst: the extreme stress of drought which has pushed the evolutionary capabilities of aroids to their limits. Indeed, it is no coincidence that most of the smallest genera of *Araceae* are predominantly either aquatics or near-xerophytes – the survivors of perhaps much larger groups that succumbed to evermore demanding habitats. In the case of those increasingly exposed to sun and dryness, devices such as prolonged dormancy, narrowed foliage, underground flowering and fruiting, and limited dispersal of seeds have ensured not only survival but marked them out as the family's pioneers in the fastest growing habitat on earth: arid land.

6
In the Shadows

Forest floor species of tropical rain forests

The sun is at its brightest at the equator but beneath its glare lies the darkest of all environments for plant life: the tropical rain forest. At most only 2 per cent of sunlight penetrates to the forest floor through the millions of green umbrellas held to intercept its rays in the upper storeys.[1] This was brought home to me on my first visit to the tropics to photograph plants when, with a 200ASA film in the camera, it was no problem to take pictures at one sixtieth of a second or faster in a village, whereas a few yards away in the forest exposures of a second or longer were needed. In this gloom, many of the forest's plants begin their lives and though the trees and climbers grow towards the bright and airy canopy, a large number of others are able to spend their entire lives in the dense, humid, breathless shade. It is this ability to tolerate such low light that has made some of the tropical aroids so successful and popular as houseplants.

—— Rain forest aroids and their habitat ——

Tropical rain forests form a broad belt around the equator, the largest areas being the Amazon basin in South America, the Congo basin in Central Africa and Southeast Asia. Days equal nights all through the year and constant high temperatures combine with regular heavy rainfall to give an atmosphere which, when inhaled, feels as if it contains more water than air. This is not just imagination, as precipitation usually does exceed evaporation and such humidity in a hot climate leads to the luxuriant growth we describe as jungle.

Rain forests vary in character from area to area according to the rainfall pattern, soil type and altitude. What they have in common is that the trees are very tall and their formation creates several distinct layers of vegetation. At ground level little can be seen of them but their trunks, to which climbers cling and around which a host of seedlings and saplings compete for light, together with a rich assortment of herbaceous plants which enjoy such shade and shelter. The ceiling of branches, lianas and foliage visible from ground level is only the lower layer of shorter trees with their burden of epiphytes. The trunks of the largest trees may not even branch for the first 20 m and the tops of these are for the most part out of sight, forming a canopy 30 m or more from the ground. Here and there the very tallest emerge from the canopy, fully exposed to sun, wind and storm, at a height of 40–50 m.

Every level of the rain forest, wherever it is, contains aroids. The American tropics have the most species and the greatest numbers of epiphytes, the Old World the most genera and richest ground flora – but whatever their distribution and habit, it is here in the rain forest that aroids are in their element and at their most numerous. They abound from the steamy lowlands to the coolest heights of the cloud forest, in dense shade on the forest floor and along the brighter forest margins and watercourses – on humus, clay, or sandy soil, on peat and limestone, on branches and thick moss, on the steepest slopes or in boggy hollows. This first chapter on rain forest aroids is concerned with those which remain at ground level: the terrestrials and lithophytes, those growing on soil and those on stones and rocks. Those which climb and perch off the ground (scandent and epiphytic species) take up chapter 7.

The diversity of species in the rain forest is legendary and, sadly, so is the rate of its destruction. The most vital and little-explored green belt on the planet is now half its former size and still it shrinks. Some of the species described in this chapter and the next may be frequently encountered in cultivation but this is no indication of how common they are in the wild or how well known their lifecycles. For the most part, the history and ecology of our tropical pot plants are 'abominable mysteries' (as Charles Darwin concluded about the origin of flowering plants as a whole). What is certain is that all rain forest species are relatively rare and are becoming scarcer as the forests are reduced or felled. Many are almost or already extinct. The aroids are no exception. While new species are being discovered every year, others are never found again, and some undoubtedly disappear without ever having been seen by scientists.

The beginning and end of *Anthurium amnicola*

A record by Robert White, Assistant Director of Las Cruces Tropical Botanical Garden in Costa Rica, typifies the situation. He describes his encounter with *Anthurium amnicola* (formerly *A. lilacinum*),[2] a small lithophyte with pretty scented mauve inflorescences, which grows on boulders in forest streams in only one area of Panama. His story goes like this:

> My first encounter with *Anthurium amnicola* in the wild in 1981 was a breathtaking event. I had been following the logging trail ever downward, enjoying the collecting but appalled at the destruction. The trail ended abruptly at a steep mud cliff. I could hear the stream gushing at the cliff's base but, try as I might, I could find no way to get down other than to sit on the edge of the cliff, push off, and slide the rest of the way. I somehow landed on my feet, but up to my knees in the icy stream. Somewhat shaken, I sat on the nearest boulder, my feet still in the water.
>
> As my head cleared and I looked to my surroundings, I realized with a start that I was sitting in the very midst of a splendid huge mat of *A. amnicola*. It was in heavy flower and the mat held numerous deep lavender blossoms. I was instantly aware of an intense minty fragrance and, leaning down, found that it came from these blossoms. Looking beyond, I saw that very nearly every boulder in the stream was covered in similar fashion . . . . a place of quiet, yet awe-inspiring beauty.

He returned the next year to find the stream bulldozed and only a few bedraggled specimens lying among the mud and overturned boulders.[3] The area was now logged. Only a few years before, this species was first discovered by Dr Robert Dressler of the Smithsonian Tropical Research Institute. Described as 'undoubtedly the finest flowering anthurium to be brought into cultivation in the last 100 years', it is today quite common in collections and has been used by Hawaiian anthurium breeders in hybridization programmes to produce compact, free-flowering plants in a wide range of colours.[4] You could say that this species was 'saved' in the nick of time – and indeed, without such intervention it might now be completely lost – but the transition from wilderness to windowsill, though fending off extinction, is a poor substitute for a wild population in its natural surroundings.

A. amnicola occupied a specialized niche and one which is unusual for the genus. Only a few other anthuriums are so far known to grow on rocks along streams. Like *A. amnicola* they have the pointed lanceolate leaves commonly found on plants with rheophytic tendencies. *A. rupicola* is sometimes found with *A. amnicola*. The recently discovered *A. sytsmae* is endemic to an area of mountains on the Caribbean slope of Panama where it is found beside fast-flowing streams and in the spray of waterfalls. Both have green spathes, the former with a white spadix and the latter with a greenish-brown. Neither have the delightful mint-and-roses fragrance of *A. amnicola*. The combination of brightly-coloured inflorescence and sweet scent in anthuriums is associated with pollination by bees.[5] (Another example is *A. armeniense*, an attractive Guatemalan species with a white spathe, pink spadix and lilac-scented white pollen.)

On the whole anthuriums enjoy humidity and the occasional soak but abhor wetness. An exception is the Brazilian *A. lindmanianum*, a robust, long-stalked species which grows in marshes. It may be seen beside pools in the botanic gardens in Munich and at Kew. Most anthuriums however prefer well-drained situations, often securing the best possible aeration by perching in trees. Some are found at ground level but usually only on steep rocky slopes or when the surface is so deep in leaf litter, fallen branches, rocks and moss, that it is almost as well-drained as an epiphytic position.

Judging from the growth habits of anthuriums which have been described by Dr Thomas Croat in his revision of the genus in Mexico and Central America,[6] a minority of anthuriums are terrestrial and most of these prefer higher elevations and steep gradients. He describes over 225 species in this region (out of a total approaching 800 in the genus as a whole), of which only 30 or so are consistently terrestrial. Then there is a significant group of species which are sometimes epiphytic and sometimes terrestrial (over 70). Generally speaking when they are found as epiphytes it is fairly close to the ground and in the conditions just mentioned. In contrast, over 100 are always epiphytic. These proportions may differ further south in tropical America, but they are unlikely to contradict the basic observation that anthuriums need extremely good drainage and are found on the forest floor only when this requirement is met.

—— Heart-shaped terrestrial anthuriums ——

Some of the finest anthuriums in cultivation are terrestrials. A number have heart-shaped leaves and pale veins; telling them apart can pose problems to amateur and specialist alike. One of the prettiest is *A. clarinervium* with dark green velvety blades, brilliant white veins and inflorescences standing well above the foliage. It has only been found on limestone outcrops in Chiapas, the southernmost state of Mexico. In cultivation it is frequently confused with *A. leuconeurum* (which is sometimes incorrectly called *A. cordatum*) and the two have even been regarded as synonymous.[7] Generally the latter is a larger plant with rather narrower leaves, less distinct veins and a wider gap between the lobes (in *A. clarinervium* the lobes meet or overlap).[8] There are grounds for suspecting that *A. leuconeurum* may in fact be a hybrid. Apparently it was collected from southern Mexico in 1860 but has not been found in the wild since. It has been noticed, however, that a cross between *A. clarinervium* and the broadly triangular *A. berriozabalense* produces offspring with a strong resemblance to *A. leuconeurum*.[9] Coincidentally, both *A. clarinervium* and *A. berriozabalense* are terrestrials from the same part of Mexico, so it could be that the original *A. leuconeurum* was a wild hybrid between the two.

Other look-alikes include the mainly terrestrial *A. magnificum* – a Colombian species which can be distinguished by its ribbed, four-sided stalks – and the epiphytic *A. crystallinum* which ranges from Panama to Peru and is probably the commonest silver-veined cordate *Anthurium* in cultivation. Its blades are not as flat as those of *A. clarinervium* and, again, it is larger-leaved and has a more open sinus (gap between the basal lobes). It hybridizes readily with the peltate pale-veined *A. forgetii*, an epiphytic Colombian species. The latter, in common with *A. crystallinum* and *A. magnificum*, has white purple-tinged ovoid berries, whereas *A. leuconeurum* and *A. clarinervium* have bright orange round berries. All have more or less velvety upper surfaces.

Three other terrestrials with heart-shaped – though not silver-veined – leaves are *A. watermaliense*, *A. papillilaminum*, and *A. corrugatum*. The first is a variable species with rather glossy sharply pointed leaves, almost triangular in outline with a very wide gap between the basal lobes. It has a green or dark purple spathe and the spadix varies from white or yellow to green or purple. The berries are orange when ripe but before this are blackish, then green. Found in Costa Rica, Panama and Colombia, it is named after Watermall, a town in Belgium where it was first sent from Colombia by an unknown collector.[10]

A. papillilaminum is a strikingly handsome Panamanian species, closely related to *A. crystallinum* but without silver veins. Its dark green, very velvety leaves are narrowly ovate and the shape of the lobes is reminiscent of a butterfly's wings. Though often seen in collections, it appears to have a limited distribution in the wild and has seldom been collected. Quite different in appearance is *A. corrugatum*, a terrestrial from Ecuador. Its shiny mid-green leaves have a network of deeply sunken veins which gives a finely puckered surface to the blade. Successful cultivation of this species is dependent upon cool nights.

Cloud forest terrestrial anthuriums

The kinds of conditions most favoured by terrestrial anthuriums are commonly found in cloud forests which extend from about 750 m to over 3000 m. Here the sun penetrates the mists for only a few hours in the late morning. Some highly scented bee-pollinated species of these zones show precise adaptations to this type of climate, only producing their perfume between about 11 a.m. and 2 p.m. when the mists clear, the air warms and insects are flying. This has been observed in *A. fragrantissimum* (which is more usually epiphytic) whose flowers develop over several weeks, emitting fragrance daily both in the male and female phases.[11] The quality and timing of scent production probably attract specific pollinators and may even have played a part in the evolution of the hordes of similar-looking but different-smelling anthuriums. Scents in the genus vary from faint to intense and from delightful to unappealing, the latter most often produced by drab-coloured or purplish spadices which may be fly-pollinated.[12]

In the cloud forests of Costa Rica is found *A. concinnatum* which can be epiphytic or terrestrial. An attractive species with narrow glossy heart-shaped leaves which are often undulate at the margins, it grows at higher elevations than any other Central American anthurium – generally at 2000–3000 m. At these altitudes the genial daytime temperature of 20–25°C plunges 30° at night to below 5°C.[13] Unperturbed by such fluctuations, this species is definitely at the cool-growing end of the tropical spectrum and keeps company with orchids such as masdevallias and odontoglossums which are notorious for failing to thrive if coddled in warmer conditions.

Palmate terrestrial anthuriums

Two outstanding terrestrial anthuriums are the palmately lobed *A. pedatoradiatum* and *A. podophyllum* which are reasonably common in cultivation. Both belong to the section *Schizoplacium*, a group of the only palmately-lobed species in the genus, which to date number seven.[14] The former reaches over 1 m tall and has leaves with up to 13 segments which are usually joined at their bases. In the subspecies *helleborifolium* several segments (usually central ones) are cut almost to the base. This form occurs in one particular locality of southern Mexico but both it and the typical subspecies grow on limestone slopes. The berries are orange and distinctly pointed. *A. podophyllum* has much more finely divided blades. Initially they

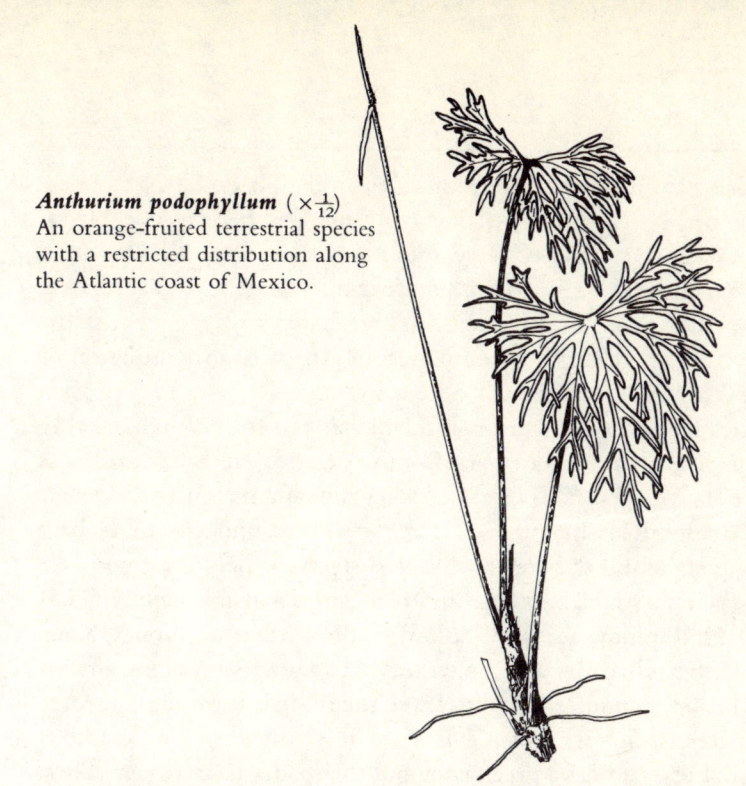

Anthurium podophyllum ($\times \frac{1}{12}$)
An orange-fruited terrestrial species with a restricted distribution along the Atlantic coast of Mexico.

are split into 5–10 lobes and then further subdivided and wavy at the margins. It is not at all common in the wild, being found in only a few places on the Atlantic coast of Mexico.

—— A ground cover *Anthurium* ——

On the whole anthuriums occur singly or in small groups and are not often seen *en masse*. An exception is *A. formosum*, one of the commonest species found from Costa Rica southwards into Colombia (and possibly Ecuador) at middle elevations (500–1500 m). It often forms large stands on shady banks at the sides of roads, with long-stalked heart-shaped blades up to 80 cm long, below which are borne inflorescences which at a glance look similar to those of a spathiphyllum, with a large erect pinkish-white to green spathe.[15] Rather similar, and also with a colonial habit, is the Venezuelan *A. nymphaeifolium*.

—— *Spathiphyllum* ——

The genus *Spathiphyllum* enjoys damp, even wet conditions deep in the forest and along watercourses. Most of the 40 or so species occur in South America but unlike other neotropical genera it has never got further into the Antilles than Trinidad (where *S. cannifolium* can be found). There are none in Africa and just three species in Melanesia. This odd distribution may be explained by the theory that at one time South America, Africa, India, Melanesia and Australasia were joined together in what is known as Gondwanaland and the genus could therefore once have spread across the whole continental mass. Successive climatic changes to the east as the continents drifted apart led to its extinction in Africa and the isolation of just a few species in Melanesia.[16]

Two species from the different continents happen to look remarkably similar. *S. cannifolium* from northern South America (and Trinidad) has broadly lanceolate leaves and fragrant flowers. The spathe is waxy, slender, and curved back with a white upper surface and green underneath. Its base clasps the long stalk and the cream-coloured spadix is held erect and free from the spathe on a stipe about 10 mm in length. *S. commutatum*, which is found in the Philippines, northern Sulawesi, the Moluccas, New Guinea and New Britain has broader and longer leaves reaching 50 cm long and up to 25 cm wide and its numerous veins leave the midrib at an angle greater than 45° (whereas those of *S. cannifolium* are at 45° or less). The spathe is slightly broader too and has a green edge but the spadix is narrower. They are very easy to confuse in cultivation.

The best known of all spathiphyllums is *S. wallisii* from Colombia. It has elliptic leaves, thin in texture and slightly wavy at the margins with 8–10 pairs of impressed veins. The long-lasting fragrant inflorescences are pure white with a green midrib on the underside. The spadix is waxy and unusually spiky because of the length of the ovaries. Commonly known as peace lilies or white sails, the plants sold under this species name are invariably cultivars and hybrids, of which the most widely grown is 'Mauna Loa'. It has broader leaves and larger inflorescences that in good conditions are produced at intervals throughout the year.

A less showy, but very attractive species occasionally available commercially is *S. floribundum*, again from Colombia. It is a smaller plant all round than *S. wallisii* and has velvety leaves and numerous inflorescences with white reflexed spathes. The flowers on the spadix are only slightly contoured and are a distinctive green and white. It grows in dense wet forest, usually along the banks of streams.

The genus *Spathiphyllum* has several interesting aspects. In its growth habit it is primitive, with a rhizome or extremely short erect stem from which sprouts dense clumps of long-stalked leaves which have a joint (geniculum or pulvinus) where the stalk joins the blade. The flat, leaf-like spathe is white in most species but invariably ages to green and in some is green and leaf-like when newly opened. The bisexual florets are equally primitive but the pollen is, in contrast, advanced and unique in the family. It caters mainly for *Euglossa* bees which are attracted by the sweet scents associated with the genus. These sophisticated insects cannot digest starch, so the pollen, unlike most in the family, is starchless.[17]

Holochlamys

Closely related to *Spathiphyllum* and similar in appearance and habit is *Holochlamys*. There is only one species, *H. guineensis*, and it is endemic to New Guinea. *Holochlamys* differs from *Spathiphyllum* mainly in having spathes which clasp the stalk and wither after flowering.

Look-alikes on different continents

It sometimes happens that although a particular genus is confined to one continent, there is another genus filling a similar niche on another continent which is remarkably alike in habit and appearance. Among tuberous aroids the obvious example is *Amorphophallus* of Asia and Africa and *Dracontium* of South America. Equally striking are the similarities between the two well-known forest floor genera *Aglaonema* and *Dieffenbachia* which, apart from the fact that the one is Asian and the other neotropical, can only really be differentiated by minute floral details. They are even about the same size in terms of number of species: *Aglaonema* having 20 or so and *Dieffenbachia* with 25–30. Both have some species with plain green leaves but their popularity as ornamentals is based on those with variegation which in the wild may be extremely variable from one group of plants to the next, and also within a population.

Aglaonema

Having said that, one of the enduringly popular aglaonemas is in fact the Chinese evergreen, *A. modestum*, which has plain green foliage and orange berries. A native of southern China and Southeast Asia, this species has long been cultivated in the east in the belief that it brings good luck. It was probably introduced into cultivation (as *A. acutispathum*) in the western world by a Mr Knaggs who came across it in Hong Kong and presented a specimen to Kew in 1885.[18] A few variegated forms are known and anyone growing them is certainly in luck as they are quite different from other aglaonemas and not at all common. 'Variegatum' is a sport with large patches of creamy white here and there; and 'Shingii', a mutation with undulating blades which have an irregular dark green border and a pale green centre. The latter has a complicated history. After its appearance among plants grown by T. O. Mahaffey, it was sold as 'Jeanne', 'Mandalay Aqua' and var. *medio-pictum* before being patented as 'Shingii' in 1964 by the company Ralston-Purina.[19] Rather similar, but in lime and bottle green, is a variety known as 'Green Goddess'.

The variegated aglaonemas which most resemble dieffenbachias are *A. commutatum* and *A. crispum* (formerly *A. roebelinii*) from the Philippines though *A. crispum* is restricted to the south of Luzon. Both are large plants reaching 1.5 m or more and have many different patterns of variegation, generally with a metallic appearance which is not characteristic of dieffenbachias. *A. crispum* has broader leaves and shorter stalks than *A. commutatum*, and sheaths which are ragged rather than smooth at the edges. The variegation consists of silvery bands down each side of the blade, leaving the centre and edges free.

A. commutatum is highly variable but on the whole its variegation follows the main lateral veins, of which there are usually four pairs (in contrast to an average of seven pairs in *A. crispum*). This species is one of the taxonomic nightmares of the family. Plants set seed without fertilization and are polyploids capable of almost infinite variety as regards variegation and considerable flexibility in leaf shape and texture – from oblong to lanceolate and from thin to leathery, with some quite pointed and twisted and others fairly blunt and flat.

Schott first described the species in 1856 and other botanists have subsequently given species status to variants which are no longer recognized. An example is Engler's *A. elegans* which he described in 1915. It

OPPOSITE]
Homalomena lindenii (New Guinea) Many homalomenas are aromatic. This species smells of aniseed when bruised; some give out a spicy fragrance at night when flowering.

OVERLEAF]
ABOVE LEFT *Taccarum weddellianum* (Brazil, Bolivia and Paraguay) The most complex leaves in the family are produced by tuberous genera and are often solitary.
ABOVE RIGHT *Philodendron pedatum* (southern Venezuela to southeast Brazil) Different coloured undersurfaces are a distinguishing feature in many aroids.
BELOW *Anchomanes difformis* (Central and West Africa) The massive divided leaf of this forest species reaches 1.5 m across on a prickly stalk 3 m tall. The wedge-shaped outer leaflets and reticulate, net-like venation are characteristic of the genus.

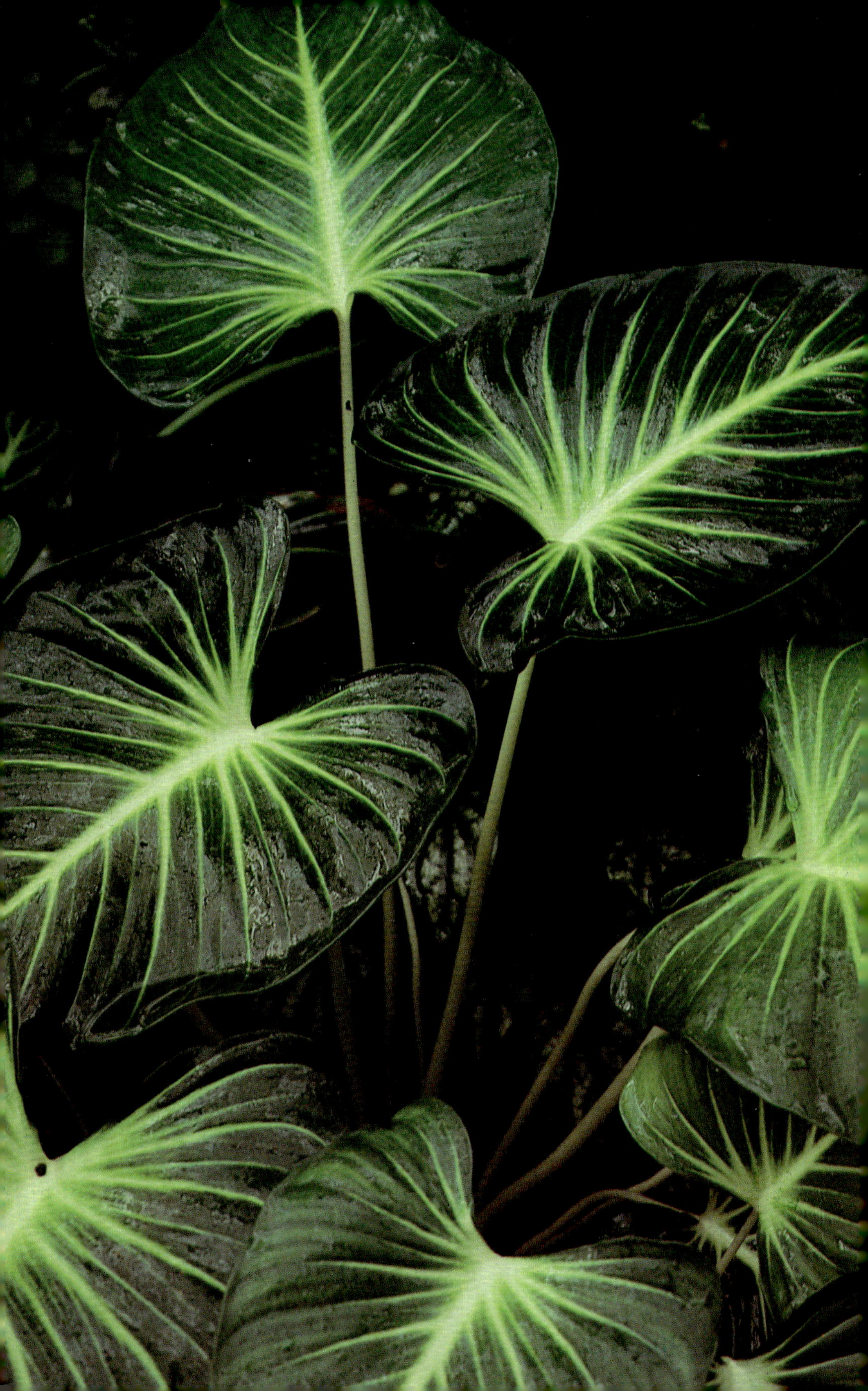

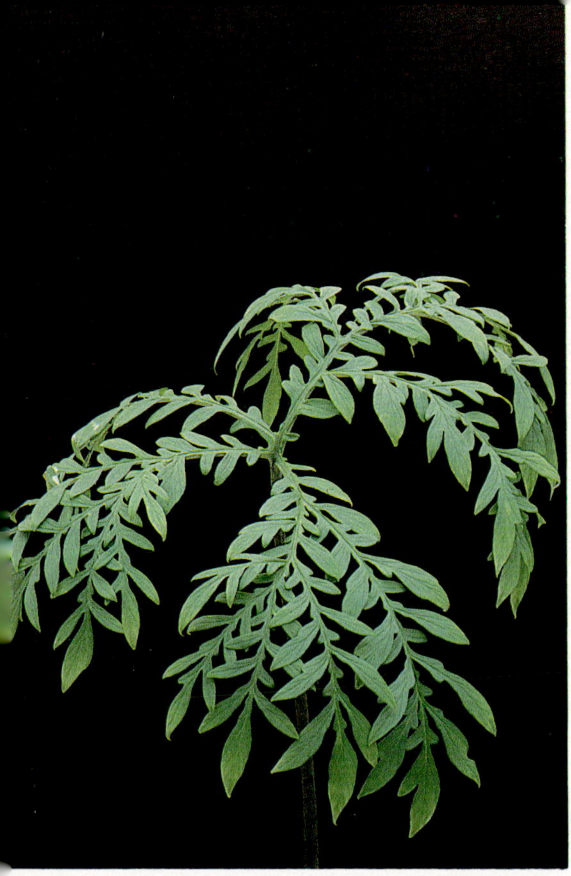

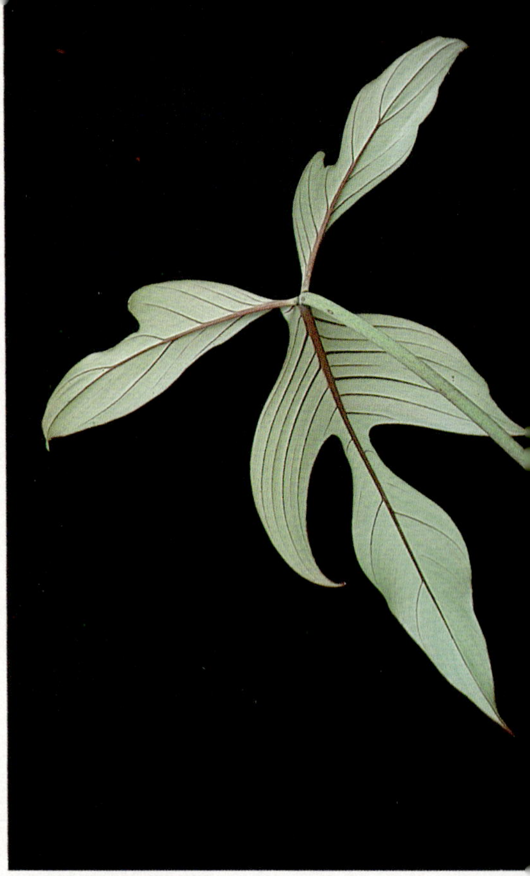

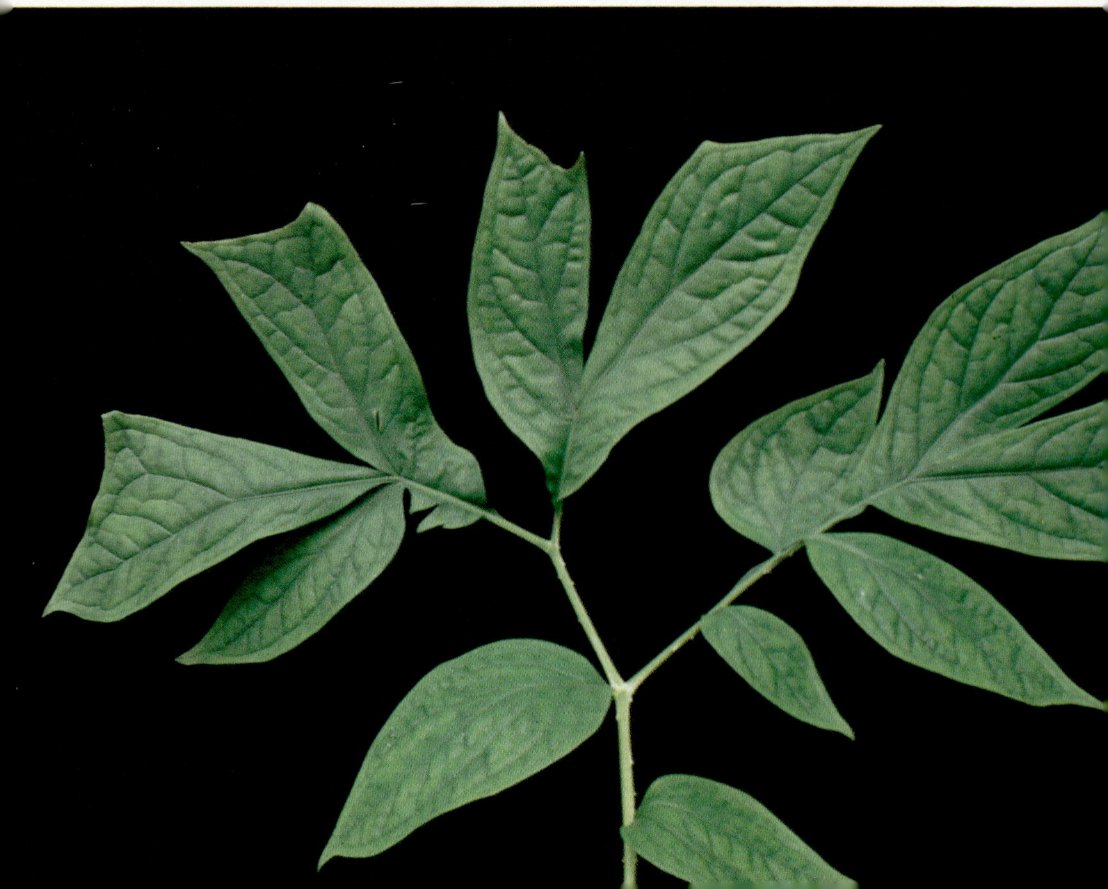

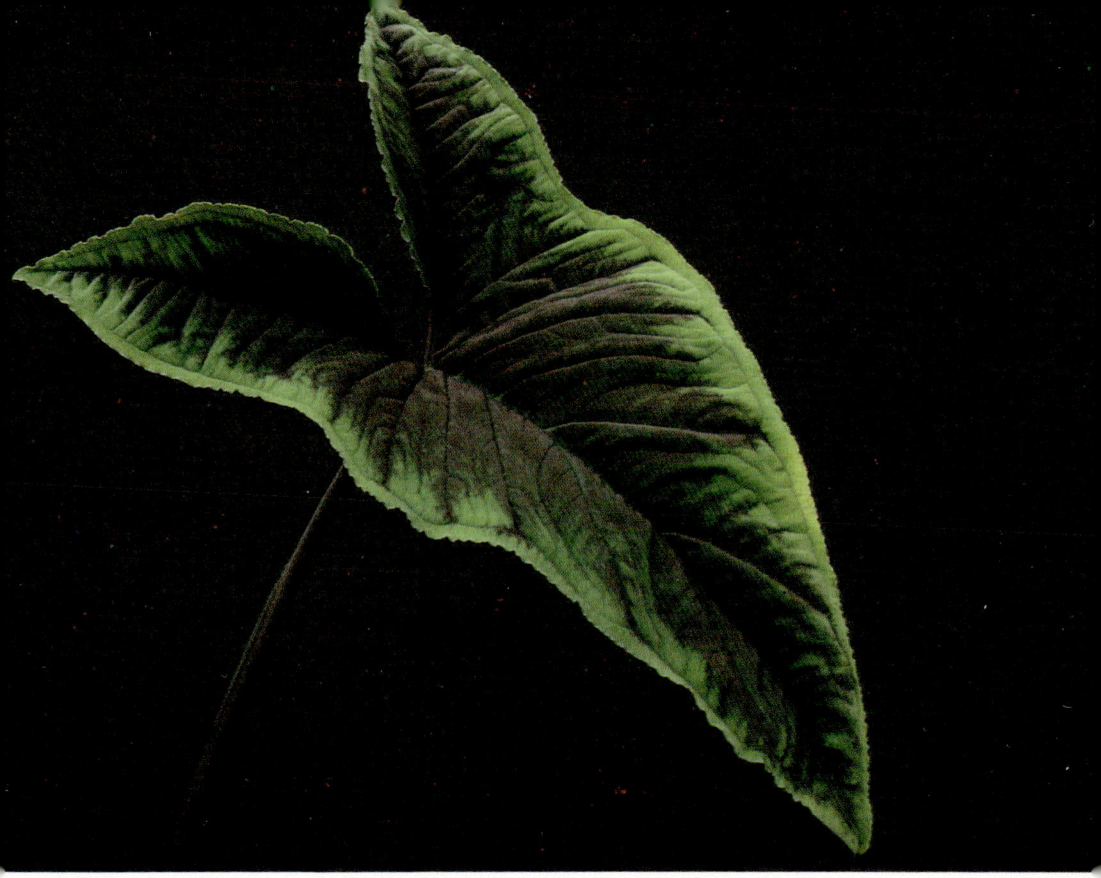

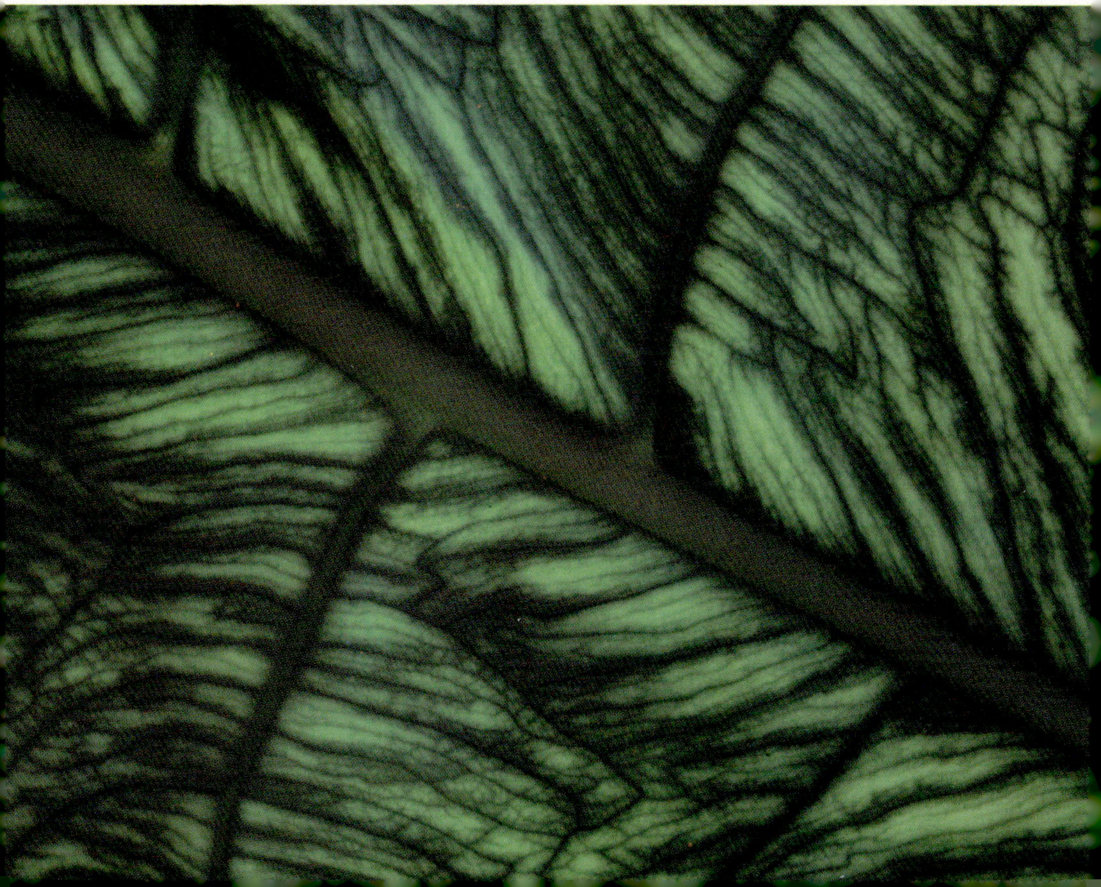

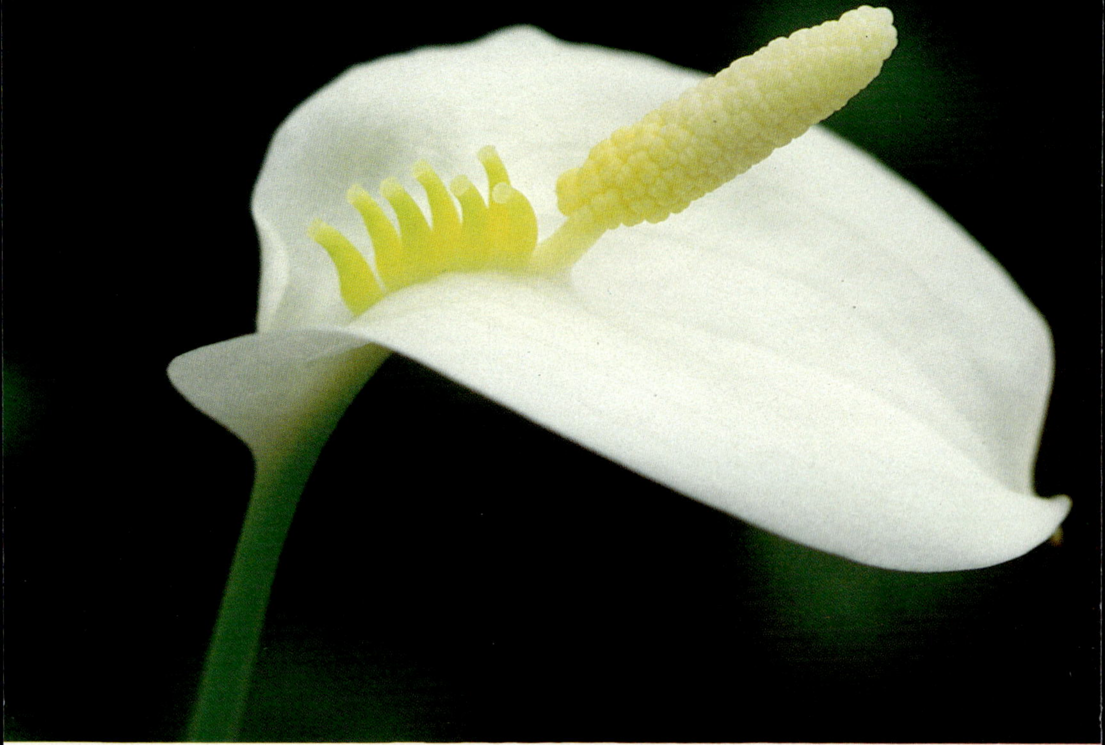
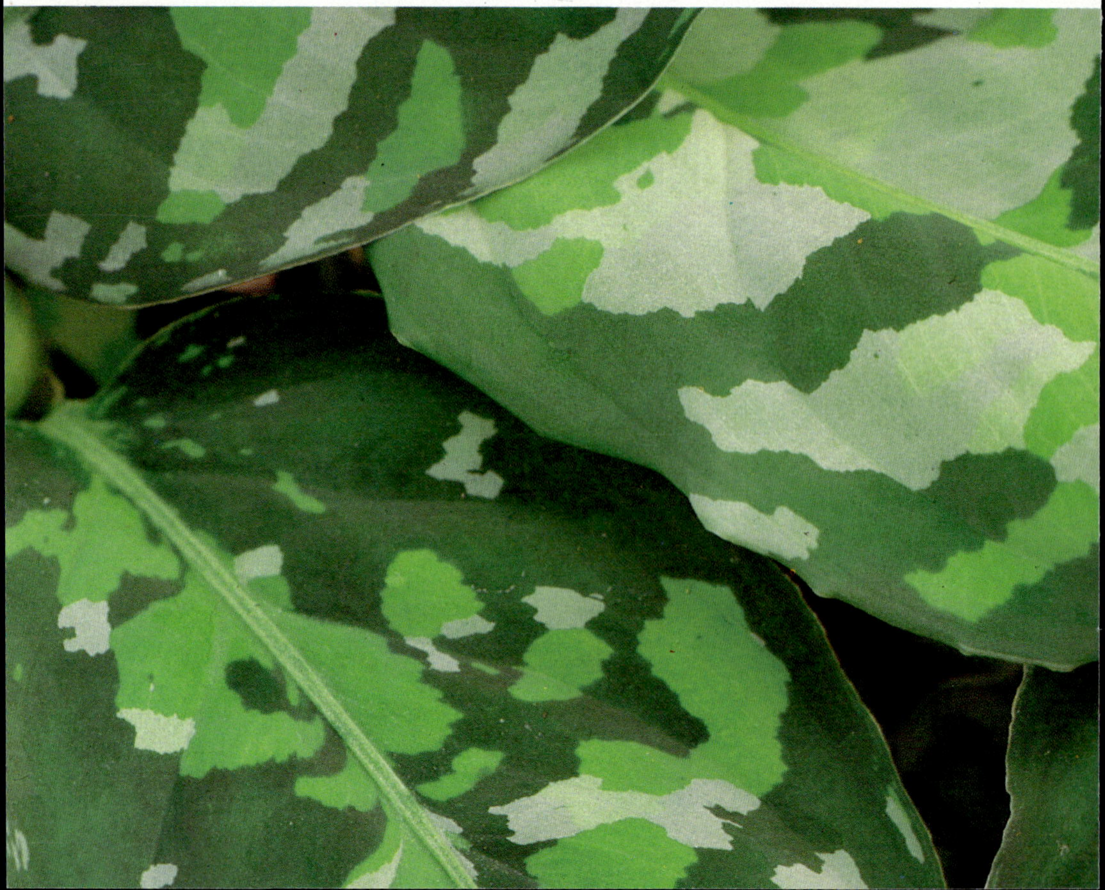

has slender leathery leaves with only a trace of variegation along the midrib. This is now considered as a minor but distinctive variant of the species (forma *elegans*).[20] Horticulturally it is of major importance though as the source of such fine cultivars as 'Pseudobracteatum' which is yellowish overall and heavily marbled in three or more different tones of cream and green.

Perhaps less easily mistaken for a dieffenbachia is *A. nitidum* which is found in Malaysia, Sumatra, and Borneo. Sometimes described as *A. oblongifolium curtisii*, this species has clusters of fairly upright shiny oblong leaves. In the wild it is rarely variegated but plants in cultivation are derived from the few with slender grey-green bars along the main lateral veins. Recently however, new boldly variegated forms have been found in southern Thailand. Some are quite different from anything that has been collected before, with broad white bands or margins. Even more surprisingly, they were growing in swamps.[21]

Also quite distinctive is *A. pictum* which is native to Sumatra. A smaller plant altogether, its slender stems sometimes branch and the piebald leaves are marked in several different shades. One of the most attractive forms is 'Tricolor' with dark green, lime green, and silver markings.

Some aglaonemas are quite unlike any dieffenbachia in appearance. *A. costatum* is a creeping rhizomatous species from Malaysia and mainland Southeast Asia. Its dark green, almost heart-shaped leaves are covered with white spots and bisected with a bold white midrib. This species was introduced to horticulture by Curtis, who discovered it on Langkawi

OVERLEAF]
ABOVE *Chlorospatha atropurpurea* (Ecuador) An outstanding species with almost black stalks and blackish variegated leaves.
BELOW *Alocasia* 'Reticulata' (Horticultural origin) The origin of this *Alocasia* is unknown. The banded stalks are similar to those of *A. zebrina* but the black-veined leaves are quite different.

OPPOSITE]
ABOVE *Callopsis volkensii* (Kenya and Tanzania) A small rhizomatous forest floor species with a peculiar spadix in which the female section of flask-shaped flowers is fused to the spathe and the cylindrical male part is held erect on a short 'stalk' (a naked portion of the spadix).
BELOW *Aglaonema pictum* 'Tricolor' (Sumatra) Most aglaonemas are variable, with some plants plain green and others with differing degrees of variegation, even within the same population. At least four different clones of this species have been named, of which this is the most colourful.

Island – just off the west coast of the Malay Peninsula – when he was collecting for Veitch's nursery. It will tolerate very low levels of light, but only really thrives in high humidity. In spite of its small stature, it has a vigorous root system and needs a deeper pot than other much taller species.

Even more unusual is *A. rotundum*. This small, slow-growing species is by no means rare in cultivation and in Southeast Asia is grown to bring good luck. However, its status and distribution as a rain forest species is undetermined. Although it was first described by N. E. Brown in 1893, it has only once been collected from the wild, in 1927 by H. H. Bartlett in the north of Sumatra.[22] *A. rotundum* lives up to its name, having almost round, dark green, glossy leaves measuring about 15 cm long and 10 cm wide. The undersurface is usually a rich burgundy and the midrib and veins are rose pink. In coloration they are reminiscent of those of the Brazilian *Calathea sanderiana* (*Marantaceae*).

Dieffenbachia

The 30 or so species of *Dieffenbachia* all belong to tropical America. They are quite common in undergrowth and clearings and along trails and streams in the rain forest. Though on a different continent, they look and behave like their Asian cousins except that they are, on the whole, rather larger plants. The common names of leopard lily and dumb cane indicate their two most outstanding characteristics – spotted leaves and extreme toxicity (even chewing a small amount can result in gross swelling of the throat and cause temporary dumbness, as described more fully in chapter 10). They are pollinated by beetles and give off a skunk-like odour when bruised. Many are bewilderingly variable, which has given rise to innumerable named forms and dubious species. Sorting this genus out is one of the many challenges to aroid taxonomists, complicated by the fact that although plants in the wild have distinctive characteristics, these are often lost when dried as herbarium specimens.[23]

In Nicholson's *Dictionary of Gardening* (1885) no less than 37 species are described but few of these are now recognized: *D. maculata, D. baraquiniana, D. magnifica, D. picta, D. amoena, D. shuttleworthii* and *D. nobilis* are generally considered variants of the commonest species in cultivation, *D. seguine*. It was one of the very first tropical aroids to be brought into cultivation in Europe, having been introduced in 1759.[24] This widely distributed and immensely variable species was at one time considered different from *D. picta* but as they were only separated by the degree of channelling in the leaf stalk (according to Engler) this is no longer regarded as sufficient to keep them apart.[25] One of the finest variants is 'Baraquiniana' (formerly *D. verschaffeltii*) which has only the occasional white spot but a long white stalk and midrib. It was found in Brazil by M. Baraquin in 1863.[26] Another fine variant is 'Jenmannii', discovered by G. S. Jenman, the Superintendent of the Botanical Garden in Georgetown, Guyana (then British Guiana) and sent to Veitch's nursery in Chelsea where it was advertised in the 1884 catalogue as 'one of the most striking new foliage plants that has come under our notice'. It is a graceful plant with slender stalks and narrow dark green blades which are blotched white along the main lateral veins. At the other extreme are the stockier, more heavily variegated 'Amoena', splashed yellow and white, and 'Memoria Corsii' with an intricate pattern of dark green veinlets, dark green blotches and cream spots on a grey-green ground. The latter may well be a hybrid, though Engler considered it a variant. 'Amoena', which means 'pleasing' was introduced from tropical America in 1880[27] and is reputedly slightly hardier than most dieffenbachias, tolerating a minimum of 13°C.[28]

A glance at the names given to the first dieffenbachias to be imported gives an idea of their magnificence as foliage plants: *imperialis, grandis, majestica, nobilis, rex, regina, splendens, triumphans,* and, of course, *magnifica*. It is almost impossible to decide which is the most outstanding. Certainly one of the most beautiful is *D. leopoldii*, a species from Costa Rica. It has brownish, white-mottled stalks, dark olive green leaves with a fine velvety sheen, an ivory midrib and, occasionally, sparse pale yellow spots.

Hybridization and mutation have further increased the range of variegation and today the most popular dieffenbachias in cultivation are chartreuse-leaved *D.* × *bausei*, the large wrinkled, heavily variegated 'Tropic Snow', compact 'Exotica' forms with almost entirely white leaves (such as 'Marianne' and 'Camille') and the robust 'Wilson's Delight' which has immaculate bottle green leaves slashed with broad white midribs.

Dieffenbachias are not easy to keep in good condition for any length of time as houseplants, being sensitive to draughts, fumes, temperature fluctuation and almost any change in conditions – as you would expect from natives of the unending warmth, humidity and stillness of the forest floor. They need about twice as much light as aglaonemas yet produce fewer leaves, and individual leaves are relatively short-lived so that the plant soon develops a bare stem. In addition, their toxicity results in many cases of accidental poisoning in the home and horticultural industry. In spite of these drawbacks they are among the most popular of all ornamental tropical plants, with an ideal combination of bold shapes, bright colours and attractive patterns and textures.

Chlorospatha

Dieffenbachias are nevertheless easy to grow compared with the genus *Chlorospatha*. For a long time this was thought to have only one species, *C. kolbii*, a spotted-stalked, pedate-leaved plant introduced in 1878[29] from Colombia (which was then known as New Granada). Now more have come to light and the genus has grown to nearer 15, and includes four rather rare species formerly described as the genus *Caladiopsis*. One of these is the extremely beautiful *C. atropurpurea* which has black stalks and green-edged black velvet leaves of a classically sagittate shape. If this species only had an easier disposition and better habit, it would be a stunning foliage plant. Unfortunately, like most chlorospathas, it is tricky to cultivate and only ever has one or two leaves at a time, with a cluster of straggly thin inflorescences appearing out of the side of a leaf stalk. Several have tripartite blades, the most distinctive being the deeply veined *C. corrugata*, a species named in 1985,[30] and the large cream-spotted *C. mirabilis*. All come from the western slopes of the Andes.

Callopsis, Ulearum and Nephthytis

Equally difficult to grow are the African genera *Callopsis* and *Nephthytis*, and the obscure Peruvian *Ulearum*. The first has only one species, *C. volkensii*, from Tanzania and Kenya. It is a small plant with heart-shaped leaves above which appear the showy inflorescences. The spathe is pure white, the spadix egg-yolk yellow and unusual in that it is partly joined to the spathe and has flask-shaped protuberant female flowers which give it the appearance of having a crest. *Callopsis* is a somewhat problematic genus and has already occupied several different positions in the family jigsaw puzzle. In some respects it is close to *Ulearum*, a rare and little known forest floor genus with two species to date, but as the latter is found only in Peru and *Callopsis* only in Africa, the similarities do nothing to solve the problem.[31]

Some botanists think *Callopsis* has links with *Nephthytis*,[32] a genus of fewer than ten species which for the most part are rhizomatous and found deep in the forest as inconspicuous members of the ground flora. One species, *N. afzelii*, seems more amenable to cultivation than most and although its plain green sagittate leaves and green inflorescences are not very exciting, it readily produces clusters of clear orange berries. More interesting is *N. poissonii* with bold triangular leaves. These African aroids need constant high temperatures (around 25°C) and humidity, heavy shade and humus-rich compost. Some species (for example, *N. hallaei* and *N. swainii*) are known to grow on very acid, nutrient-poor soil and do best in a mixture of peat and sphagnum moss.[33]

Endemics of Madagascar and the Seychelles

Madagascar is renowned for its numbers of animals and plants found nowhere else and the aroids there are no exception. Three endemic genera make up the tribe *Arophyteae*: *Carlephyton* with three species; *Colletogyne* with one; and *Arophyton* (which now includes the monotypic genera *Humbertina* and *Synandrogyne*) with seven. None were known to Engler and some have been described only recently.[34] In flower and pollen structure they are among the most advanced of all aroids. They are mostly tuberous with a dormant period during the dry season and generally grow in pockets of humus on or between rocks in the forests. Most of the species occur on a particular type of rock, whether gneiss, granite, basalt or limestone. The very rare *Colletogyne perrieri*, which has a purple-spotted white spathe and red-dotted spadix, is only known from one limestone region where it is found in association with *Carlephyton diegoense*. *Colletogyne* and *Carlephyton* have one or a few heart-shaped leaves, but in *Arophyton* they may be heart-shaped, hastate, trilobed or pedate, according to the species. *Arophyton* is also distinguished by having some rhizomatous species, including *A. buchetii* which is usually epiphytic on *Pandanus*.

Colletogyne perrieri ($\times \frac{1}{5}$)
The rare monotypic genus *Colletogyne* has been found in only one limestone region of Madagascar. It has a purple-spotted white spathe and a red-dotted spadix.

Further north in the Indian Ocean are the Seychelles with their unique endemic aroid: *Protarum sechellarum*, whose leaves have radiate elliptic leaflets. This monotypic genus has few clear affinities. Formerly it was a tribe of subfamily *Aroideae* but has recently been moved to *Colocasioideae*.[35]

Xanthosoma

The tribe *Caladieae* includes not only the obscure genus *Chlorospatha*, but the well-known *Xanthosoma* and *Caladium*. They are very closely related, one of their main differences being that in *Chlorospatha* and *Xanthosoma* the pollen grains are in clusters of four (tetrads) and in *Caladium* they are shed singly. Not surprisingly, some species have caused confusion. *X. lindenii* was so named by Engler but after Michael Madison's revision of *Caladium*[36] must now go by the new name of *C. lindenii*. Though mostly plain green in the wild,[37] a form of this Colombian species is one of the most brilliantly variegated of all aroids, with fresh green and white banded hastate leaves. Widely grown as a houseplant in the United States, it is virtually unknown in Great Britain.

Also difficult to classify is *X. striatipes*. It has two main areas of distribution in South America: in the north from Pará (Brazil) to French Guiana, Surinam and Guyana; and in the south of Brazil from Minas Gerais west into Paraguay. The two populations vary in leaf shape and seed size and, most significantly, the ovary of southern plants is more like that of a *Caladium* than of a *Xanthosoma*. This species grows in seasonally wet places at the edges of forests, forming a chain of tubers, each corresponding to the amount of growth made in a season. In this respect it is typical of the general tendency for xanthosomas to grow in rather wet places, often in bright light and not necessarily in the depths of the forest.

Xanthosoma has 57 species, all in South America. Those that do grow in the dim understorey or ground flora of the rain forest display some interesting features: trunk-like stems which raise the terminal head of leaves that much nearer the light, pale fruit-scented inflorescences pollinated by ruteline scarab beetles which fly at dusk and spathes which fill with rain after flowering to encourage insect visitors to depart. In addition, some species are downy or bear bulbils on the stem, or both. There are very few aroids which are downy and the majority belong to this genus. Bulbils on aerial parts are rather more common, though still unusual, but the combination of the two features is quite unique. An example is *X. pubescens* which is not only downy but produces in the region of 900 minuscule bulbils, each less than 1 mm, at each node. Its relative *X. viviparum* (which is not downy) has two distinct modes of growth. Some populations rarely flower, instead bearing about 400 bulbils in special crescents of tissue on the stems. Other have larger leaves and flower quite frequently but bulbils are few.[38] No detailed studies have been made, but it may be that reproducing by bulbils takes less energy than flowering and is resorted to when the plant is low in resources.

Recently the first pedate-leaved downy *Xanthosoma* has been described. *X. plowmanii* is an attractive tuberous species with light green leaves, sometimes marked white along the midribs. It grows on granite outcrops in moist forest in a small area of southern Brazil and was first collected by Timothy Plowman in 1980. Only when a specimen flowered five years later was it possible to describe it as a new *Xanthosoma*.[39]

——— A would-be carnivorous aroid? ———

The most unusual *Xanthosoma* is *X. atrovirens* from Ecuador. It has two very peculiar forms, one with flaps of tissue like small leaves on the underside of the leaf ('Appendiculatum'), and the other with long-tailed pouches at the apex of each leaf ('Variegata Monstrosa'). Similar excrescences to those of 'Appendiculatum' are found on certain taro (*Colocasia esculenta*) cultivars such as *Piko lehua apii*, but the pouched form is probably unique in *Araceae*.

What purpose could these pouches serve? Many plants in the rain forest have downward pointing 'drip tips' which assist in shedding rain from the surface of the leaf. If the leaf stays wet for any length of time, evaporation from the pores (stomata) is hindered and the plant has a circulatory problem as then it cannot draw in water and nutrients from the ground. Wet leaves also impair photosynthesis and make the leaf surface prone to colonization by algae and other epiphytes. Drip tips are even more important for species which hold their leaves almost horizontally in order to intercept as much light as possible.[40] But instead of shedding water, *X. atrovirens* 'Variegata Montrosa' must actually collect it.

It has been said that in spite of the intense competition, the tropical rain forest is the soft option for plants in that it does not take much to merely survive where there are no stressful seasons, prolonged droughts or cold spells to weed out the weak and raise the threshold of survival adaptations. In such an environment even freaks stand a good chance and the occasional freak may eventually turn out to have a feature which comes in useful. And so the monster of one epoch becomes the new species of the next.[41]

The connection is inescapable: the pouches of *X. atrovirens* 'Variegata Monstrosa' look like the pitchers of insectivorous plants. Without fossil evidence we cannot be sure how carnivorous families such as *Sarraceniaceae* evolved but one theory is that they arose as a result of scyphogeny, the freak cupping of leaves. Then, as water and debris such as dead insects collect in the sacs a nutrient solution forms which provides a foliar feed for the plant. In poor soils a plant with this deformity would therefore have an advantage. With this scenario it is amusing to surmise that this *Xanthosoma* could be a proto-insectivorous aroid!

Homalomena

There are only about 10 *Homalomena* species in South America. The rest – some 100 – occur in Southern Asia and the southwest Pacific region. They can vary greatly in appearance. Several of the South American homalomenas are downy. At Fairchild Tropical Garden in Florida I photographed an unidentified Colombian species[42] with heart-shaped blades covered in the finest hairs and stalks which were both downy and spiny. Just recently, a new species taller than a man was found by a botanist in Sumatra. With blades up to 80 cm long and 50 cm wide on stalks over 1.5 m long, it is not surprising that he named it *H. megalophylla*.[43] In contrast, the Bornean *H. minutissima* has leaves measuring only 5–10 cm in length.

In some Southeast Asian rain forests homalomenas are the commonest aroid and many species can be found within a short distance in clumps or small colonies on the forest floor or on rocks. One such area is Harau in West Sumatra, a valley formed by a fault and walled with sheer cliffs over 40 m high, down which waterfalls plunge as spray into pools way below. At least one large cordate *Homalomena* species grows – in association with the remarkable forked fern *Dipteris conjugata* – all the way up the cliff in the constant spray of the falls. At the bottom, virtually in the falls, are mats of a small species with the lanceolate leaves so often found on plants living close to or in water. And nearby, on a dry ledge under the cliff are several plants of a tiny elliptic-leaved species. I had the impression that homalomenas are perpetually in flower as almost all had dead, dying, open, and emerging inflorescences. One plant of *H. cordata* in Java had at least 20 inflorescences. They had waxy white spathes which opened just enough to expose the slightly longer white spadix which was covered to the tip with male florets and frequented by very tiny flies.

Very few homalomenas are in cultivation. The best-known is *H. wallisii* from Colombia which looks like a squat dieffenbachia, with oval leaves heavily splashed with yellow and neat clumps of foliage at ground level. *H. lindenii* is also seen occasionally in collections. It is found wild in New Guinea and has long-stalked olive green heart-shaped leaves and yellow midrib and veins which are almost luminous when wet. In common with many homalomenas, the plant smells strongly of anise.

──── *Schismatoglottis* ────

Similar in appearance and habit to *Homalomena*, and thought by most taxonomists to be closely related, is the genus *Schismatoglottis*. They are about the same too in terms of numbers of species and distribution, *Schismatoglottis* having close on 120 species, mostly in tropical Asia but a few in the American tropics. Both genera have rather thin cordate or lanceolate leaves, often with the veins deeply impressed and a clump-forming habit. One significant difference is that in *Schismatoglottis* the leaves are more frequently variegated and the upper part of the spathe sheers off after flowering. In fact, the genus name means 'split-tongue'. Again, few are now seen in cultivation, although a wide variety are listed in 19th century manuals,[44] but those that are show an exquisite, though subtle style of colouring. One of the finest is a plant known to collectors as *S.* 'Silver Heart' but whose identity is unknown, which has blades of a uniform silvery blue. Better known is *S. neoguineensis* from New Guinea, which has medium green heart-shaped leaves sprinkled haphazardly with uneven-sized pale green blotches. The leaves of *S. picta* (sometimes called *S. calyptrata*) are also very decorative. They are mid-green and cordate and the shape is echoed in a broad outline of a narrow heart which looks as if it has been sprayed on the centre of the blade with silver paint.

──── *Alocasia* ────

Even the greatest admirers of the genus *Alocasia* must concede that there is nothing subtle about these plants. These are the glossies of the family, their shining shields dominating the scene wherever they grow. There are some 70 species, all from tropical Asia, where they are found here and there in the ground flora of the rain forest. Some grow as lithophytes in colonies on limestone rocks, with rhizomes exposed and only their roots penetrating the humus and rock fissures. Many of the mountains in Southeast Asia where these plants grow are extremely steep and rugged, composed of heavily eroded porous limestone overlaid with a loose layer of leafmould.

The true terrestrials are more scattered, occurring as individuals or small groups in dense forest or more open areas along watercourses and sometimes in marshy places in full sun.

Alocasias have long-stalked, often peltate leaves and either rhizomes or tubers (or both) and may undergo dormancy, the reasons for which are not understood. In some species it seems to be triggered by a drop in temperature and humidity, but others (for example *A. magnifica*) become dormant regardless of conditions. This element of unpredictability is further compounded by the fact that — at least in cultivation — some species seem to be short-lived, dying suddenly but often leaving offsets at the end of stolons. Basically though their origins dictate that they need constant warmth and humidity, plenty of moisture but excellent drainage and, for the most part, shade.

Identifying alocasias is no easy task and this is another example of a genus which is badly in need of revision. Many species are naturally very variable and the difficulty of assessing similarities and differences is increased in cultivation because of the large numbers of hybrids and the way that characteristics can change in response to conditions.

⎯⎯ The copper alocasia and its kin ⎯⎯

A. cuprea is unmistakable though and has been a firm favourite in cultivation since it was sent to Veitch's nursery from Borneo by Thomas Lobb in 1859.[45] Lobb was one of the most discerning Victorian plant collectors. A man of few words, he rarely described his acquisitions. When he did comment that something was 'very pretty', it was understood to mean exceptionally fine and deserving the utmost care. Whether he made this remark about *A. cuprea* is unrecorded, but it is certainly one of his most outstanding introductions. The foliage of *A. cuprea* looks exactly like burnished metal and a skilled metal worker could make a perfect imitation of its unique leaf. The oval blades are about 25 cm long and have their back lobes joined nearly all the way so that the peltate structure is accentuated. The areas along the join, the broad raised midrib and the arching lateral veins are heavily shaded (almost black in some plants), emphasizing the depressions. The underside is a rich purple. The overall finish is coppery though there are some variations in colour between specimens and between leaves of different ages. Some are predominantly green and some more blackish but on the whole this is a remarkably unvarying species. The only complication is what is usually referred to as 'the green form'.

Described by Schott as *A. alba* and variously known as 'Green Shield', 'Green Cuprea' and 'Green Velvet', some collectors regard this as a separate species, others as a variety.[46] One grower believes it to be a sport, as it apparently appeared as an offset in a pot of *A. cuprea*.[47] Whatever the truth about its origins, there are a number of obvious differences between the two. The green form has rather larger softer leaves of a bright lime green shaded with olive green. Differences have also been recorded in the inflorescences and tubers. Perhaps the most noticeable thing, apart from the colour, is the fact that the green form grows much more easily. I was surprised to see it in the gravel at Jerry Horne's nursery in Miami, behaving, if not looking, like a weed.

Rather like a miniature form of *A. cuprea* is a recently introduced species from the Philippines which was promptly christened 'Quilted Dreams' and 'Hawaii' by horticulturists, but is now described as *A. bullata*.[48] This is a lithophyte and forms dense colonies, 10–20 cm high, of puckered leaves which, like its larger relative, have dark shading around the midrib and primary veins.

Along the same lines as *A. cuprea* and 'Green Cuprea' is the much rarer Bornean species *A. guttata* var. *imperialis* which has a bluish-grey surface and blue-black shading round the midrib and main veins, and back lobes not joined. Apparently the Czar of Russia was so impressed by this plant that he asked the botanist N. E. Brown to name it after him: hence the '*imperialis*'. '*Guttata*' refers to the purple-spotted stalks.

—— Silver veins and purple backs ——

Probably the most variable of all alocasias is *A. longiloba*. Extreme variability is often associated with a wide distribution, in this case from Thailand and the Malay Peninsula to Borneo and several other islands of Indonesia, including Java. David Burnett, in his review of the genus in cultivation,[49] gave up the attempt to describe it and made the telling observation that some of the differences in forms of this species are more significant than between it and other species. The leaves are sagittate but can be anything from 15 to 55 cm in length on mature plants, and to 70 cm in the variety 'Magnifica'. In some forms the lobes are spread wide apart so that the shape is nearer hastate. Generally the upper surface of the blade is a glossy dark green, with a purple shiny reverse. Midrib and primary veins are silvery grey.

Also very variable is *A. lowii*, first introduced by Hugh Low from

Borneo in 1862. Again the blades are large (up to 50 cm) and often very broad (30 cm or so in 'Grandis'). Many plants which have been grown as *A. putzeysii*, *A. thibautiana* and *A. korthalsii* are now considered to be varieties of *A. lowii*. (*A. korthalsii* does in fact exist but is a rare species from Sarawak with olive-green blades, midrib and main veins in brilliant silver and an unmistakable glossy brownish-red lower surface which is unique in the genus.)[50] All have confusingly similar silvery venation and purple undersurfaces and in particular resemble the form of *A. lowii* known as 'Veitchii'. But whatever their genetic conformation, they are assets to any collection. The magnificent blue-grey leaves and silver veins and veinlets of *A. thibautiana*, as it was named on its arrival from Borneo in 1878, led to its being described as 'by far the finest of the genus' in a genus where superlatives abound.

The largest of the silver-veined purple-backed alocasias is *A. watsoniana* which was brought into cultivation by Sander in about 1880. It was first found in Java but also occurs in Sumatra and possibly the Malay Peninsula too. Its leaves are often over 90 cm long and 50 cm wide and are characterized by concentric wrinkles radiating from the y-shape which is formed by the midrib dividing into the lobes. This puckering of the blade is also seen in some clones of the best known of all the alocasia hybrids: *A.* x *amazonica*. Again silver-veined and purple-backed, this cross was made in the 19th century between *A. lowii* and *A. sanderiana*, inheriting the colour and basic leaf shape from the former and the lobed margins from the latter.

The kris plant

A. sanderiana has leaves which stop you in your tracks. It looks more like a creation of Art Nouveau than of Mother Nature. The blades – some 40 cm long – are highly glossy, very dark green, narrow and deeply (often angularly) lobed, with midrib, primary lateral veins and lobed margins drawn in a dazzling silvery-white. The undersurface is green in the species but purple in the variety 'Gandavensis'. A description of it was published in William Bull's *Retail List of New, Beautiful, and Rare Plants* of 1884: 'A remarkably handsome and truly grand aroid introduced from the Eastern Archipelago, and forming one of the finest of the variegated stove plants yet introduced to Europe.' Its common name of kris plant refers to the Indonesia dagger known as a kris, which has the blade scalloped down both sides. This species is found only on the Philippine island of Mindanao and is now very scarce in the wild.[51]

Another Philippine species with the same blend of primitive and sophisticated in its effect is *A. micholitziana*. The form 'Maxkowskii' (also known as 'African Mask', 'Green Velvet', and 'Green Goddess') has broader and darker leaves than the species. They are the texture of finest velveteen, precisely marked with a stylized pure white outline of the midrib and main laterals.

Alocasia sanderiana ($\times \frac{1}{5}$)
This remarkable aroid from the Philippines is now scarce in the wild and is protected by CITES legislation. Hybrids of this species have sinuous more shallowly lobed blades – the best-known being
A. × *amazonica*.

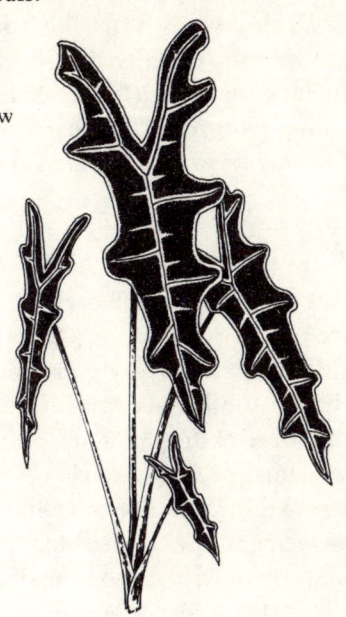

—— Outsize alocasias ——

Alocasia excels both in beauty and size. Two huge arborescent species are particularly common in tropical countries: *A. macrorrhiza* and *A. odora*. The former probably came from Sri Lanka originally but has been introduced all over the tropics as a food crop. It is very variable but commonly has bright green leaves some 1.2 m in length which are not peltate on mature plants. The leaves are held almost erect in a crown at the top of a trunk 1–2 m tall. Of the many named forms, the most interesting foliage is produced by 'Variegata'. Its piebald leaves have areas of dark green, grey-green, and ivory: each leaf different and occasionally one that is predominantly white. *A. odora* is very similar to *A. macrorrhiza* but has consistently peltate leaves held stiffly upright. In its homelands of tropical Asia it can be seen growing in full sun.

Another very large species is *A. portei* (formerly *Schizocasia portei*) from the Philippines. It bears deeply divided leaves up to 2 m long, the lobes of which have frilly margins. The leaf stalks are as long again, giving the plant a final height of over 3 m. New leaves are erect but older ones arch to a horizontal or slightly drooping position. Mature plants are usually surrounded by offsets and form imposing colonies.

The largest-leaved of all alocasias is *A. robusta* from Borneo. It was first collected in the early 1960s and formally described in 1967.[52] The light green leaves reach 2 m long and 1.2 m wide. It is unusual in having a purplish-black spathe limb. This feature is shared by the rare but much smaller *A. scabriuscula* from Sarawak, which has the added distinction of producing bright red tubers.

A half-hardy alocasia

In comparison with the other species described, *A cucullata* is rather unexciting in appearance, with plain green peltate heart-shaped leaves a mere 15–30 cm in length, marked only by a prominent midrib and main veins. It does however have a number of interesting features: one being that it is by far the hardiest member of the genus, unscathed by temperatures of 4–5°C and only slight damaged by a touch of frost. Another is that although it has a widespread distribution (from China and Japan to Burma, India and Sri Lanka) it is remarkably unvariable. Since its introduction to horticulture in 1826 it has remained rather on the sidelines but has much to recommend it, not the least that it forms compact plants with many long-lived leaves.

The *Zebrina* group

Although most alocasias are admired for their leaves, a few are grown for their beautifully patterned stalks. *A. zebrina* and *A. wenzelii* are almost indistinguishable but for the fact that the latter is rather short-lived, larger and faster growing – a cultivated seedling being on record as having produced leaves 1.3 m long in 13 months.[53] *A. zebrina* takes much longer to reach barely half the size and is slow to reproduce in cultivation. It was first collected by John Gould Veitch in 1862 when he was searching for *Phalaenopsis* orchids on the Philippine island of Luzon. Now, due to over-collection, this Philippine endemic is on the verge of extinction.[54] Both *A. zebrina* and *A. wenzelii* have elegant non-peltate arrow-shaped leaves which point upwards and long strikingly banded stalks which are green to almost white with horizontal zig-zag mottling in dark brown. Similar is 'Tigrina Superba' which has slender bluish pointed leaves whose margins curl under. The veins on the undersurface are black and the stalks are usually banded but may be plain blackish-green.

A. 'Reticulata' has obvious similarities to the plants described in the last paragraph but its origins are unknown. It has not been recorded in the wild and may be a sport. Apparently it was first sold in the 1970s by Fantastic Gardens in Florida. It has the same banded stalks as its presumed relatives but the blades are quite unique, basically lime green with every vein and veinlet blackish-green and the primary laterals occurring at almost right angles to the midrib.

Xenophya

No description of *Alocasia* is complete without mentioning the genus *Xenophya*. There are only two species, both with very distinctive leaves: *X. brancaefolia*, a large plant reaching 2 m, whose leaves are pinnately divided; and *X. lauterbachiana*, with dark stalks often mottled like those of *A. zebrina*, and upright flat lanceolate leaves, regularly lobed at the margin and purple beneath. They are from lowland rain forest in New Guinea. Formerly the genus was known as *Schizocasia*[55] and in the future it might well become a section of *Alocasia*.[56] The two genera do seem to have only the minutest floral differences and their closeness has recently been demonstrated by a hybrid made by Lawrence Garner of Miami between *X. lauterbachiana* and an alocasia.[57]

Alocasias are the epitome of exotic tropical foliage plants. They, like many other aroids in cultivation, exhibit qualities which we generally associate with elaborate man-made hybrids, such as immense size, bold shapes and showy colours. As we admire these magnificent specimens it is hard to believe that their wild relatives were just part of the tangle of infinitely variable vegetation which covers the dim and fragile floor of the rain forest. Some are so popular as to be commonplace and we take them for granted, growing them expertly in our homes and gardens but knowing practically nothing about their natural lives – let alone whether they still exist in the wild.

7
Towards the Light

Tropical climbers and epiphytes

If there is one place on earth which has made me wish for wings, it is the tropical rain forest. Not to glide above its expanses of tree-tops which from on high look like varicoloured broccoli, but to flit from plant to plant in the cavernous greenery below their crowns. For a creature with two legs (and, in my case, a rucksack of camera equipment and a tripod as well) it is the world's most uninviting obstacle course: an assemblage of mud, ooze and wetness interspersed with roots and rocks concealed beneath loose leaf litter, trip wire lianas and barbed wire thorns, outsize spiders' webs (always at face level), inordinate numbers of insects who seem particularly irritated by human endeavour and (in the Old World at least) leeches which wave at you as you pass and stick to you like iron filings to a magnet. And if you take these hazards in your somewhat faltering stride, there is still the sweltering heat and torrential rain which present further barriers to botanizing in what may be paradise for plants but is infernally difficult for anything but a bird or a butterfly to get about in.

It is not therefore surprising that we know so little about life in the rain forest. Even the most dedicated tropical botanist, professional or amateur, cannot hope to achieve the same detailed patient observations that have been lavished on the world's temperate flora. And chances are that, being an earthbound human, he or she will tend to concentrate efforts on the most accessible species at or near ground level and will probably show an understandable preference for working at higher elevations where conditions are less oppressive. As a result our knowledge of tropical plants is biased towards those growing within easy reach in the least inhospitable regions.

——— The upper storeys of the rain forest ———

The most inaccessible of all zones in the rain forest are the upper storeys, from where the trees first branch to where they brush the sky. Life at the top is richer in many respects than down in the shadows. Every tree is laden with epiphytes, draped in climbers and skeined with lianas. Here most of the forest's leafing, flowering and fruiting takes place, providing food for many of its creatures which, in spite of their numerous swarms and flocks, still constitute less than a quarter of one per cent of the total living material, a mere drop in this ocean of greenery. And what falls from these heights – leaves, petals, dead wood, rotten fruit – feeds the bulk of them: the decomposers in the soil which outnumber and outweigh in their billions, the larger, more familiar creatures.[1]

Getting to grips with plant life at these heights and densities poses serious problems. Straightforward collecting can be done in a random fashion from fallen branches or rather more systematically by following behind the plough, so to speak, and gleaning specimens as logging takes place. The more determined have also been known to fell trees for themselves or shoot down branches with high-powered rifles. But none of these techniques gives any insight into lifecycles and relationships. This can only be done by painstakingly climbing up, braving attacks by the ant hordes which relentlessly defend their plant territories and taking other formidable risks to life and limb. The first westerner to climb into the canopy was probably Sir Francis Drake who, in 1573, scaled a 'giant and goodlie tree' in the Serrania del Darien, Panama, in order to view two oceans: a sight which inspired his voyage round the world. Only in the last 60 years though has any progress been made in exploring the canopy itself and so far, even with a battery of military and mountaineering skills, very few aerial walkways and tree-top laboratories have been established.[2] If anything, the oceans and outer space are better known.

In the face of such difficulties it is inevitable that we know most about species which have been observed at length only when they have been removed from their natural habitat. It is a commonplace of tropical botany that many species have never been seen flowering, fruiting or even – in the case of some tuberous aroids – in leaf, in their wild state and almost all the information about them has been gained during subsequent cultivation. Aspects such as population dynamics, pollination and fruit dispersal are for the most part unknown and will only be revealed through multi-dimensional rain forest exploration.

A case of mistaken identity

Bearing in mind the problems of collection and observation and the relative rarity of many species, it is no surprise that some species – especially those with quite different juvenile and adult forms – have led botanists a merry dance. The most notorious so far is *Rhaphidophora celatocaulis* (as the juvenile form of *R. korthalsii* was known for many years), which was sent to Veitch's nursery from northwest Borneo by the collector F. W. Burbidge. For about 40 years it went under an assortment of names: N. E. Brown called it *Pothos celatocaulis* and Engler and Krause decided it was *Monstera latevaginata*. These were not far from the mark compared with its horticultural label of *Marcgravia paradoxa* which is not even a monocot genus, let alone a member of *Araceae*. It was a plant growing at the German University in Prague which finally gave the game away when its small almost stalkless leaves, which overlap and press closely to the bark of a tree, at last saw the light and completely changed their habit to long-stalked, deeply cut adult foliage.[3] Only then could the two entirely different forms be seen as one and the same species: *R. korthalsii*.

—— Shingle plants ——

The juvenile form of *R. korthalsii* is known as a shingle plant because the blades are held close to the climbing surface and overlap methodically like tiles on a roof. This habit is also found in *Scindapsus* and *Monstera* (section *Marcgraviopsis*) but is not unique to *Araceae*. It is also known in such genera as *Ficus (Moraceae)* and *Hoya (Asclepiadiaceae)*, as well of course as in *Marcgravia (Marcgraviaceae)*. It is obviously a very compact way of growing. Keeping the foliage clamped to the bark ensures that as little water as possible is lost from the leaves as they do not waft about. This is further improved by having pores (stomata) only on the undersurface of the leaves, so what does escape remains trapped around the adhesive climbing roots.

—— Towards the darkness ——

Seedlings of *Monstera* species in sections *Marcgraviopsis* and *Echinospadix* begin life, as other climbers, by germinating on the forest floor but instead of putting out leaves and roots, they produce a runner (stolon).[4] This cord-like growth is leafless and rootless, bearing only tiny scale leaves at the nodes, and its development is fuelled by reserves in the large seed that is typical of this kind of climber. It proceeds along the ground or under the leaf litter and can reach several metres in length, becoming green when exposed to light and white if concealed. In either position though, it can 'see' where it is going, being attracted to dark areas above ground – a phenomenon known as skototropism.[5] As soon as it reaches a vertical or sloping dark surface, a tree trunk or rock face, it puts out its first pair of flat opposite leaves, sprouts adhesive adventitious roots, clamps on to its host and changes orientation completely – towards the light. Snaking through the crowds to where it can grow unimpeded, it avoids the intense competition that conventional seedlings have to face.

⎯⎯ Variegated juveniles ⎯⎯

Not all climbers with juvenile forms have the shingle habit though. Some have a variegated juvenile phase, usually with the blade a different shape from the adult form. A few adopt both strategies: the juvenile foliage of *Monstera dubia* is both silver-variegated and a shingle plant. What advantage pale variegation might confer is puzzling, as cutting down the photosynthetic area would seem to be a disadvantage in heavy shade. It might be that this is offset by the advantage of making vulnerable juvenile foliage look completely different from the adult, enabling young plants to play hide-and-seek with butterflies and other insects looking for the right leaf on which to lay their eggs.

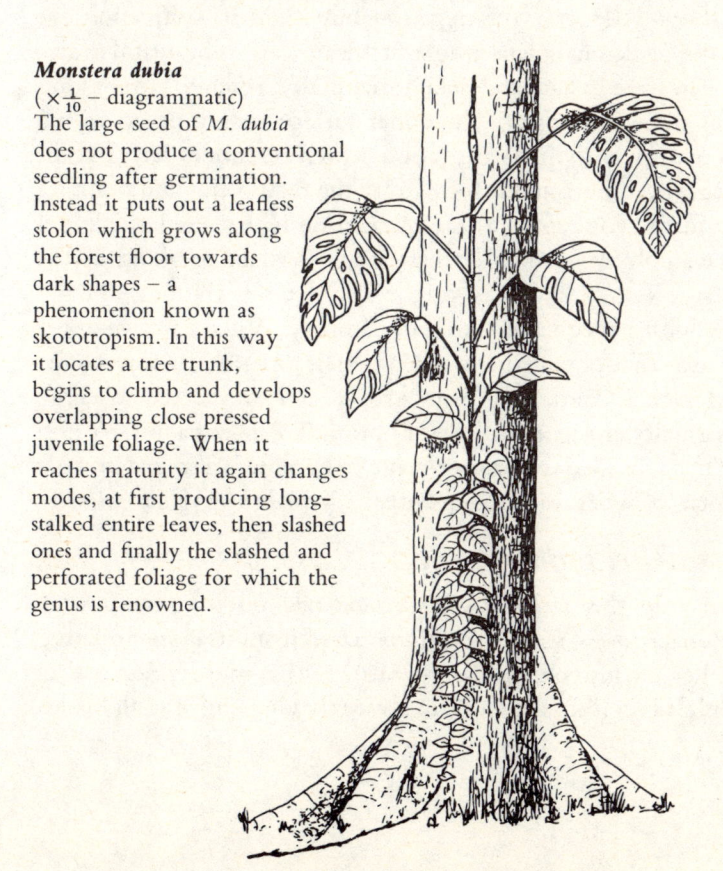

Monstera dubia
($\times \frac{1}{10}$ – diagrammatic)
The large seed of *M. dubia* does not produce a conventional seedling after germination. Instead it puts out a leafless stolon which grows along the forest floor towards dark shapes – a phenomenon known as skototropism. In this way it locates a tree trunk, begins to climb and develops overlapping close pressed juvenile foliage. When it reaches maturity it again changes modes, at first producing long-stalked entire leaves, then slashed ones and finally the slashed and perforated foliage for which the genus is renowned.

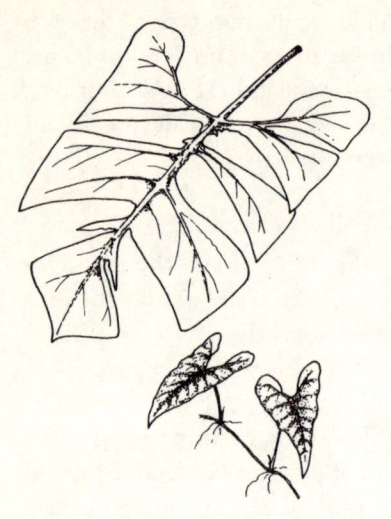

Cercestis mirabilis
(formerly ***Rhektophyllum mirabile***)
This West African climber has two stages of growth:
Below: ($\times \frac{1}{20}$) A juvenile mode with sagittate leaves, which in some plants is variegated with a white fern-like pattern.
Left: ($\times \frac{1}{10}$) A mature stage, in which the foliage is larger and broader, with irregular slashes and wedge-shaped lobes.

One of the most distinctive juvenile forms is seen in some clones of *Cercestis mirabilis* (formerly known as *Rhektophyllum mirabile*). In the wild many plants show little or no variegation but those in cultivation are derived from the occasional plant whose first leaves are as beautiful as any in the family. They are hastate and held horizontally, reaching 30 cm long on stalks about the same length. The upper surface is dark green with a most unusual pattern of white variegation which resembles fern fronds, each one fitting into the segments formed by the main veins and pointing towards the midrib. The result of this variegation is that the large leaves look more like a colony of small feathery ones. Perhaps breaking up the outline of a large leaf in this way might also deter herbivores. This is another species with a chequered taxonomic history. When a well-marked juvenile form was first obtained from West Africa in 1887, it was named *Nephthytis picturata*. Eventually a plant grown at Kew started to climb (young plants usually creep on the forest floor until reaching a suitable host tree) and produced divided fenestrate foliage – revealing its true identity as the juvenile form of what was subsequently called *Rhektophyllum mirabile*.

Rhektophyllum and *Cercestis*

Until very recently the genus *Rhektophyllum* had but two species (*R. mirabile* and *R. camerunense*) and was differentiated from its closest relative, *Cercestis* (another exclusively African genus), by having divided and/or perforated adult leaves (those of *Cercestis* always having entire adult blades, though almost any shape from oval to trilobed). However, the flowers of the two genera are regarded by some botanists as being virtually identical so it has been decided to 'sink' *Rhektophyllum* and transfer its species to *Cercestis*.[6] There are therefore about 10 species of *Cercestis* now but none are common in cultivation, apart from its new acquisition, *C. mirabilis*. All *Cercestis* species are scandent and at intervals put out long leafless shoots (flagella). Some species are widespread in the forest and provide useful materials (see chapter 10).

Culcasia

A still larger African genus is *Culcasia* with 27 species, most of which are climbers and seldom seen in the western world. They have simple, unlobed, entire leaves but otherwise resemble *Cercestis*, apart from the fact that they never put out flagella. Some are most unusual: *C. rotundifolia* (a rare species from Gabon) is a small slender climber with round leaves that give it the appearance of a shingle plant. A number of culcasias have medicinal properties and other practical uses (see chapter 10).

Rhaphidophora, Epipremnum and *Scindapsus*

The genera *Rhaphidophora*, *Epipremnum* and *Scindapsus* are chiefly distinguished by the number of ovules they have, ranging from many in *Rhaphidophora* to a few in *Epipremnum* and only one in *Scindapsus*. As far as outward appearances go, they are all climbers and mainly Asian in distribution.

Apart from *R. korthalsii* which is sometimes available in its juvenile shingle form (and is an interesting plant to grow on surfaces in a greenhouse) few of the 120 or so species of *Rhaphidophora* are seen in cultivation, though the glossy palm-like leaves of the adult *R. decursiva* are very ornamental.

Likewise *Epipremnum* only has a couple of its 15 species which are cultivated to any extent. The commonest is *E. pinnatum* 'Aureum' (often called *E. aureum*) which is among the most popular of all variegated aroids for pots, hanging baskets and, where the climate permits, for landscape use as dense ground cover or a bold climber. When planted outdoors it will carpet the ground with juvenile foliage but starts to climb and abruptly changes leaf size as soon as it reaches a vertical surface, though the variegation remains. *E. pinnatum* 'Aureum' has also run the gamut of genera, having previously been put into *Pothos*, *Scindapsus*, and *Rhaphidophora*. Commonly known as golden pothos, hunter's robe or devil's ivy, it has bright yellow and green marbled leaves, oval when juvenile and measuring about 15 cm, but becoming much larger (to 80 cm) and irregularly slashed as the plant matures. It originated in the Solomon Islands but is now grown all over the tropics. Flowering is not unknown in cultivation[7] but mature plants at least 25 years old failed to produce inflorescences in Florida.[8] Extremely popular too is the form 'Marble Queen' which is exactly the same but with white variegation. It entered horticulture in 1935, having originated as a sport in the nursery belonging to Erich Gedalius.[9]

Scindapsus is a larger genus, with some 36 species, but again only one is widely cultivated. *S. pictus* was introduced from Borneo by Thomas Lobb. In the *Gardener's Chronicle* of 1859 it was described as having leaves which rival *Anoechtochilus* orchids in beauty. The comparison is apt, though the orchids' leaves are pencilled in gold or silver, whereas those of the aroid are silver-smudged, with the finest silver line round the margin. Its small leaves are a lop-sided heart shape, very dark green with a crystalline lustre. In the horticultural trade it is often called the satin pothos. This delicate species can be found straggling along the forest floor in parts of Southeast Asia. I found many isolated plants, mostly under half a metre in length, in dense shade in the rain forests of western Sumatra, but never saw one climbing. It is likely that some juvenile climbers can persist for many years, making little growth unless a gap in the canopy appears. Then they, and other immature plants which have been biding their time, undergo a great spurt of growth in the rain forest space race. When *S. pictus* does this its leaves do not change significantly other than in size, reaching 15–20 cm long.

The pothoid group

For all that the houseplants *Epipremnum* and *Scindapsus* are sometimes labelled pothos, they do not look much like the genus which goes under that name. *Pothos* is one of three genera in the subfamily *Pothoideae*. It has about 50 species but the other two, *Pothoidium* and *Pedicellarum* have only one each. All are Asian in distribution (though *Pothos scandens* has reached Madagascar) and rather primitive florally but specialized as regards their climbing habit. Like most aroid climbers, the leaves are produced alternately on opposite sides of the stem, giving two vertical rows which fit neatly against the host tree. Some species have flattened leaf-like stalks (phyllodes) which emphasize the back-to-the-wall and ladder formations characteristic of root climbers. In *P. repens* the leaf-like stalks are much larger than the blades and obviously play a far greater role in photosynthesis. In fact the blades are reduced to mere tabs on the end of each long phyllode. Another specialization of pothoid climbers is that the position of the leaves can be constantly adjusted according to light levels by flexing a joint (pulvinus or geniculum) which is situated where the stalk meets the blade. Leaves with a joint are especially common in climbing and epiphytic aroids. They can also be seen in monsteras and their relatives and in all anthuriums.

Pothos and its kin are not showy plants. Their modest foliage is complemented by equally demure inflorescences which in many cases are difficult to recognize as belonging to *Araceae*. Generally the spathes are little different from bracts and the spadices scarcely more noticeable. In some species they resemble catkins and are produced at the end of hanging shoots, and in others they are no more than tiny bobbles which stud each axil. At first glance, some species appear to have more than one spathe. *P. seemannii*, which is found scrambling in damp ravines in southern China, has globose spadices the size of peas, each subtended by several bracts

Pothoidium lobbianum (×$\frac{1}{3}$)
The pothoid group – *Pothos*, *Pothoidium* and *Pedicellarum* – are primitive florally but have a specialized ladder-like climbing habit. In this monotypic genus, the leaf stalks are flattened into phyllodes, performing the functions of leaves, and the uppermost inflorescences have no spathes but are protected during development by foliage leaves.

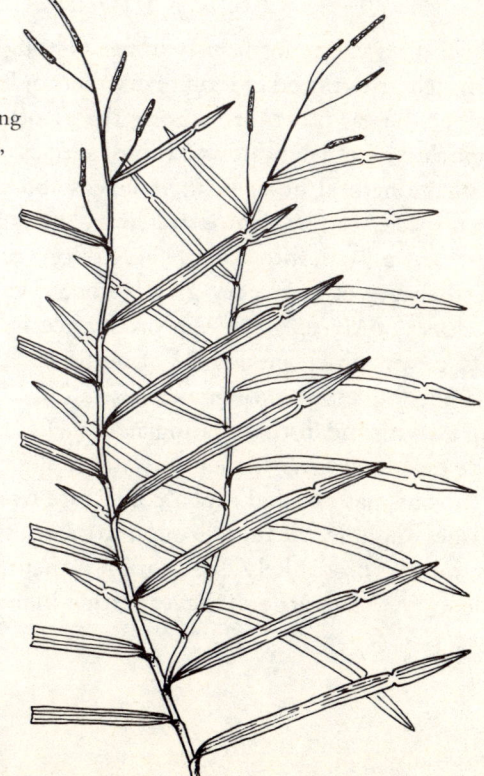

looking little different from the spathe. Species in the *remotiflorus* and *beccarianus* groups have thin spikes of scattered flowers which are quite unlike the fleshy densely packed spadices of most aroids. The uppermost inflorescence on each shoot of *Pothoidium lobbianum* does not even have a spathe and instead is enclosed during development by a young leaf.

Most peculiar of all is *Pedicellarum paiei* which is unique in *Araceae* in having stalked (pedicellate) florets, unlike all other aroids in which the flowers sit flat (sessile) on the spadix.[10] This monotypic genus is a rarity: so far only two plants have ever been examined by botanists, both dried herbarium specimens collected in Sarawak – one in the 1920s and the other in 1962.

Because of their unprepossessing appearance pothoid climbers are rarely seen in cultivation, yet they have a certain charm and curiosity value in collections. When pressed as herbarium specimens the segmented foliage of *Pothos* is reminiscent of mistletoe or even of seaweed. The commonest is *P. scandens*. It has a wide distribution in forests from Madagascar to India, Sri Lanka, Indo-China and the Philippines. Various forms have been recorded, some densely branched with numerous tiny grass-like leaves, each with a characteristically flattened stalk. It bears disproportionately large red berries.

—— Holey monsteras ——

Aroid climbers vary enormously in size. Though in theory all are capable of unlimited growth and indefinite numbers of leaves, some are diminutive in stature whereas others are among the giants of the family. One genus which includes both delicate and robust species is *Monstera*. Many of the 24 species have natural holes in their leaves and the huge variation in size between species can be demonstrated by the fact that a leaf of one of the smallest can easily fit into a single hole of one of the giants. The numbers, sizes and shapes of hole vary greatly too, even within the same species. Some clones of *M. obliqua* have more surface area devoted to holes than to leaf tissue, yet others are devoid of perforations.

This phenomenon, known as fenestration, is caused by areas of tissue ceasing growth and distintegrating at an early stage in the formation of the leaf; the earliest causing slashes, the latest holes. The possible advantage of these cut-outs has inspired much conjecture over the years. Perhaps holes and slashes might let the rain through to the roots, or reduce tearing in high winds, but the most likely explanation is that they assist cooling, causing turbulence in the air around the leaf rather like the blades of a fan.[11] Apart from one species of *Aponogetonaceae* fenestration is found in no other monocot family but *Araceae*. Within *Araceae* slashes and perforations occur in *Dracontioides* and in young leaves of *Dracontium* and *Anchomanes*, and most commonly in climbers, particularly in *Monstera, Epipremnum, Rhaphidophora, Cercestis* and the small Southeast Asian genera *Anadendrum* and *Amydrium*. The last two genera are rather isolated within this group as although they share these features, they lack others characteristic of monsteroid genera, notably the needle-like cells (trichosclereids) which are thought to be a defence against herbivores. *Amydrium* has the additional feature of reticulate venation: the foliage of *A. medium* is slashed and perforated like a monstera but has an obvious network of veins.

The best known monstera with holes and slashes is of course *M. deliciosa* (formerly *Philodendron pertusum*), a Central American species. Under its common name of Swiss cheese plant this is one of the most familiar aroids in cultivation, instantly recognizable even by those who have never heard of monsteras, let alone aroids. Taking over from the aspidistra of Victorian times as the indispensable complement to interior decor, it has become the archetypical indoor plant of the 20th century. Ironically, it is not at all common in the wild and when it was first collected in Mexico in 1832 and sent to Munich it received no acclaim. About ten years later, further collections were made in Mexico and Guatemala and sent to Copenhagen and Berlin respectively and it was cuttings from these plants that began the exponential growth of *M. deliciosa* as one of the greatest horticultural successes of all time.

It is not only the foliage of *M. deliciosa* which is outstanding. Though juvenile leaves are simply heart-shaped it soon starts to produce larger leaves with cut-away margins and when it reaches maturity, with profusely holed and split blades up to 90 cm long and 75 cm wide, it also bears large cream inflorescences. After flowering the thick waxy spathe withers and the spadix enlarges to produce juicy berries. At first these are protected by irritant trichosclereids in the bluish-green outer layer of the ovaries, but when ripe this breaks away. The yellow fruits are edible and taste of fruit salad but unfortunately fruiting does not take place in houseplants. Though many rapidly outgrow both pot and room, they are still only youngsters and far from reaching maturity or their potential height of 20 m and more.

Although *M. deliciosa* is most familiar as an ornamental indoor plant it is also grown outdoors in tropical countries and, because one of its outstanding features is toughness, in the subtropics too, where it frequently survives near-freezing winter temperatures. Not only is it the hardiest monstera, it is also the only species which grows on rocks or soil as readily as up trees. This feature is exploited in landscaping, where *M. deliciosa* is used as a ground cover shrub. In this kind of situation it flowers just as freely.

The very first monstera to be described was *M. adansonii*. A detailed account was given by the French botanist Charles Plumier as early as 1693. His observations were made in Martinique and he called it *Arum hederaceum, ampliis foliis perforatis*.[12] Linnaeus named it *Dracontium pertusum*, which in a horticultural list of 1777 was translated as 'perfoliate-leav'd dragon'.[13] In contrast to *M. deliciosa*, this species grows like a weed in Central and South America but has never made any impact as an ornamental plant. It is extremely variable (so much so that Schott made over 20 species out of the specimens at his disposal), has a widespread distribution and is highly adaptable, taking to telegraph poles as readily as trees.

Some species of *Monstera* are thought to be pollinated by tiny stingless *Trigona* bees. They are attracted to a sticky substance which oozes from sterile flowers at the base of the spadix. The bees collect this for food and for binding together various materials used in nest building.[14] The pollen itself is starchy and therefore unlikely to be of use to them as food, as bees cannot digest starch, but it falls on to them as they busily scrape up the syrup and is transferred to the next inflorescence in their search for further supplies.

—— Snakes and ladders ——

When a juvenile climber begins its ascent of a vertical surface, it cannot sense what lies ahead – whether there is one or 100 metres above, or whether the surface is sufficient to support it. Yet as it progresses, node by node, it somehow detects the nature of the substrate and if it proves unsuitable will abandon that particular route towards the light. The escape strategy takes the form of changing its growth pattern completely and turning tail by letting go of its host at the next node and developing a naked whip-like shoot (flagellum) which elongates and dangles until it reaches the ground. Then it grows horizontally for as far as 20 m – still without roots or foliage – searching for another tree. When it finds one, it does another abrupt about-turn and begins to climb, producing leaves and adhesive roots again. This versatility is seen in species such as the small and often lacy-leaved *M. obliqua* and the huge *M. acuminata*, which has pendent oval leaves heavily perforated with slit-like holes. The former rapidly outgrows the

Monstera tuberculata ($\times \frac{1}{8}$)
Not all monsteras have slashed and perforate leaves. This unusual species has short-stalked simple adult foliage which resembles the juvenile leaves of other species. It also has completely pendent inflorescences.

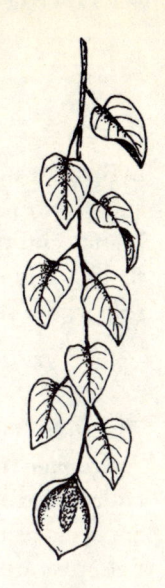

saplings and shrubs it favours as hosts and often needs to drop off and start again. The latter, like many monsteras, will only grow on the main trunk of a tree, and when it reaches the first forks sends out escape shoots.[15]

Variations of this behaviour are found in *M. dubia* which uses the technique to go up and down the same tree, thus utilizing the available space to greatest advantage. Its hanging shoots are not however devoid of foliage but bear thicker smaller leaves to cut down on water loss.[16]

The oddest monstera of all is *M. tuberculata*. It also climbs up a tree and then dangles, but its growth form hardly changes. Throughout all stages it produces the short-stalked oval overlapping leaves which are characteristic of several juvenile monsteras and maturity is signified only by the production of an inflorescence at the end of a hanging shoot. It is thought that this species has arisen by neoteny,[17] a type of evolutionary development whereby a juvenile becomes sexually mature. The inflorescence is one of the few truly pendent ones in *Araceae*. It has a concave white spathe which opens widely and is almost round apart from the pointed apex – rather like one half of a shell. The spadix lies down the centre, its surface covered in long spiky styles which give it the appearance of an immature fir cone.

Aerial roots

The root system of aroid climbers is highly specialized. Usually two types of root are produced: short ones, generally put out at right angles to the stem, which have adhesive root hairs and attach the plant to the climbing surface; and long feeding roots which absorb water and nutrients. Initially the long ones may dangle, often at great length from the plant's position high in the upper storey of the rain forest, but are capable of penetrating soil and fissures if these are encountered. Once inside dark moist soil they branch and form a dense network. Aerial roots of this type are extremely flexible and strong and when anchored in the soil they play an important role in the structure of the forest, acting rather like guy ropes.

—— *Heteropsis* ——

The South American genus *Heteropsis* contains some 13 species. *H. spruceanum* is named after the English botanist Richard Spruce, who collected aroids along the Rio Negro in Amazonas, Brazil, in the 1850s. In appearance it is rather like a *Pothos*, with slender pointed lanceolate leaves in flat rows down either side of the stem. The roots are used for lashing together poles in the construction of buildings and are said to be as strong as nylon parachute cord.[18] For use in basket and wicker work the outer layer (cortex) is stripped off and the inner fibres are woven while fresh. Spruce sent rolls of roots back to Kew and also collected other araceous cordage species such as *Philodendron goeldii*, which at one time was also named after him (as *Thaumatophyllum spruceanum*). He described the uses of these plants in some detail and commented that 'The only limit to the length of the roots is the height of the trees on which the plants grow, for when perfect they always reach earth or water, hanging down like so many bellropes.'[19]

The inflorescences of *Heteropsis* have spathes which drop off during flowering. The exposed spadix is visited by bees and plant bugs (*Hemiptera*). The berries hang at the end of the stems and (at least in *H. integerrima*) give off the smell of fermenting or decaying fruit which is often associated with dispersal by bats.[20]

—— *Rhodospatha* ——

Another little-known genus of South American climbers is *Rhodospatha* which is characterized by thin-textured oval or elliptic leaves that in most of the 23 species are pink when new. The spathes, as the genus name suggests (*rhodon* meaning 'rose'), are often pinkish too. Juvenile plants have been seen to form large areas of ground cover in the forests in the northern Andes.[21] They are rarely seen in cultivation nowadays but in Victorian times *R. blanda*, a Brazilian species, was available and *R. picta*, with very dark green leaves marbled bright yellow-green, was grown by the famous horticulturist William Bull, though its origin was unrecorded.

Stenospermation multiovulatum ($\times \frac{1}{6}$)
Most species of the tropical American genus *Stenospermation* are epiphytes. Typically they have finely veined simple leaves, crook-shaped peduncles, deciduous spathes and downward-pointing spadices of bisexual flowers.

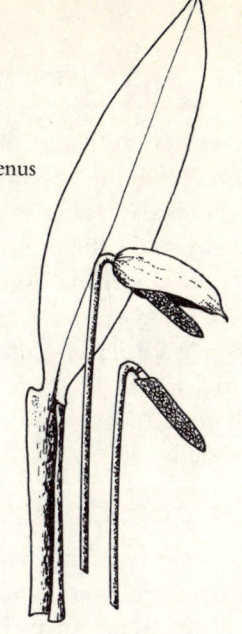

––– *Stenospermation* – the complete epiphyte –––

The South American genus *Stenospermation* has about 20 species and is closely related to *Rhodospatha*. Noticeable differences are the non-climbing predominantly epiphytic habit, horizontal rhizome and erect stem with thick leathery leaves with fine lateral veins. Most species have unusual crook-shaped inflorescences: the flower stalk (peduncle) being bent over at the top so that the cream-coloured spadix points downwards at an angle of 45°. The pale spathe drops off at the start of flowering. If it were not for the deciduous spathes, which often lie on the leaves like fallen rose petals, *Stenospermation* would probably have more potential in horticulture.

––– *Alloschemone* –––

One of the rarest South American climbers is *Alloschemone occidentalis*, the only known species in the genus. At one time this genus was dropped and its species added to *Scindapsus*[22] but after examination of plants in the wild it was reinstated. It has been found only three times: in 1830, 1934 and most recently in 1978 by scientists on the Projeto Flora Amazonas, whose aim was to gather material for a new flora of the Amazon region. The expedition began in Manaus and one of the members came across a postcard of the Río Negro in which the distinctive lobed leaves of the adult plant could be seen at the top of the photograph. They were therefore hopeful of finding it and they did, but although there were plenty of plants in one area, none were flowering or fruiting.[23] The seeds are still unknown but a cutting of that collection is now growing well in Munich Botanic Gardens, so fruits may one day be produced in cultivation.[24] The juvenile foliage of *A. occidentalis* is lanceolate. As the plant ages the leaves have a pre-adult stage in which they have just one gaping incision which extends right to the midrib. The mature leaf is entirely different again, with deeply divided leaves produced only at every fifth or sixth node. They are cut almost to the midrib into 4–6 lobes and have a pronounced joint (pulvinus) where they join the stalk.

A goosefoot out of step

Syngonium is the only genus of climbers in the subfamily *Colocasioideae*. Its closest relatives are therefore *Xanthosoma*, *Colocasia* and *Alocasia*, to which outwardly it bears little resemblance. There are 33 species of *Syngonium* throughout the New World tropics,[25] the most familiar of which is *S. podophyllum*, the goosefoot plant. As a juvenile this is an immensely successful houseplant, but like most syngoniums undergoes successive changes in leaf shape as it grows, and takes on quite a different character when mature. At first the leaves are sagittate to hastate, sometimes with a grey-green band down the centre. When pre-adult they develop three leaflets and progressively increase the number to 11 and lose the variegation. Since its introduction around 1870 the most highly variegated clones have been selected and those in cultivation today have very pale blades when young. Cultivars with cream foliage only faintly green at the margins are the most popular now, with 'White Butterfly' leading the field.

Epiphytic and climbing anthuriums

The tropical American genus *Anthurium* has two outstanding claims to fame: it displays a diversity of foliage unmatched anywhere else in the plant world and is the largest genus of aroids by far, with over 700 species. In fact, about half of all aroids in the New World are anthuriums. They are at their most numerous and diverse in northern South America but extend from the north of Mexico to southern Brazil and northern Argentina, as well as to the islands of the Caribbean. When Engler revised the genus in 1905 he described 58 species in Central America. Today, 219 have been determined by Dr Thomas Croat in the course of his remarkable work on *Anthurium*,[26] with others collected but insufficiently known and a good many probably still undiscovered.

The anthuriums in Panama alone now number over 150 and more than 80 of these are endemic. The reasons for this proliferation of species probably lie with the terrain, as many are isolated on mountain tops, and also with the fact that the berries are dispersed by rain forest birds which frequently have small rigidly defined territories. One such montane endemic recently described by Dr Croat is *A. globosum* which characteristically has a fat, often spherical spadix above an almost round spathe which is held horizontal and slightly cupped beneath it. This shape of spadix is unusual in the genus. Generally it is slender, tapered and tail-like, as the name *Anthurium*, meaning tail flower, suggests. *A. globosum* belongs to the large section *Calomystrium* and like most species in the group has heart-shaped leaves. In common with many anthuriums, it grows both epiphytically or terrestrially at higher altitudes in the rain forest.[27]

⸻ Scarlet spathes ⸻

Best known of all species in the section *Calomystrium* is *A. andraeanum*, popularly called wax flower on account of its brilliant red glossy spathes which are undoubtedly the finest imitation of shiny plastic in the natural world. It grows as an epiphyte in the wet mountain forests of northern Ecuador and southern Colombia; its local Ecuadorean name is *cresta de gallo*, meaning cockscomb – another good description of its flamboyant inflorescences.

This species was one of the most exciting horticultural discoveries of the 19th century. When the collector Edouard André found it in Colombia in 1876, he sent it to Jean Linden's nursery in Belgium and described it as 'without doubt one of the most beautiful, if not the most beautiful, of all my discoveries in South America'.[28] In 1882 a plant reached Kew and details were published in Curtis's *Botanical Magazine* by J. D. Hooker who commented, a trifle disparagingly, that it was 'one of the gaudiest plants that have of late been introduced into cultivation'. Two years later it was advertised in more glowing terms in William Bull's *Retail List of New, Beautiful, and Rare Plants* as 'one of the most striking and remarkable flowering plants', and soon every major nursery in Europe stocked it. Among them was Sander and Son, and it was from here that the Minister of Finance of the Republic of Hawaii, Samuel Mills Damon, bought a plant on a visit to London in 1889. Presumably some hybridization had already taken place, as his purchase had shell pink inflorescences (two reds can produce red, pink, or white offspring). On his return to Oahu he entrusted it to his Scottish gardener, Daniel Macintyre, who diligently propagated the plant and cross-pollinated its offspring to produce over 80 hybrids which laid the foundations of the Hawaiian anthurium industry and established this aroid as the symbolic 'heart of Hawaii'.

It is well known that many birds are attracted to bright colours, especially scarlet, and that bird-pollinated flowers are invariably brilliant reds, oranges or yellows. In spite of the eye-catching colour of the spathes and the contrasting yellow and white banded spadices, there are no records of birds pollinating *A. andraeanum*. Quite probably the spathes play a role in seed dispersal, serving to attract birds after flowering, as the spathe persists until the rather inconspicuous yellowish fruits are ripe.[29]

A. andraeanum is mostly valued for its cut-flowers and does not do very well as a houseplant. On the other hand, *A. scherzerianum*, a Costa Rican species which also has large red inflorescences, makes an ideal indoor specimen, tolerating low temperatures and low humidity, both of which are anathema to its cousin. In spite of their superficial similarity these two species are very different on closer inspection and not closely related. *A. scherzerianum* has brittle, leathery, elliptic leaves, minutely dotted on both surfaces with dark glands, whereas those of *A. andraeanum* are heart-shaped and gland-free. The spathe is similarly broadly oval, but not distinctly cordate at the base or puckered and glossy like that of *A. andraeanum*. The spadix is also different – orange to red and usually coiled upwards, whereas that of *A. andraeanum* is straight and downward pointing. Lastly the fruits are red to orange instead of yellowish. Perhaps most significantly from a grower's point of view, they inhabit different zones of the rain forest, *A. andraeanum* being found as a scandent epiphyte at 400–1300 m and

OPPOSITE]
Anthurium gracile (South America and the Antilles) A self-pollinating or apomictic epiphyte – one that does not require fertilization for reproduction – which lives in association with ants; like orchids, its roots have a thick outer covering known as velamen.

OVERLEAF]
ABOVE *Philodendron squamiferum* (Surinam and French Guiana) The stalks of this climbing species are clad in rubbery projections which at a distance resemble moss.
BELOW *Philodendron rugosum* (Ecuador) A recently discovered species which has the upper surface of the leaves minutely etched by numerous fine veins.

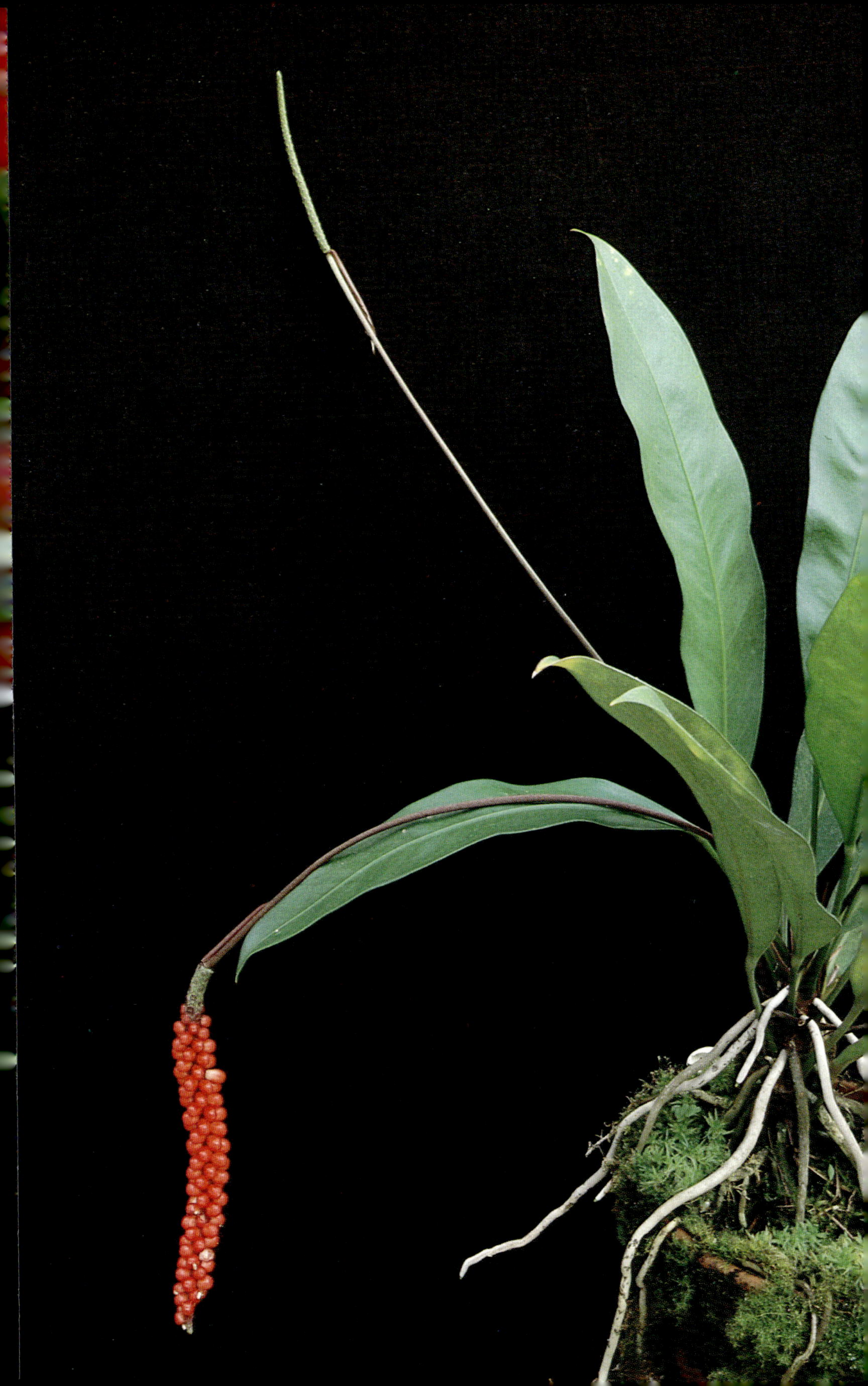

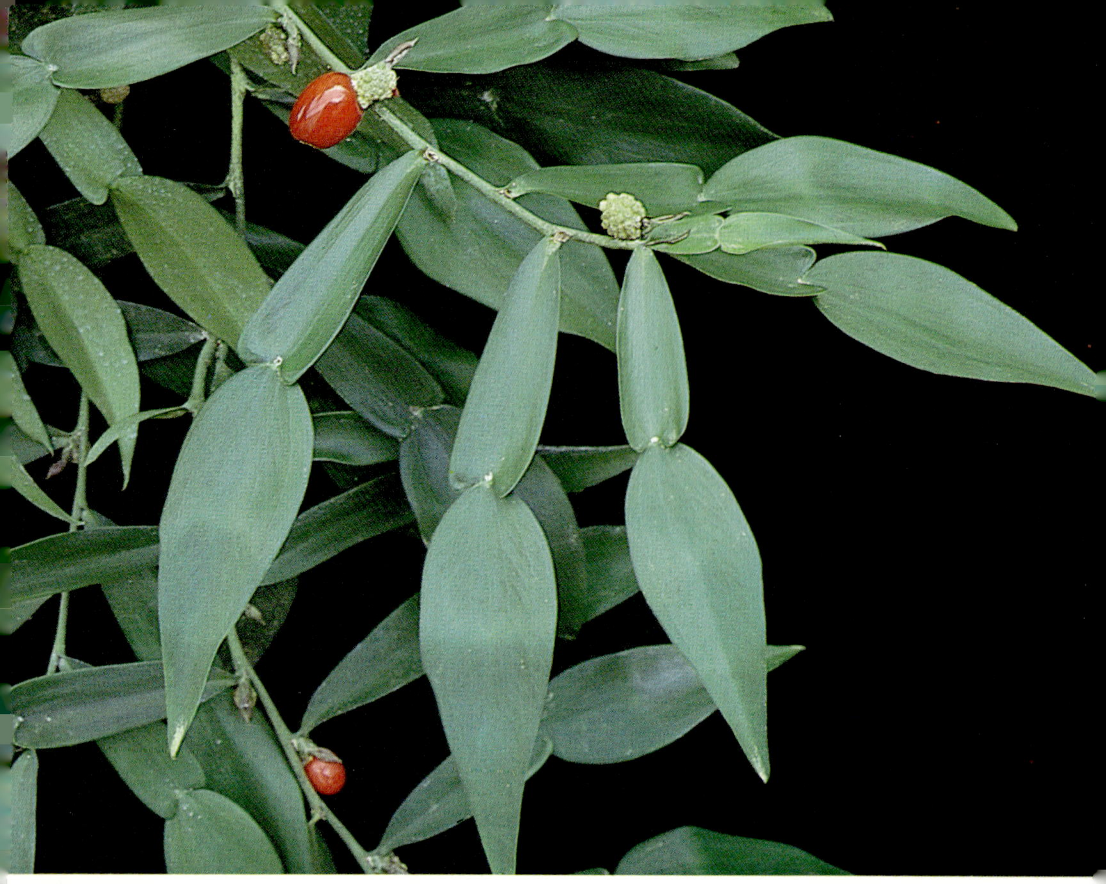

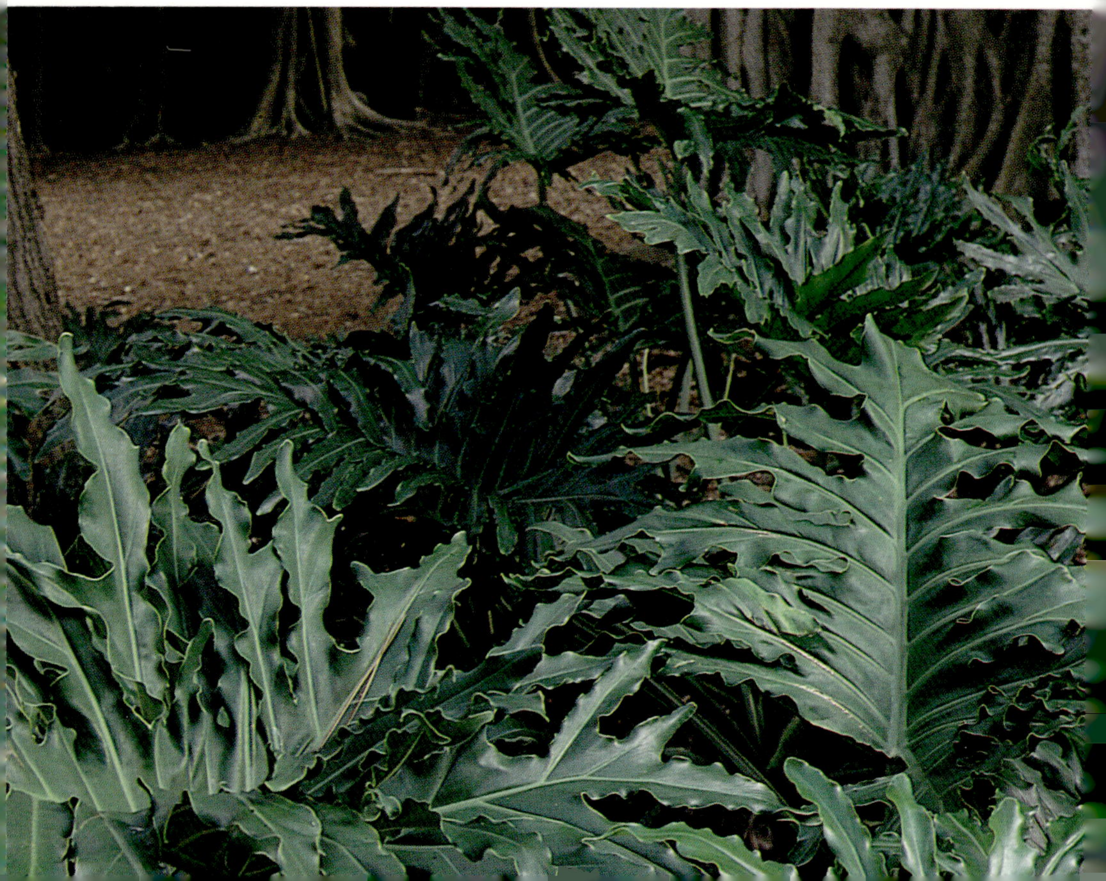

A. scherzerianum with more of a clump-forming habit from 1300 m–2100 m, sometimes epiphytically and sometimes at ground level – hence the latter's ability to thrive in cooler conditions.

Inevitably, many attempts have been made to cross these two scarlet anthuriums, the only ones in the genus, but all have failed. *A. scherzerianum* belongs to the very different section *Porphyrochitonium* whose members are characterized by, among other things, glandular dotting on leaf blades which are never heart-shaped. However, it has been crossed with some members of its own group, the most interesting being *A. wendlingeri*, a species from Costa Rica and Panama. This is an epiphyte with smooth, almost velvety pendent strap-shaped leaf blades up to 80 cm long and long pale spadices which hang down and curl extraordinarily, like pigs' tails. The spathes are lanceolate, greenish to purplish and also point downwards. The hybrids are midway between the two, with slender oval orange-grey spathes and quite long upward-pointing coiled spadices. As one would expect, a greater range of shapes and colours showed up in the second generation.[30]

Pendent anthuriums

Epiphytic anthuriums have several different habits of growth: vine-like and creeping; rosulate (rosette-forming) or bird's nest (named after the epiphytic bird's nest fern, *Asplenium nidus*); and the pendent strap-leaved mode as seen in *A. wendlingeri* and *A. punctatum*. Both latter species belong to section *Porphyrochitonium* but apart from the pendent leaves are rather

OVERLEAF]
ABOVE LEFT *Anthurium andraeanum* (southwest Colombia and northwest Ecuador)
The striking green and red cultivar is known as 'Mickey Mouse'.
ABOVE RIGHT *Anthurium spectabile* (Costa Rica) A pendent epiphytic species with blades up to 1.5 m long.
BELOW LEFT *Alocasia micholitziana* (Philippines) This fine form of the species goes under various names: 'Maxkowskii', 'Green Velvet', 'Green Goddess' and 'African Mask'.
BELOW RIGHT *Dieffenbachia seguine* 'Nobilis' (tropical America) Dumb cane is one of the most popular of houseplants but its deadly chemistry causes many cases of accidental poisoning each year in the home and in the horticultural industry.

OPPOSITE]
ABOVE *Pothos scandens* (India, Southeast Asia, China, Madagascar and the Comores)
A shrubby climber with pea-sized inflorescences, disproportionately large berries, and leaves which have flattened leaf-like stalks (phyllodes).
ABOVE *Philodendron bipinnatifidum* (southeast Brazil) A variable arborescent species which often grows in full sun.

different in appearance, the foliage of *A. punctatum* being ridged by deep-set veins. It also has white rather than reddish berries.

One of the finest pendent anthuriums is *A. spectabile* from Costa Rica. It has huge oblanceolate leaves which can reach over 1.5 m long, suspended from stalks 15–50 cm in length which are square in cross section. The blades are glossy and conspicuously hatched with rows of raised primary lateral veins which leave the midrib with military precision at 45°. Surprisingly, it belongs to the section *Pachyneurium* which consists of species with involute vernation (both leaf edges rolled at the margin when in bud), most of which are rosulate in habit.

⸺ Wallis's treasures ⸺

The most striking pendent anthurium must surely be *A. veitchii*. A specimen growing in the Palmengarten in Frankfurt at the turn of the century lived up to its reputation in Victorian times as 'the noblest inhabitant of European stoves'.[31] It had over 20 oblong leathery leaves approaching 1 m long and 30 cm wide, heart-shaped at the base and regularly ribbed down their length by arching, deeply sunken primary veins. The flower stalks are 15–50 cm long, slightly shorter than those which support the massive blades, and bear pale green spathes and greenish cream spadices whose male florets release purple pollen. The fruits are red. Engler placed it in section *Calomystrium*.[32]

A. veitchii and the equally spectacular but very different *A. warocqueanum* were both discovered by Gustav Wallis. He entered the service of Veitch and Son in 1870 and in 1872 was sent to New Granada (present-day Colombia). Two years later he returned with these magnificent plants, both from Colombia, which have few rivals for their foliage anywhere in the plant world. *A. warocqueanum* is an epiphytic creeper; in a well-grown specimen (and it is not easy to grow) the pendent elongated heart-shaped leaves reach up to 90 cm long. They are a rich velvety green with white veins. It has been put in section *Cardiolonchium* because of its velvety cordate blades. This is however disputed by John Banta, who in his breeding programme with anthuriums has discovered incompatibilities within sections that suggest features such as this may not necessarily indicate a close relationship.[33] He finds berry colour of greater significance. (Those of *A. warocqueanum* are red whereas those of other species in the section are orange or purple.)

Creeping anthuriums

In complete contrast to the majestic anthuriums are the small creepers. Several are quite distinctive. *A. radicans* makes a dense growth of puckered heart-shaped leaves and clusters of pale inflorescences with short spiky spadices. Its combination of characteristics is unique and it stands alone in section *Chamaerepium*. It can however be crossed with species outside its section. With *A. dressleri* (a purple-berried terrestrial species from Panama and Colombia with typical *Cardiolonchium* velvety cordate blades), it forms superb compact plants with spreading clusters of deeply veined heart-shaped leaves.

An even smaller and more puckered creeper is *A. clidemioides* from lowland rain forests in Costa Rica. At a glance it could easily be mistaken for a species of *Dioscoreaceae* or *Melastomataceae* with its vine of almost stalkless heart-shaped leaves. It is in the section *Polyphyllium* which also includes *A. flexile*, a Central American climber. Superficially they would seem to have little in common, as the latter has unremarkable smooth elliptic leaves but they are the only two anthuriums with roots between the nodes and the only ones in which new leaves are not protected by cataphylls. Instead, each new leaf emerges from a sheath in the stalk of the previous leaf.[34]

Berried treasures

The commonest anthurium of all is a small creeper, *A. scandens*. It grows just about everywhere an anthurium can grow, from Mexico across to the West Indies and down to southern Brazil, in all kinds of rain forest from sea level to 2700 m. Genetically it is extremely complex and this accounts for its variable appearance. It is very similar to *A. trinerve*, which is considered by some botanists to be yet another variant. The latter is confined to rain forest in the Amazonian region and is differentiated by its persistently erect white spathe, white spadix, prominent pistils and terrestrial habit. Both species are apparently self-pollinating and not crossable with any others. *A. scandens* has rather plain elliptic leaves and tiny green inflorescences but is worth growing for its berries alone. They are large for the size of the plant, and may be a pearly white, pale lilac or purple.[35]

Two other anthuriums with attractive fruiting spadices are *A. bakeri* and *A. gracile*. The former is a Central American epiphyte of wet forests with slender gland-dotted matt leaves some 20–50 cm long and brilliant scarlet pointed berries. *A. gracile* is rather similar with bright red berries but is smaller and has a purplish-brown rather than cream spadix and no glandular dotting. This is another species which is unique genetically, and self-pollinating, so readily bears fruit. It also has white roots which in the wild are often infested with ants[36] and are apparently covered, like those of orchids, in a layer of velamen, a thick outer skin that absorbs water from the atmosphere.

——— Bird's nest anthuriums ———

Some of the most distinctive epiphytic anthuriums are of the bird's nest type. Nearly all have involute vernation and belong to section *Pachyneurium*. Among the few exceptions are *A. hacumense* and *A. hookeri* which, because their leaves are convolute in bud (one side wrapped round the other like a Swiss roll) and gland-dotted on the lower surface, belong to section *Porphyrochitonium*. *A. hacumense* is peculiar also in having a stipe almost 30 cm long which often exceeds the length of the spadix. A more orthodox rosulate species is the huge *A. salviniae* which occurs in wet forests from Mexico to Colombia. It has broad oblong short-stalked leaves up to 1.8 m long and pendent inflorescences with narrow purplish-green spathes and lavender spadices. The spadix can reach over 40 cm long, which gave rise to the synonym, *A. enormispadix*. Because of its dimensions, it is less common in cultivation than the very similar but smaller *A. schlechtendalii*, a species found from Mexico to Costa Rica. In cultivation it is

Anthurium salviniae ($\times \frac{1}{20}$)
Bird's nest anthuriums are specialized epiphytes which can tolerate dry exposed positions in trees and on rocks. The vase of leaves funnels debris and rainwater to the dense mass of roots at the base, forming a damp compost heap which sustains the plant.

sometimes described as *A. tetragonum* and is frequently confused with *A. crassinervium* (which is restricted to northeast Colombia and northern Venezuela). Both *A. schlechtendalii* and *A. crassinervium* and a number of other very similar bird's nest anthuriums are known as the '*Anthurium crassinervium* complex'. They occur from Mexico to Paraguay and have in common leathery leaves which often undulate at the margins and red or purplish-red berries. Otherwise they are variable in appearance and adaptable as to habitat, taking to the trees in moist forests and occurring as smaller plants on rocks in drier localities.[37]

One of the most handsome and manageable-sized bird's nest anthuriums is *A. superbum* from seasonally flooded forests in Ecuador. It has about ten very dark green shiny leathery leaves some 50 cm long which are furrowed by the main veins and held stiffly upright. The midrib is prominent and sharply angled and young leaves are purplish underneath. Additional attractions are the pink roots and mauve berries.

Perhaps the oddest and most distinctive of the birds' nests is *A. reflexinervium*,[38] a small species from Peru which has such heavily puckered leaves that it looks like a rather elegant heartless Savoy cabbage.

Bird's nest anthuriums have a highly specialized growth habit. The broad upright, sometimes overlapping leaves form a funnel which greatly increases the catchment area for rain. The vase also serves to collect debris so that the plant forms its own compost heap which releases nutrients as it decomposes. The dense root mass grows both downwards to anchor the plant to tree or rock and upwards to retain falling organic matter. For this reason, these species are also known as litter-basket epiphytes. Their efficiency at water and nutrient retention means that they can colonize forests with a pronounced dry season, especially if their leaves are thick and leathery too. Some Mexican endemics such as *A. schlechtendalii* subspecies *jimenezii*, *A. nizandense* and *A. halmoorei*, and *A. affine* in northeast Brazil not only endure extremely dry spells but can grow on exposed rocks.

—— Philodendrons ——

Philodendron, with over 340 species described to date, is the second largest genus after *Anthurium*. Though less diverse in foliage, it is more varied in its habits and habitats. It includes arborescent species as well as climbers and epiphytes and frequents wet places in addition to rain forests and dry scrub. The most familiar however are the climbers and epiphytes of tropical rain forest, from which we have derived some of the most popular ornamental aroids of all. They are pre-eminent in northwest South America but extend from Mexico through Central America, across to the West Indies and down the whole of South America, reaching just into southern temperate regions around Montevideo. The generalization may be made that philodendrons tend to be more diverse at lower altitudes than anthuriums and prefer higher positions in the canopy.

The name *Philodendron* means tree-loving and the majority of species are found growing up trees though many will also grow on rocks or the ground in certain situations. Like most climbing aroids, philodendrons germinate in soil but may later die off at ground level and become semi-epiphytic, clinging on with short clasping roots and sustaining themselves through long feeder roots which dangle earthwards and eventually penetrate soil or crevices in rocks.

Philodendron seedlings have heart-shaped leaves but subsequent growth may go through a succession of sizes and shapes until the final adult stage is reached. *P. radiatum* is a good example: beginning with entire blades and becoming increasingly incised so that mature foliage is pinnatifid (cut almost to the midrib). As in *Anthurium*, almost the whole range of aroid leaf shapes can be found in this genus, from simple oval, elliptic and linear, or basic sagittate, hastate and cordate, to the more complex tripartite, lobed,

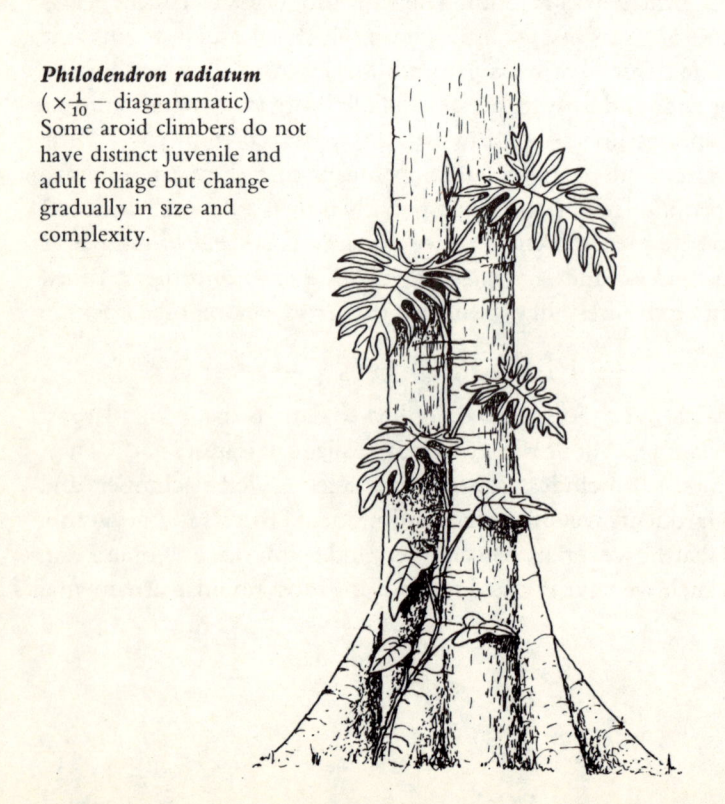

Philodendron radiatum
($\times \frac{1}{10}$ – diagrammatic)
Some aroid climbers do not have distinct juvenile and adult foliage but change gradually in size and complexity.

pinnately lobed and pedate. The veins are arched and neatly parallel to one another, in contrast to those of anthuriums in which the main laterals are connected by an irregular network of subsidiary veins. This can clearly be seen by comparing *P. melanochrysum* with *A. warocqueanum*. Both have rather similar leaves: elongate heart-shapes with pale veins and a velvety texture, but the main laterals of the *Philodendron* are quite simple, whereas those of the *Anthurium* branch irregularly into smaller silver veins.

Unlike anthuriums, but like monsteras, philodendrons are able to put out long slender leafless shoots to explore the space around them for more suitable conditions if they experience limitations to growth. Some species, such as *P. fragrantissimum* and *P. linnaei*, regularly alternate modes of growth so that rosettes of foliage occur at intervals along leafless sections of stem.

Philodendron inflorescences show features which indicate highly specific pollination strategies. They are mostly borne on short stalks, singly or in clusters of up to 11, and have waxy persistent spathes which are tubular at the base, hooded above and often bicoloured. In some species the base and upper portions contrast and in others the inner and outer surfaces differ. In most cases the paler colour is white or green and the darker shade red or crimson. The spadix is usually white, more or less the same length as the spathe and has a zone of sterile florets (staminodes) between the male and female flowers. Many species are fragrant and the release of scent is often precisely timed to coincide with the habits of their pollinators: scarab beetles belonging to the families *Rutelinae* and *Dynastinae*.[39] Other insects pay visits but have not so far been implicated in pollination. Wasps of the *Chalcidae* family lay eggs in the ovaries of many species which results in galled inflorescences.[40] *Trigona* bees have been observed collecting the resin which is exuded during flowering from either the spathe or the spadix, a feature which is unique in *Araceae*. However, the resin's real purpose is to glue pollen to pollinating beetles. Resin is also found in stems and leaves. It may be colourless, yellow, orange or red, but turns brown when exposed to the air.[41]

—— Ant plants ——

In most philodendrons the nectaries – glands which secrete a sweet sticky substance – are outside the flower (extra-floral). They may occur on spathes, stalks, sheaths, and on the underneath of leaves. *P. leal-costae* has them on the outside of the spathe and *P. martianum* has so many on the sheaths that the surface becomes streaked with nectar. These secretions attract ants whose aggressive presence deters insect pests which would otherwise devour the plant. Species in other families are known to entertain ants for their own benefit and some even provide special inflated organs for the ants to live in. No aroid has yet been found which goes to such great lengths but a number can be considered myrmecophiles, that is, living in close association with ants. The obvious example is *P. myrmecophilum* (also known as *P. deflexum* and *P. megalophyllum*) of the Amazon basin. It perches in trees, along with bromeliads, epiphyllums and peperomias. The ants feed from extra-floral nectaries at the top of the leaf stalks and set up home near this food supply, transporting soil grain by grain from the forest floor to build a nest around its roots. The plant not only benefits from having them as bodyguards, but also utilizes the nutrients in the nesting material.[42]

—— Climbing philodendrons ——

Most of the philodendrons popular in cultivation are climbers: some delicate and others robust, but each having its characteristic level in the strata of rain forest vegetation, from low creepers to high-rise lianas. The best known of all is *P. scandens*, the sweetheart vine, which, as ornamentals go, must be regarded as the tropical equivalent to ivy (*Hedera helix*). As a houseplant it grows almost as easily, given a little more warmth, and in the tropics will quickly curtain a wall or fence with innumerable neatly pointed cordate leaves, much the same as ivy does in temperate areas. In fact, an early but incorrect name was *Arum hederaceum*. Like *Anthurium scandens* it is extremely common and widespread in the wild, ranging from Mexico and the West Indies (from where it was introduced to horticulture in 1793)[43] to eastern Brazil.

Because of its variability, wide distribution and differing juvenile forms, *P. scandens* has acquired a number of names over the years and is sometimes still referred to as *P. oxycardium* or *P. cordatum*. Several subspecies are now recognized and their juvenile forms are frequently used as ornamentals. The commonest is subspecies *oxycardium* from Mexico which has its new leaves flushed brown and the loveliest is the Panamanian subspecies *scandens* forma *micans*, whose foliage is a velvety dark green with reddish undersides. All forms are capable of great vegetative spread and climb high into tree tops. Flowering is rare. When it does occur, greenish-white inflorescences are borne at the end of pendent shoots which are sent from tree top to ground level.

Silver-variegated philodendrons include the lovely *P. ornatum* (sometimes called *P. sodiroi* or *P. asperatum*), an average-sized climber with bluish-green heart-shaped leaves which are heavily smudged with silver grey in the juvenile stage. The fine foliage has the added attractions of reddish veins on the undersurface and purplish stalks which are verrucose (covered with fine wart-like protuberances). Adult blades are similar in shape but reach 60 cm or more and have few, if any, silver markings. Also silver-blotched is *P. mamei*, differing in broader, finely furrowed blades with less variegation. It is endemic to Ecuador and slow-growing, with internodes so short that it appears to form clumps.

Most connoisseurs would agree that the finest cordate philodendron is *P. melanochrysum*. It is often still referred to as *P. andreanum*, the name originally given which commemorates its first European collector, Edouard André, who found it in Colombia in 1886.[44] '*Melanochrysum*' means 'black-gold' and this describes the minute gold sparkles which the velvety blackish-green blades reveal when sunlit. The elongated heart-shaped foliage is further enhanced by the salmon flush of new leaves and the contrasting white midrib and main lateral veins.

Extra texture

Its chief rival in beauty is *P. verrucosum* which has thinner broader leaves with lightly undulating margins and indented main veins, around which the green of the blade fades to almost white. They are brilliant satiny lime green when young, with the areas between the main veins flushed olive above and pink below. The stalks and outer surfaces of sheaths and inflorescences are densely covered in soft bristles.

An interestingly textured epidermis is a feature of several ornamental philodendrons. *P. squamiferum* (from Surinam and French Guiana) is a distinctive species with blades cut into five flaring lobes and dark red stalks clothed in bristles which resemble moss. The Central American *P. pterotum* is one of several species in which the stalk is flattened and has winged margins.

A recently discovered species, *P. rugosum*,[45] has a fascinating leaf texture. The leathery leaves have broadly heart-shaped basal lobes which meet or slightly overlap and the surface is given a leather-like grain by the countless fine impressed veins and veinlets. It grows at quite high elevations (1000–1700 m) in one area of Pastaza in Ecuador and is reported to be a tough adaptable plant in cultivation.[46]

Shapes galore

As well as a variety of colours and textures, every conceivable kind of lobing distinguishes *Philodendron* climbers. *P. radiatum* is incised two-thirds of the way to the midrib five to ten times on each side of the blade when mature (juveniles have fewer shallow indentations and are often sold as *P. dubium*). *P. lacerum* is similar but cut only halfway, as is *P. angustisectum* (formerly *P. elegans*) which is pinnatisect to within a few centimetres of the midrib. The unusual five-lobed leaf shape of *P. squamiferum* is also seen in the juvenile stage of *P. pedatum* (formerly known as *P. laciniatum*). Adult leaves develop additional lobing though still retain five main divisions. *P. pedatum* is found from southern Venezuela to southeast Brazil. The cultivar 'Florida' has irregular white variegation. Some areas remain dark green, others are marbled green and white, with pure white lobes here and there.

The elephant's ear climbers

Perhaps best known of all climbers in the genus are a group of species with shiny sagittate leaves known popularly as elephant's ears: *P. erubescens*, *P. imbe* (also known as *P. sanguineum*) and *P. ilsemannii* (often called *P. sagittifolium*). The first is unmistakable with its purple stems and stalks, pink sheaths, dark green blades with maroon undersides and purple and red inflorescences. It originated in Colombia but is now grown worldwide and has numerous colourful cultivars with names like 'Red Emerald' and 'Red Duchess'. Most dramatic are ones such as 'Brandy Wine', 'Black Cardinal' and 'King of Spades' which have dusky purplish black foliage. In complete contrast, the cultivar 'Golden Erubescens' is clear yellow with pink undersides and 'Prince of Orange' has orange new leaves which change to yellow and finally pale green as they develop.

P. imbe from eastern Brazil is very similar but with narrower leaves and less red coloration. Leaf undersurfaces are faintly reddish, stalks and sheaths are merely dotted red and the inflorescences are green and cream with a red flush inside the tube. It can also be distinguished by having only about three main veins which leave the midrib almost at right angles.

Quite different again is *P. ilsemannii* which is irregularly variegated with white. Each leaf is different – some basically white with sparse green spotting and others with large green areas which occasionally cover one half of a blade. Young leaves are an exquisite creamy pink. It was first described in 1908 but its origin is unknown. Engler regarded it as the juvenile phase of *P. sagittifolium*[47] but other sources suggest a link with the variegated form of *P. cordatum*.[48] In addition to these species, many elephant's ears in cultivation come under the name *P. domesticum* (also known as *P. hastatum*). No one seems to know where this kind came from either. It differs from *P. imbe* in having five to six main laterals and a slightly flattened ridged (rather than round) leaf stalk.

Rosette-forming philodendrons

Bird's nest philodendrons grow in the crowns of trees, often in exposed positions high in the canopy. *P. wendlandii* is a Central American species with dark green narrowly oval to oblong leaves. The short stout stalks have flat upper sides and continue into broad midribs which taper but remain pronounced. The spathes are pale reddish green at the base with hoods that are cream outside and dark red within. It is an important species in hybridization, producing compact upright-leaved plants. *P.* 'Wend-imbe' has bright green leaves with undersides faintly marbled with red and dazzling inflorescences with pure white spadices and red spathes lined with yellow.

Unusually long-stalked, semi-pendent inflorescences are borne by *P. insigne*. They have an immaculate white outer spathe, a ruby red interior and a white spadix which secretes resin during flowering.[49]

Coming up with the same solution

Plants perched high up in trees dry out very quickly indeed. Epiphytic orchids have got round dehydration during spells of rainless weather by having leaf stalks modified to store water and these so-called pseudobulbs help keep the plant turgid when moisture is at a premium. *P. martianum* has come up with a similar design. It has shiny leathery ovate leaves about 45 cm long, supported by swollen spongy stalks. The internodes are very short so the leaves form a clump which is basically rosulate though in heavy shade they may splay out towards the light. The inflorescence is cream. This species was first called *P. cannifolium* because of a vague resemblance to the leaves of cannas. A better description might have reflected the interesting likeness to those of orchids.

Self-heading philodendrons

Outstanding among philodendrons are those with well-developed stems topped by a crown of foliage. In many species the stem is rigid, stout and unbranched and acts as a trunk. Others branch forming a thicket with several heads of leaves and may even hang from trees suspended by their tough aerial roots, as *P. goeldii* often does. Species with an arborescent habit make imposing plants for landscape use in the tropics and when young are excellent non-climbing pot plants. They cannot however be propagated easily from cuttings as they effectively have only the one growing point, lateral buds being present but difficult to stimulate into growth.

P. bipinnatifidum is the arborescent *Philodendron par excellence*. The juvenile stage with barely undulating leaves is a very popular indoor plant. As it grows the blades become more lobed and wavy until they reach a metre in length on stalks as long again – the whole plant measuring some 2 m tall on a trunk about 10 cm in diameter. The inflorescences are normally solitary with green spathes averaging 20–30 cm long which are maroon or green at the base outside and cream inside. Some populations are pollinated exclusively by *Erioscelis emarginata*, a scarab beetle of the family *Dynastinae*, but others appear to have a wider variety of visitors.[50] They are attracted by the strong resinous fruity scent which the flowers emit at dusk. During flowering the spadix raises its temperature over 20° C above that of the surrounding air to dissipate this odour.[51] The berries are white, sweet and edible when ripe and are used as food by local people, who also cut the thick roots for cordage. It is a variable species with a wide distribution through northern Argentina, Bolivia, Paraguay and southeastern Brazil. Some confusion exists over the correct name, as in the past botanists (including Engler) have recognized both *P. bipinnatifidum* and *P. selloum*. Some have used the two names to differentiate distinctive populations[52] but recent studies suggest that they are best regarded as a single polymorphic species.[53]

P. bipinnatifidum is adapted to high light levels. In rain forests it tends to grow epiphytically to avoid heavy shade but the arborescent habit offers a move away from dependence on trees for a leg up towards the light. Species with this habit, unlike climbers and bird's nest epiphytes, can very easily stand on their own roots and with the makings of a trunk and a dense crown of leaves are not overshadowed by everything else around them. So inevitably *P. bipinnatifidum* and its kin are mainly found in areas where the tree cover is thinner and the light higher – the edges of waterways, swamps, coastal scrub or outcrops of rock – and growing tree-like in the light can cast their own shadows.

8
The Titans

Giant tuberous species of the tropics

The island of Sumatra is bisected by the equator and down the western side runs a chain of active volcanoes with rugged limestone cliffs and chasms whose contours are deceptively softened into green mounds of vegetation. The rain forests which clothe these mountains are some of the oldest in the world and 60 million years of uninterrupted growth has nurtured the evolution of a tenth of all known species. Human constructions and cultivation scar them, often extensively in the south. Roads are cut through, with villages at intervals like beads on a string, and open sunny paddy fields reach to the very edges of steep jungle-clad slopes. But for all the felling the abiding impression of the island is that the dark and silent forest still predominates, a fearsome presence which guards its secrets well.

—— A botanist in wonderland ——

It was in 1878 that one of its most remarkable secrets was first revealed to science: the existence of an aroid with an inflorescence taller than a man. The discovery of this species (*Amorphophallus titanum*), which must rank as one of the greatest highlights of natural history exploration, was made by the Italian botanist and explorer Dr Odoardo Beccari. He gave an account of the event in a letter to his friend, Marchese Bardo Corsi Salviati:

> It was on the 6th of August 1878 at Ajer Mantcior in the Padang Province of Sumatra, that I found the leaves of this extraordinary plant. Shortly afterwards, being at Kaju Tanam, a place not far from Ajer Mantcior, and there informed that *Amorphophallus* was common in the surrounding hills, I offered a large reward to anyone who would bring me a flower. This promise produced a more speedy effect than I could have hoped for, for on the 5th September, towards midday, I had the satisfaction of possessing a flower of this marvel. The single flower (or more correctly, inflorescence) with the tuber (from which it springs directly) form together so ponderous a mass, that for the purpose of transporting it, it had to be lashed to a long pole, the ends of which were

placed on the shoulders of two men. To give an idea of the size of this gigantic flower, it is enough to say that a man standing upright can barely reach the top of the spadix with his hand, which occupies the centre of the flower, and that with open arms he can scarcely reach halfway round the circumference of the funnel-shaped spathe, from the bottom of which the spadix arises.'[1]

Beccari sent tubers and seeds to his friend in Florence, but the tubers were detained by customs in Marseilles under legislation aimed at preventing the introduction of *Phylloxera* (insects which attack grape vines) and they rotted.

——— The Sumatran Titan reaches Europe ———

The seeds fared better. Some germinated in the Marchese's garden, though presumably did little else and of these, one which was sent to Kew flowered ten years later (in 1889) after skilled cultivation by Mr W. Watson.[2] His protégé had put on over 25 kg (57 lb) in weight and rewarded him with an inflorescence 2 m (6 ft 9 in) tall. Until this happened, Dr Beccari's description was understandably greeted with some disbelief — traveller's tales, like fishermen's, often containing a certain element of exaggeration. Even J. D. Hooker, the director of Kew, had his doubts and he was obviously relieved when a second westerner (the traveller H. Forbes) reported having come across the plant: 'unimpeachable as is Dr Beccari's character for scientific accuracy, it is satisfactory to have had . . . a full confirmation of his account'.[3] Forbes's descriptions were no less spectacular: he recorded a tuber measuring over 2 m in circumference which took 12 men to bear its weight.[4]

——— A shocking sight ———

Plant hunting was of course at its height in the Victorian era and new species were brought from the farthest corners of the earth to enthrall a discerning audience who enjoyed the curious as much as the beautiful. Reports of this plant therefore created an unprecedented stir, but inevitably many were not only surprised but shocked by the details. For this was the period when governesses protected their charges from the indecent sight and smell of the stinkhorn fungus (*Phallus impudicus*) by ordering a daily routine of destruction in the autumn as it emerged in shrubberies where young ladies were likely to pass by. Regardless of this concern for propriety, the botanists of the day refused to beat about the bush and gave both fungus and the aroid names which graphically described their appearance: *Amorphophallus* meaning 'shapeless phallus'.

It is not hard to imagine then that the lifesize painting which Beccari commissioned in 1881 to commemorate his extraordinary discovery must have caused a few raised eyebrows. It measured over 5.5 m by almost 4.75 m (18 ft × 15 ft 6 in) and showed a massive leaf growing out of the ground and two Sumatrans carrying the flowering plant lashed to a pole. Apparently it was later presented to Kew and once graced the ceiling of the Wood Museum. Then it disappeared. William Blunt, in his book *In for a Penny*, made the plea: 'If Puritans contrived its removal, surely in this permissive age it could be reinstated.'[5] I am sure it could – only it has never been seen again!

Vital statistics

The Titan arum's statistics are unbelievable. The tallest inflorescence reported was 3.3 m[6] and the average is around 2 m. It has a fairly short stout spotted stalk (25–35 cm long and 10 cm in diameter) and a few large grey-green spotted bracts. At first the spadix is almost completely enveloped within the concertinaed ribs of the spathe but as the bud develops its growth accelerates greatly. When open, the spathe is like an upturned bell, fluted and frilled towards the margin, pale green with whitish spots on the

Amorphophallus titanum
Right: ($\times \frac{1}{25}$) The inflorescence of the Titan arum or *bunga bangkai* (corpse flower) of Sumatra is the largest in the family, reaching 2 m or more.
Far right: ($\times \frac{1}{15}$) It is pollinated by large beetles which are lured into the spathe by the powerful smell of decaying flesh. They climb freely over the flowers but are unable to negotiate the smooth spathe walls or overhanging bulge of the appendix and are thus detained while the separate phases of female and male flowering take place. When flowering is over the inflorescence begins to collapse, enabling the pollen-covered beetles to gain a foothold and escape – perhaps to a newly opened Titan arum.

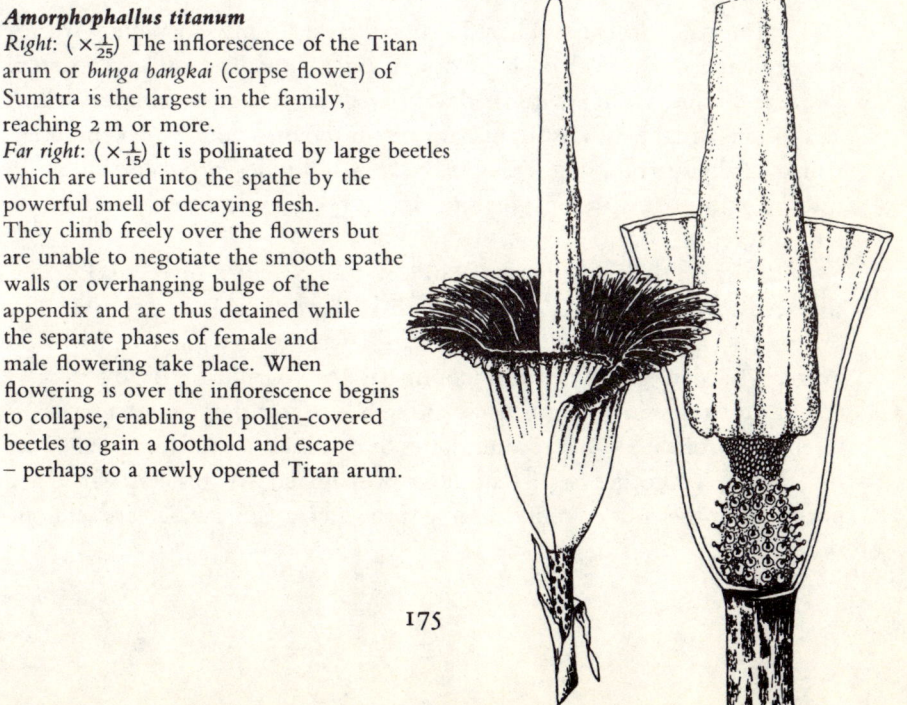

outside, and inside a rich dark crimson fading to greenish-yellow at the base. It measures about 1.25 m in height, 1 m across, and over 3 m in circumference. The spadix appendix is a pale, rather dirty yellow, creased, and 1–2 m tall. It is essentially hollow inside apart from a web of white fibres. The flowers are hidden in the vase formed by the lower spathe. Male florets number anything from over 450 to nearly 5000[7] and are cream in colour, closely packed in a 6–8 cm ring just below the point where the spadix bulges to form the sterile visible portion. Below them is a wider light green band from which the 700 or so female florets protrude. Each is 2 cm long and has a bulbous pink base containing the ovary, a narrow purple style and a yellowish stigma. When these are ready to receive pollen, the yellow rind of the appendix emits an overpowering, nauseating stench which gives the plant its Indonesian name of *bunga bangkai*, meaning corpse flower. The smell is 'perhaps the most powerful and disgusting odor produced by any plant'.[8] Observers have noticed that it comes in waves and at close quarters acts as a floral tear gas, making the eyes smart and water. One chemical thought to be responsible is butyric aldehyde.[9]

The inflorescence is produced after a dormant period of three or four months and the single leaf develops later. If the flowers are pollinated, olive-sized fruits 3–4 cm long are formed in a cylindrical cluster about 50 cm in length. They turn bright red when ripe and look very much like an Alice-in-Wonderland version of the familiar arum berries of temperate hedgerows.

The leaf is a marvel in itself and at first glance resembles a sapling tree. A mature one has a smooth dark green stalk over 2 m tall and at least 20 cm in diameter at the base. It is patterned with irregular pale green blotches which at a distance make it indistinguishable from the moss and lichen encrusted trunks of the surrounding trees. At the apex it divides into three and again into several more divisions, making an umbrella of leaflets up to 15 m in circumference.

Both the single leaf and solitary inflorescence are produced, at different times, by a hemispherical tuber which is the largest in the world. The greatest weight recorded is 75 kg[10] and over 50 kg (the weight of an average-sized woman) is not uncommon. One corm dug up by Beccari was almost 1.4 m in circumference and so heavy that the two men carrying it fell and broke it.[11] One account states that numerous buds are formed on the tuber.[12] Another says that it is propagated by offsets which are produced at the ends of runners some three metres long which are sent out by the parent tuber before it dies.[13] However, it is also thought that *A. titanum* does not produce offsets at all and only an annual thickening of the base of the corm takes place.[14] Certainly, it is not successfully propagated from adventitious buds.[15] In the literature there is likewise some disagreement about whether the tuber shrinks, grows, or stays the same after flowering or developing a new leaf.[16] It would seem likely that the production of an inflorescence would deplete the tuber, but as soon as flowering was over and the new leaf emerged, weight would rapidly be regained. The first tuber to flower at Kew was recorded as losing 9 lbs during flowering.[17]

There was once a rumour that this monstrous inflorescence is pollinated by elephants. This does conjure up the most delightful cartoon image of jumbo flapping from one elephantine bloom to another, dipping a pollen-covered trunk into the voluminous spathe, but the truth is more mundane. Instead the fearful smell attracts dung and carrion beetles, particularly the large *Diamesus osculans* and *Creophilus villipennis*.[18] Once at the bottom of the spathe they are trapped – unable to scale the smooth spathe walls and prevented from escaping up the spadix by the steep overhang formed directly above the zone of male flowers by the bulge of the appendix, from which they repeatedly drop off back into the vase. Only after the pollen has been released do the tissues break down sufficiently for them to climb out, receiving a coating as they make their final crawl over the male flowers.

Anyone who has been to the equatorial tropics will have vivid memories of the rain. Going outside during a Sumatran downpour is like walking into a waterfall and so it would seem that a plant which produces a vase-shaped inflorescence is asking for trouble. This occurred to observers of the specimen which flowered in New York Botanic Garden in 1937 and a bucket of water was poured into the spathe to test its water-holding capacity. They were relieved to find that it seeped away rapidly through a 1 cm gap at the base of the overlap. In addition, the overhanging bulge of the appendix may shelter the flowers, especially the males which are directly under it. When flowering has finished the spathe collapses and twists, effectively sealing the chamber so that the fertilized female flowers are kept dry.[19] As the fruits develop, it rots away.

—— In captivity ——

The Titan arum is difficult to cultivate – not because, as was also once rumoured, it eats its grower – but because it is prone to rotting, does not reliably increase in size and fails to produce seeds or offsets. Botanic gardens in Europe have flowered it only 11 times and all but the first flowering at Kew have been from imported mature tubers. The best record, not surprisingly, is held by Bogor Botanic Gardens in Java where it grows outdoors and flowers approximately every two years. But even there it has never set seed. The reason for this is because the female flowers open first and by the time the male florets are producing pollen they are no longer receptive. Therefore there has to be another inflorescence in the vicinity whose male flowers coincide with its female flowers. Fruits are apparently rare in the wild too. Plants grow in small groups in the forest and a mature tuber flowers only every second or third year at the most, lasting but a few days. In addition, flowering can take place at any time of the year, so all-in-all chances are against two plants in the same area producing inflorescences whose male and female ripeness coincides. Perhaps the massive inflorescence is necessary to produce a sufficiently strong smell to attract pollinators to blooms growing far apart.

Whenever the Titan arum has flowered in cultivation it has attracted an enormous amount of attention and is by far the greatest superstar of the botanical world. Every report of the Titan arum flowering in cultivation makes fascinating reading. At Kew in 1891 an artist named Miss Smith had the unenviable task of remaining in close proximity to the inflorescence in order to draw it. The atrocious stench made her suffer 'prolonged martyrdom that terminated in illness'.[20] The flowering at Kew in 1926 received such publicity that 'so great was the crowd . . . that a special cordon of police was required to maintain the queue at the entrance to the conservatory'.[21] Onlookers were eager to be photographed in its presence in spite of the smell. At one of the early Kew flowerings they face the camera without trace of a grimace but in 1956 when it bloomed in the botanic gardens of the Dutch city of Leiden, such decorum was a thing of the past and the observers are all holding their noses, the looks of admiration evidently tinged with nausea.[22] In Java the story is the same: at the 1970 flowering, which lasted two days, 34,000 came to see the bloom.[23]

Here is the scene in the New York Botanic Garden in 1937:

Within a week (of the flower bud's appearance) news photographers had begun their invasion of the tropical aquatic house (in a steamy temperature of 96°F) where the plant was being grown. The daily press kept the public informed of the progress of the gigantic bloom, which was increasing an average of four inches in height a day. When the first definite announcement was made that it was opening (about 3 pm on June 7) the invasion became a bombardment of the greenhouse, with photographers from half a dozen motion picture companies setting up apparatus on a platform built in the pool and around the margins, cameramen from nearly every newspaper and press service in town as well as from many magazines taking pictures from all angles, and reporters from the major papers and weekly magazines making observations and asking more questions than staff members had heard in all the previous years together. . . . The public were kept outside, but when the ventilators were opened a trifle at night to keep down the temperature in an effort to prolong the life of the bloom, enough of the widely publicised stench crept out into the atmosphere to satisfy the curious as to the quality of the smell.[24]

This plant had been imported from Sumatra in 1932 as a tuber weighing almost 27.25 kg. The spectacular event so impressed people that when a second imported specimen flowered two years later it was designated as the official flower of the Bronx. The Borough President commemorated it with the following announcement:

> *Amorphophallus titanum* (giant krubi) is the world's largest flower. The only place in the Western Hemisphere where this giant krubi has bloomed is the Bronx. I, therefore, feel it proper that the Bronx should adopt it as its official flower. I, therefore, proclaim the *Amorphophallus titanum* (giant krubi) as the official flower of the Bronx, as its tremendous size shall be symbolic of the largest and fastest growing borough of the City of New York. There are many other sweeter-smelling flowers, but none as large and distinctive.[25]

It would not do to pass comments on what else the inflorescence might symbolize but it must be observed that it is not in fact the largest flower in the world. This title is held by the flower of the leafless, stemless parasite *Rafflesia arnoldii* which measures 1 m across. Coincidentally it grows in the same habitats as the Titan arum but is slightly easier to find flowering (though neither is easy). A further coincidence is that it too was collected by Beccari and it was once known as *Rafflesia titan*. *A. titanum* is not of course a flower but an inflorescence (a spike of many small flowers). But alas, it is not, technically speaking, the largest inflorescence either, as the talipot palm (*Corypha elata*) – again from Sumatra – produces one that is an astounding 6–8 m long with a preposterous 60 million individual flowers. (Not surprisingly it takes over 30 years to manage such a display and then promptly dies.) However, it differs in structure from that of *Amorphophallus* in that the flowers are borne on numerous branches. The *Guinness Book of Plant Facts and Feats* gives *Puya raimondii* as the largest known inflorescence but, again, it is branched. So what we are left with is that *A. titanum* is the world's largest unbranched inflorescence. To most people though it looks to all intents and purposes like a flower because the individual florets are not seen. And – technicalities apart – it is certainly the most impressive single bloom you can ever see.

—— In the wild ——

A. titanum grows only in western Sumatra on steep limestone hillsides overlaid with loose humus, often in secondary forest, at 120–365 m above sea level. Plants tend to occur in colonies (though not necessarily very close together) but no studies have been made of population sizes and densities or the approximate ages of plants occurring in the same locality. It enjoys shade but is said to flower more readily in open situations – though it is probable that only inflorescences near habitation and in clearings are noticed. When I explored deeper into the forest I was impressed by how difficult it would be to spot a developing inflorescence, even of this size (an open one would be much easier because of the smell). Light levels at the forest floor are reduced by as much as 90 per cent and the density of vegetation is such that visibility is down to a few metres.

Both the inflorescences I found were indeed in the open, though recently and unnaturally so: the first at the side of a main road where the surrounding vegetation had been hacked down to improve accessibility for viewing the bloom; the second on a small area of extremely steep hillside which had been cleared for bringing down chunks of limestone. Unfortunately, both were also vandalized: the first before I arrived and the second after I had photographed the bud.

There is still a great deal that is unknown about the Titan arum because it is so difficult to cultivate for any length of time and hard to find flowering in the wild. No one knows just how rare it is, how long an individual plant in the wild lives or how long it takes a wild seedling to reach flowering size (though Beccari's seedling flowered after 11 years and one young plant in cultivation with a tuber weighing only 7 kg produced an inflorescence

85 cm in height).[26] Neither have any detailed studies been made of its metabolism, though one observation states that the spadix appendix of a plant in cultivation was 4°C above ambient[27] and that the bud grows at 4–20 cm a day.[28] And the difficulties of observing plants in the wild mean that details of pollination, fruit development and dispersal are very sketchy indeed. One of the few accounts of wild plants was made by Dr H. A. Schrage, a surgeon in Sumatra who went orchid hunting in the forests. Although he came across *Amorphophallus titanum* often enough, he only twice found it flowering and once saw a hornbill feeding on the berries.[29]

The greatest mystery of all though is why this giant aroid has a tuber and becomes dormant. Its environment is one of the most unvarying on earth with every day of the year the same length and an equatorial climate with constant high humidity, an average daily temperature of 28°C and an evenly distributed annual rainfall of 310 cm. As one writer put it: 'The hot season is slightly hotter and not quite so wet as the wet season, while the wet season is slightly wetter and not quite so hot as the hot season.'[30] The 'cool' season is December to February but the temperature drops only a degree or two with the increased cloud cover; the driest months are March, April and May though it still rains most days. It is known that tropical rain forest species may be sensitive to extremely small variations in conditions such as these but, nevertheless, wild populations have been observed to have plants in all stages at any one time and it is therefore difficult to detect a response to these slight changes.

What then triggers dormancy and what is its purpose? Obviously the production of an inflorescence which is large and fast-growing, raises the temperature of its tissues and emits an extremely strong smell needs considerable reserves of carbohydrates (or possibly lipids) and this may only be possible by amassing and storing nutrients in a tuber. And possibly the climate and habitat were once less equable and the tuberous growth and dormancy evolved in response to a pronounced dry season. Until detailed studies are made, these and other features of the Titan arum must remain speculative.

A. titanum must be one of the greatest wonders of the living world yet, strangely, it is not well known: tourist literature on Sumatra never fails to mention *Rafflesia* but in spite of the fact that the Titan arum's image appears on the 500 rupiah note, few people – tourist or local – have heard of it, let alone seen it.

Another Sumatran giant

Even greater obscurity surrounds the contender for the heavyweight title. *A. decus-silvae* is also endemic to Sumatra and is known to produce inflorescences 4 m tall (one was recorded at 4.36 m) and have tubers comparable in size to those of *A. titanum*.[31] However, two-thirds of the inflorescence consists of a spotted stalk the thickness of a man's leg and the actual spathe and spadix, though huge, are smaller and less impressive than those of the Titan arum. Little is known about this species other than the few records of dimensions. It was first described in 1920 as two different species, *A. decus-silvae* and *A. brooksii*.[32] These are now considered (by Mitsuro Hotta and Josef Bogner) one and the same[33] but as the names appeared almost simultaneously, no one has yet published a decision on which should have priority – though the evidence favours the former.[34]

The genus *Amorphophallus*

Amorphophallus is a genus of around 100 tuberous species found in tropical Asia and Africa. They vary greatly in size and shape and many are poorly known, probably quite rare, and difficult to cultivate. As well as having beautifully patterned stalks and fine floral details, they produce some of the most fascinating inflorescences in the family: textural marvels whose

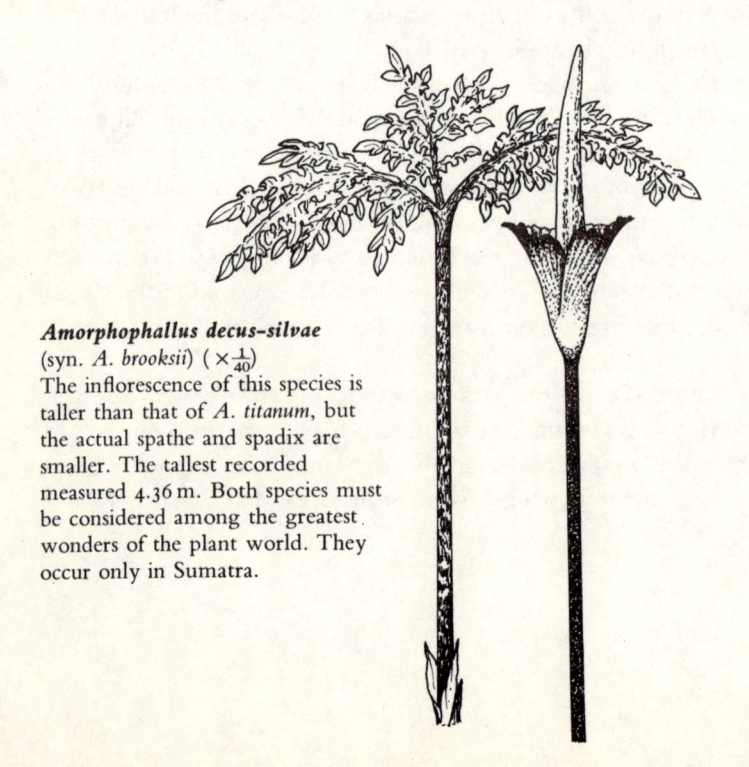

Amorphophallus decus-silvae
(syn. *A. brooksii*) ($\times \frac{1}{40}$)
The inflorescence of this species is taller than that of *A. titanum*, but the actual spathe and spadix are smaller. The tallest recorded measured 4.36 m. Both species must be considered among the greatest wonders of the plant world. They occur only in Sumatra.

combination of bold forms and convoluted surfaces would bring delight to any sculptor.

In such a visually surprising genus it is quite impossible to name the strangest but certainly one of the most unusual is *A. pendulus* from Sarawak, whose inflorescence is like a rather larger, coarser whiplash arisaema in shape. The spathe is dark purple and hooded and from the dark interior droops a long pale yellow string-like appendix which reaches up to 36 cm in length. The inflorescence is borne on a stalk 20–40 cm tall which is the same blackish-purple and, like the tubular lower half of the spathe, is speckled with circular light green spots. The leaf is equally fine, with a densely spotted reddish stalk and dark purple-tinged leaflets which often have silvery variegation along the midribs. This is a recently described species which was first seen in flower by botanists at the Royal Botanic Gardens, Kew, in 1968.[35]

Equally bizarre is the hairy amorphophallus, *A. cirrifer*, a species first described when it flowered at Kew after several dormant fist-sized tubers were sent from Thailand by a British doctor in 1922.[36] Though the foetid inflorescence is quite repulsive it also has an air of the ridiculous in that the stout tapered blackish-purple appendix is held horizontally, giving it, somewhat incongruously, the appearance of having been subject to strong winds. The surface of this extraordinary appendage is unpleasantly dimpled and sprouts the occasional 1 cm-long hair-like process in the same livid purple. These 'hairs' also occur sparsely among the flowers which are enclosed by the tube of the spathe. Though not one of the giants of the genus – the spathe blade measuring only about 11 cm by 7.5 cm and the appendix up to 40 cm in length and 3 cm across at its swollen base – it is sufficiently gross to induce amazement and disbelief in any observer.

As is characteristic of amorphophalluses, *A. cirrifer* flowers after several months of dormancy and shortly afterwards produces a single triple-forked umbrella of leaflets supported by a fleshy mottled stalk. Though the genus is acclaimed for the most gigantic, grotesque, evil-smelling inflorescences in the plant world, its decorative leaves are seldom given due recognition. This was not always the case however. One 19th century dictionary of gardening recommends them as 'strikingly effective in subtropical bedding' and lists several suitable species including the newly discovered *A. titanum*![37] Perhaps Victorian gardeners were more imaginative than those of today but, for whatever reason, *Amorphophallus* leaves no longer feature as 'dot' plants in the bedding schemes of parks and gardens.

For the record, if interest should revive in their potential as foliage plants, the best species for use outdoors in temperate climates would be *A. konjac*, formerly known as *A. rivieri*.[38] Though chiefly grown as a crop in eastern Asia for its edible tubers (see chapter 9), it is also occasionally sold as a dormant tuber under the name of devil's tongue or leopard palm, as a curious and ornamental plant.

The curious part is of course the large inflorescence, which is the colour of raw liver and has 'a reek reminiscent of dead rats'.[39] It can reach over 50 cm on a stalk 1.5 m tall. Its ornamental value lies in the spotted leaf stalk and complex array of leaflets. The tubers reach 9–13 kg and put out stolons at the ends of which offsets develop. It is a reasonably hardy species which propagates easily[40] and is found wild in China, southern Japan, the Philippines and Nam-Phan (Cochin China).

In a genus renowned for its nightmarish blooms, *A. paeoniifolius* (formerly *A. campanulatus*) is surely one of the ugliest. It is also huge. For some years the star aroid of the Marie Selby Botanic Gardens in Florida was thought to be *A. titanum*.[41] It grew well and in 1983 when four years old weighed 34 kg and produced a densely spotted stalk 2 m tall with a crown of leaflets 2.45 m across. The next year it flowered and revealed its true identity as this species.[42] All parts of the inflorescence are convoluted: the squat spadix is wrinkled, warted and dome-shaped and sits in the centre of a very frilly spathe, emitting 'an abominable and putrid odor'.[43] It is commonly known as elephant yam and native to Sumatra and Malaysia. Its large tubers are used as food in Southeast Asia (see chapter 9).

It may perhaps be a slight exaggeration to call *A. prainii* attractive, but its pleasantly-coloured fungus-like inflorescence is certainly less repulsive than most of the genus. The tuber is up to 25 cm across and produces a short stout mottled stalk and a shallow recurved pale primrose spathe. The inner hollow of the spathe is dark maroon and from it arises a bright yellow spadix with a swollen creased appendix which ends abruptly – as if cut off – and is the same delicate shade as the spathe. This is quite a common species of lowland rain forest in Malaysia. Its local name is *likir* and there are records of the tubers being collected both for cooking and for their juice, which if used raw and combined with extracts of other plants, such as *Rauvolfia perakensis* (*Apocynaceae*), is said to make a dart poison strong enough to kill a tiger or rhinoceros.[44] Apparently the *likir* juice speeds the absorption of toxins such as reserpine, an alkaloid found in *Rauvolfia*, which acts as a tranquillizer and cardiac depressant.

Dracontium

The American equivalent of the Asian Titans is the genus *Dracontium* which has a dozen species or so. Again they are tuberous, producing a solitary divided leaf with a mottled stalk, though generally it is rough and finely patterned whereas in *Amorphophallus* it is more distinctly blotched. The unpleasant-smelling inflorescence appears after dormancy, with or just before the emergence of the new leaf, and is generally dark green and purple, hooded or boat-shaped with the purple or brown spadix much smaller than the spathe. The fruits contain one or two seeds 5–15 mm long. The seeds are unusual, being kidney-shaped, corky and crinkled. They float on water but the plants are not associated with aquatic habitats and the details of dispersal are unknown.[45] Dracontiums are rain forest plants but tend to grow singly here and there rather than in groups.

One of the largest dracontiums is *D. pittieri*. It grows in wet lowland forests of Costa Rica and its complex leaf on a stout purple-patterned stalk is 3–4 m tall. In this species, leaf and inflorescence are sometimes seen together.[46] Similar in size is *D. gigas*, a Nicaraguan species. The illustration in Engler's *Pflanzenreich*[47] shows a man standing beside a leaf which is over three times his height. Allowing for a little artistic licence it is still a giant. As with all dracontiums, however, the inflorescences are small compared with those of amorphophalluses, though the leaves are equally majestic.

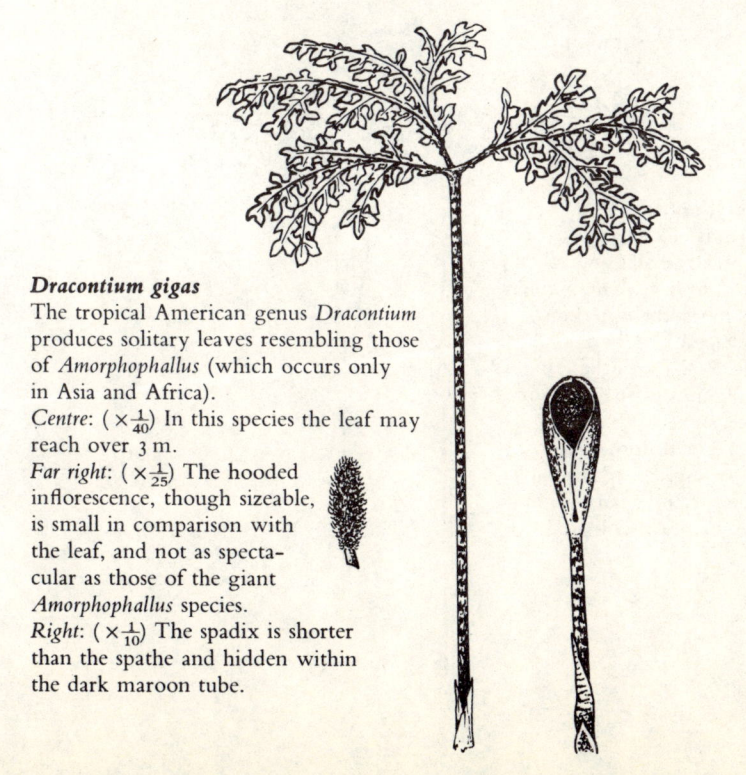

Dracontium gigas
The tropical American genus *Dracontium* produces solitary leaves resembling those of *Amorphophallus* (which occurs only in Asia and Africa).
Centre: ($\times \frac{1}{40}$) In this species the leaf may reach over 3 m.
Far right: ($\times \frac{1}{25}$) The hooded inflorescence, though sizeable, is small in comparison with the leaf, and not as spectacular as those of the giant *Amorphophallus* species.
Right: ($\times \frac{1}{10}$) The spadix is shorter than the spathe and hidden within the dark maroon tube.

Anchomanes

Giant tuberous aroids are found on the African continent too. One species, *Anchomanes difformis*, is one of the most spectacular species to be seen in the tropical aroid house at the Royal Botanic Gardens, Kew, where it forms a thicket of robust spiny stalks over 3 m tall which spread out into a canopy of wedge-shaped leaflets. Now and again pale peach-coloured inflorescences up to 30 cm long are borne on prickly stalks about 75 cm tall. The lower margins of the spathe slightly overlap around the female part of the erect spadix and are held free of the upper portion which is covered to the apex in a dense mosaic of squarish male florets. This species grows in rain forest, riverine forest and savannah woodlands in West, Central and East Africa and may have greenish, purplish or reddish-brown inflorescences.[48] Its heavily wrinkled rhizomes are about 30 cm in diameter and extend horizontally, more or less on the surface, for several metres. They are used in the region as part of a treatment for sores caused by leprosy.[49]

Anchomanes has only about five species and differs from *Amorphophallus* and *Dracontium* in mostly having spiny stalks and tuberous rhizomes with 'eyes' and roots from many points (rather than semi-globose tubers which root from the crown). The rhizomes spread and plants therefore tend to

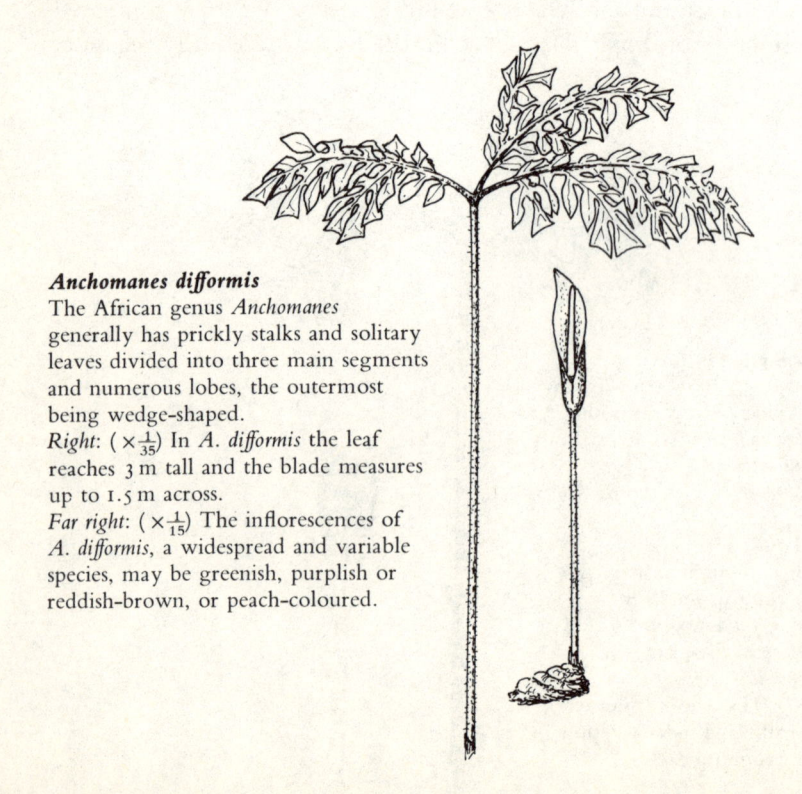

Anchomanes difformis
The African genus *Anchomanes* generally has prickly stalks and solitary leaves divided into three main segments and numerous lobes, the outermost being wedge-shaped.
Right: ($\times \frac{1}{35}$) In *A. difformis* the leaf reaches 3 m tall and the blade measures up to 1.5 m across.
Far right: ($\times \frac{1}{15}$) The inflorescences of *A. difformis*, a widespread and variable species, may be greenish, purplish or reddish-brown, or peach-coloured.

form colonies. They undergo an annual dormant period, with inflorescences appearing before or at the same time as the leaves.

All the genera mentioned in this account of the Titans belong to the subfamily *Lasioideae*. This group of closely related genera also includes the monotypic *Dracontioides* which reaches 2 m tall; *Urospatha*, of which *U. sagittifolia* reaches similar heights; and *Cyrtosperma* with the edible swamp taro, *C. merkusii*, often exceeding 3 m.

Any account of these giant tuberous aroids is bound to concentrate on their statistics. In some cases their dimensions are just about all we know of them. Of course, the size is itself impressive: the first thing we notice, the last thing we forget and the most easily conveyed in descriptions. But mere measurements do little to communicate the experience of seeing these remarkable species – especially in the wild – and tell nothing of their lifecycles, which may be equally astounding. Statistics of a far more vital nature, concerning population dynamics, pollination biology, dispersal and vegetative reproduction for example, still await discovery in the rain forests which dwarf even these Titans.

9
An Acquired Taste

Aroids as food plants

The oldest cultivated crop in the world is an aroid: taro (*Colocasia esculenta*). It has been grown in tropical and subtropical Asia for over 10,000 years.[1] It is thought that the ancient irrigation systems of terraced rice paddies seen today may well have been constructed originally for taro long before rice came on the scene – and that rice may have first come to notice as a weed in the flooded taro patches.[2] Grains were brought into cultivation in Neolithic times, from about 10,000–7000 BC, and yams before this, in the Mesolithic, when it would appear that taro was already domesticated.[3]

Because of its long history of cultivation in so many places the origin of taro is hard to determine. The most likely beginnings were in eastern India, from where it spread to southern China, Formosa (which it reached around 9000 BC), Japan and the Philippines and then through New Guinea and the islands of Indonesia. It also went westwards to the Middle East and the eastern Mediterranean and was recorded in Egypt over 2500 years ago. From there its cultivation was adopted right across the continent of Africa and in the 19th century eventually reached the New World when it was loaded aboard slave ships on the West African coast to feed their human cargo.[4]

Development of taro as a food crop

Like most aroids *Colocasia esculenta* tends to be acrid in its raw state and is usually only edible after thorough cooking, when the needle-like crystals of calcium oxalate and other irritants have been broken down and rendered harmless. A few cultivars, such as the Indonesian *Lampung hitam*, have such low acridity that they can be eaten raw. Wild plants are often extremely acrid and the first culinary experiments must have resulted in some very unpleasant experiences, as eating even the tiniest amount raw causes an intense discomfort consistent with having thousands of microscopic needles stuck into lips, mouth and throat. It says much for human tenacity and observation that over the centuries as many as 1000 varieties of *C. esculenta* have been developed worldwide, all varying in colour, size, flavour, texture, acridity, disease resistance and suitability to differing climatic conditions and cultivation techniques. The explanation of this immense variability lies in its genes. *C. esculenta* has anything between 22 and 42 chromosomes per cell and during cell division they behave erratically.[5] Thus new permutations, and hence varieties, are formed without recourse to cross-pollination and seed production. As a consequence, taro has always been propagated vegetatively and growers have never bothered with breeding, relying instead on a keen eye to spot interesting mutations. Selective hybridization is in fact rather difficult as many clones seldom flower and even more rarely set seed. Some, such as the Indonesian cultivar *Loma 2*, have never been known to flower.[6]

It must therefore largely be left to scientific hybridization and mutation breeding programmes to raise new varieties of taro with significantly increased yields, better food value, and lower acridity but so far modern plant breeding methods have made little or no impact. The vast majority of edible aroids currently being grown (of which taro is the most important) are very ancient, little documented, and, in many areas, fast disappearing as traditional diets and cultures succumb to western influence. Currently an estimated four million tonnes are grown each year, mostly on a subsistence basis.[7] This quantity is consumed by some 400 million people. The potential for crop improvement is enormous. Even now, one square kilometre of taro can feed 5000 people for a year.[8] With superior cultivars and the introduction of new cultivation and processing methods, it could undoubtedly feed many more and to a higher nutritional standard than at present.

The first step in improving any crop is to assess all its characteristics. This means collecting and recording every variety and learning to appreciate the needs and tastes of those who grow it. Taro is virtually unknown in the West, though increasing amounts are being imported each year to satisfy the demands of immigrant communities and those interested in novel foodstuffs. In the tropics it is a different story. Here taro is ubiquitous and in a number of places – New Guinea, the Solomons and other islands of Oceania – is the staple food. Many countries with subtropical regions grow it too, from Japan (the Japanese displayed it at the Health Exhibition in London in 1884 as 'Japanese potatoes')[9] to New Zealand, where it was introduced by the first Maori immigrants.[10]

—— Varieties and names ——

Wherever *Colocasia esculenta* is grown, there are local varieties and names.

The best known are: taro and kalo in the Pacific islands; eddo and dasheen in the West Indies; old cocoyam in Africa; keladi and talas in Southeast Asia; arvi in India; and imo in Japan. The many varieties do nevertheless fall into two main categories; the hardier, more tolerant eddo type (sometimes referred to as *C. esculenta* var. *antiquorum*, or *C. esculenta* var. *globulifera*), which has a small main corm surrounded by relatively large and numerous cormels; and the dasheen type (*C. esculenta* var. *esculenta*) with a large corm, up to 30 cm long and 15 cm in diameter, and fewer smaller cormels.

Appearance

In spite of the large number of varieties, plants of *C. esculenta* are similar in general appearance and it needs a trained eye to register the more subtle differences in growth habit, shape and colour of leaves and corms. On average they reach 1 m tall with a whorl of large long-stalked blades which are peltate, heart-shaped, about 50 cm in length and held more or less vertically. Some cultivars have a reddish blotch at the 'navel' where the blade joins the leaf: examples being *Pula tusi tusi* from Samoa and the Hawaiian *Kumu*. Other striking variants are the purple variegated leaves of *Uahi apele* (a Hawaiian variety prized for medicinal use), and the many *Piko* cultivars of Hawaii which, instead of peltate foliage, have their blades lobed to the navel or *piko*. One of these, *Piko lehua apii*, is most unusual in having blades ruffled with strange ridges and frills on the underside. A plot of this variety looks at first glance to be suffering from a disease but in fact the deformed leaves are quite healthy.

Colocasia esculenta ($\times \frac{1}{10}$)
Taro is widely grown as a root crop in the tropics and is the staple food of the Hawaiian Islands. There are two main kinds: *Far right, above*: dasheen, with a large central corm and relatively small cormlets. *Far right, below*: eddo, with a smaller main corm and larger more numerous cormlets.

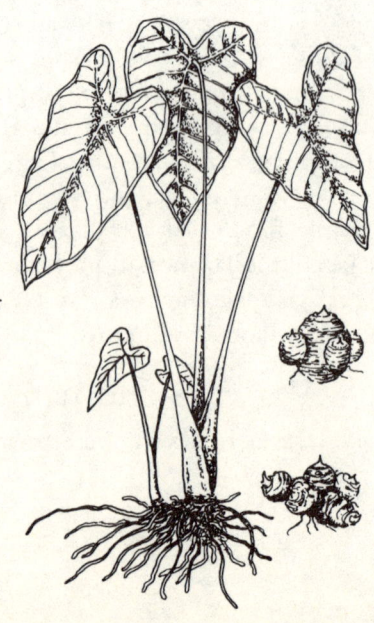

When taro does flower it produces clusters of two to five fragrant inflorescences in the leaf axils on stalks 15–30 cm long. The spathe reaches 20 cm and is pale yellow with a greenish basal tube. The spadix is 6–14 cm long and has a sterile appendix which is longer in eddo types. The scent is fruity – rather like mango – and pollination is probably by fruit flies.[11] The berries are green when ripe and rarely contain seeds.

Taro has a dense but shallow root system arising from the lower parts of the corms. The corms themselves are brown-skinned cylindrical or spherical stem tubers which show scars where the leaves were attached. The inner flesh is starchy and may be anything from white or yellow to pink, orange, purplish or grey in colour and varies in texture when cooked from mealy to glutinous.

—— Cultivation ——

Taro is an adaptable crop and there are varieties which thrive in conditions as different as equatorial wetlands to cool Himalayan foothills. It needs an average daily temperature of 21–27°C and is essentially water-loving but intolerant of stagnant conditions. Cultivation therefore usually involves a certain amount of levelling and ditching in order to control water levels and flow. It can be grown under dryland (also known as upland) culture in areas with as little as 1.75 m of rain per year provided it is irrigated. On the whole taro does best in heavy slightly acid soils (pH 5.5–6.5) which are rich in humus. In many countries it is the first crop after the forest has been cleared and burnt.

The main ways of propagating taro are by cormels, suckers and cuttings. The latter, known as *huli* in Hawaii, consist of the lower portions of the stalks attached to a few centimetres of corm: in other words, what is left after cutting off the main corm and the leaves during harvesting. Pacific islanders generally prefer this method and it usually gives higher yields. Whatever the method, planting wetland taro is just as laborious as rice, involving the turning over and levelling of heavy mud by water buffalo and hand implements in most parts of the world or by rotavators where affordable. Weeding the crop for the first few months until the leaves are large enough to smother competition is also arduous, wading in thick mud and bending over for long hours, and can account for half the total labour time. No less demanding is the harvesting, when the corms have to be loosened from the mud with poles before being sliced off from the foliage.

Pests and diseases

As with any crop, a number of pests and diseases can affect taro and are a serious hazard where a single crop is grown. In most parts of the world though, taro is still part of traditional husbandry which relies on small plantings, regular rotation of crops and interplanting, to encourage healthy plants. Taro may follow several years of rice or serve as a nurse crop for young cocoa, or accompany maize, legumes and okra. Often it is planted at the edges of rice or sugar cane and is common on marginal land and in fish ponds.

The worst problems are rots, caused mainly by *Pythium* and *Sclerotium* organisms, which lead to distintegration of the corms. Equally serious is taro leaf blight, a fungal disease caused by *Phytophthora colocasiae*. It is quite common in the Pacific region and may reach epidemic proportions among overcrowded plants when the weather is very humid and cloudy. These conditions mostly afflict wetland taro. A number of cultivars are resistant and are of particular importance in breeding programmes to produce varieties which can maintain vigour, even in intensive cultivation. Dryland taro may be badly affected by root-knot nematodes (*Meloidogyne* species). All taro may suffer from dasheen mosaic virus which is spread by aphids and is commonest in intensive cultivation. It does not kill the plants but stunts growth and reduces yields. The disease is systemic, but affected plants may not show the symptoms (feathery pale areas along the veins) for some time. There is no cure. Among insect pests, the most damaging is the taro leafhopper (*Tarophagus proserpina*) which lays eggs in the stalks and can ruin a crop. The most satisfactory control is the Philippine egg-sucking bug (*Cyrtorhinus fulvus*) which is introduced to prey on the pest.

OPPOSITE]
Symplocarpus foetidus (eastern North America) Skunk cabbage, polecat weed and swamp cabbage are just some of the common names for this wetland species. The squat cowrie-shaped inflorescences heat up enough to melt their way through ice and snow; the stout rhizome is used in American and European herbal medicine.

OVERLEAF]
ABOVE LEFT *Colocasia esculenta* Known as *keladi* in Indonesia, the tubers of *Colocasia esculenta* are here on the way to market in Bogor, Java. The leaves are also eaten.
ABOVE RIGHT *Colocasia esculenta* Lehua is an ancient variety of Hawaiian taro once reserved for royalty on account of its red-skinned, pink-fleshed tubers.
BELOW *Colocasia esculenta* Traditional taro patches (*lo'i*) are still to be seen in the Hawaiian Islands. These are on the Keenae Peninsula, Maui.

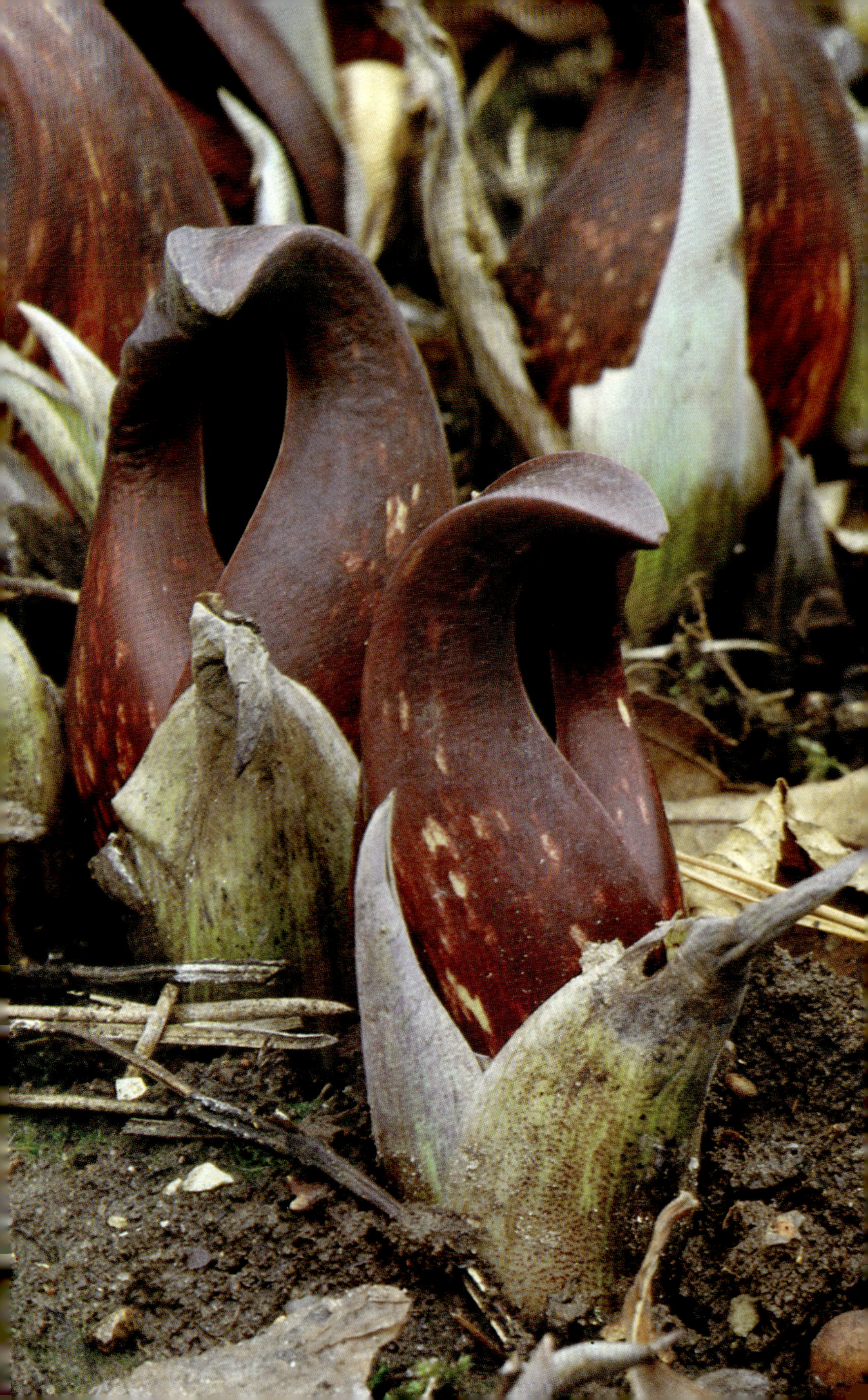

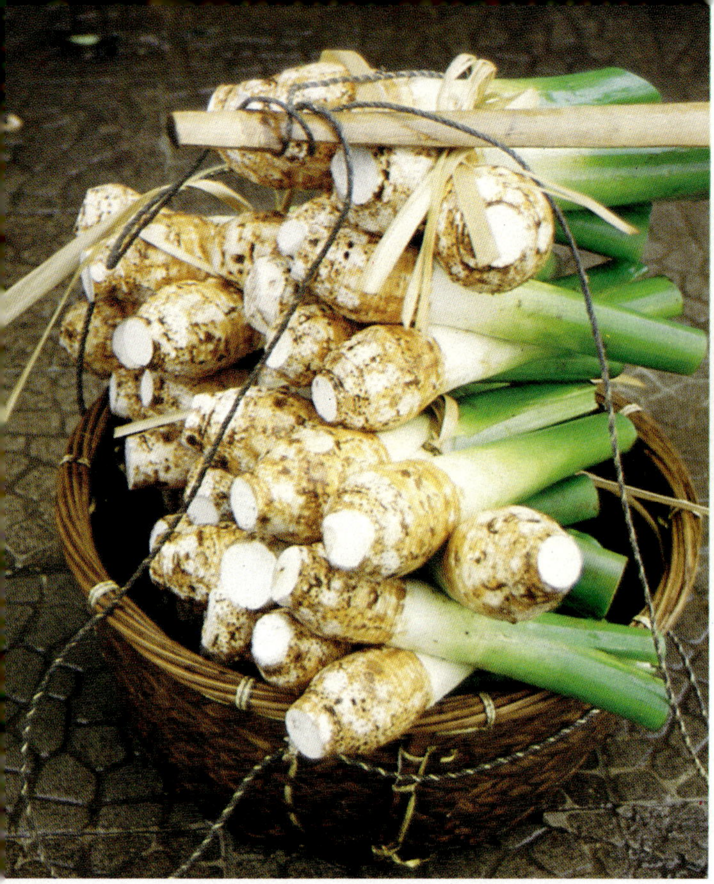
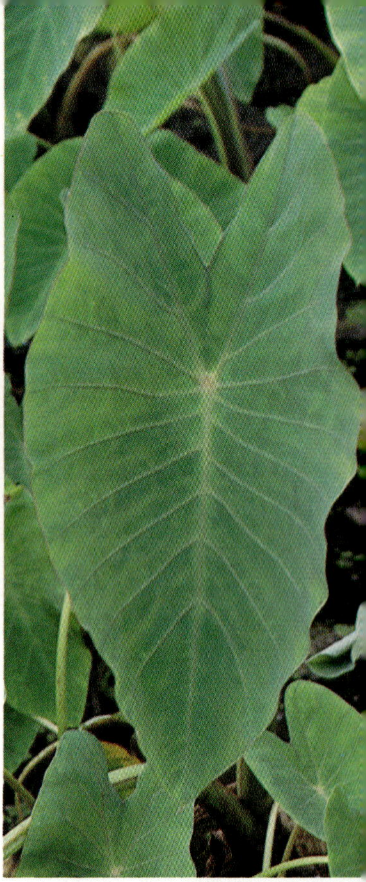
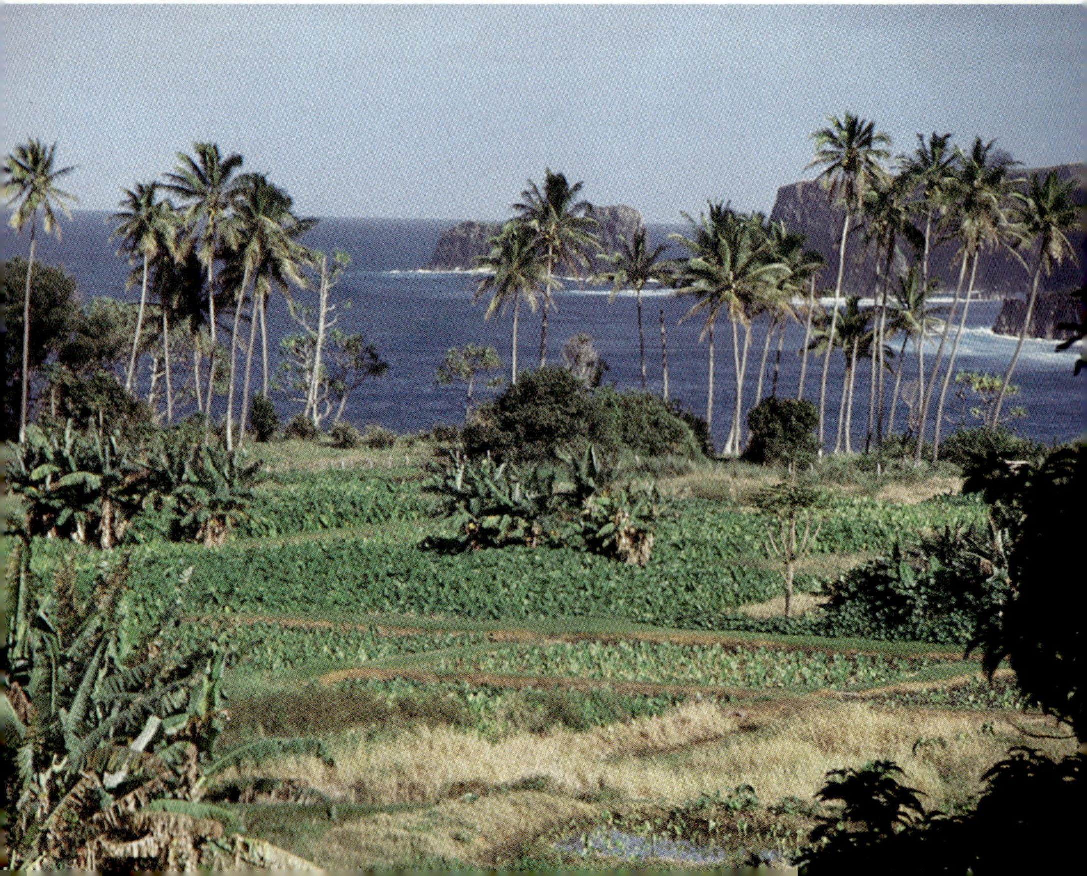

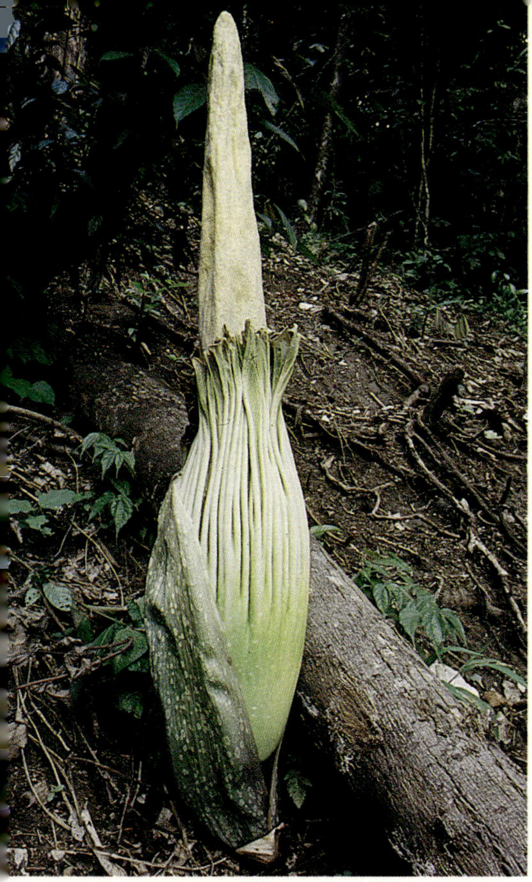
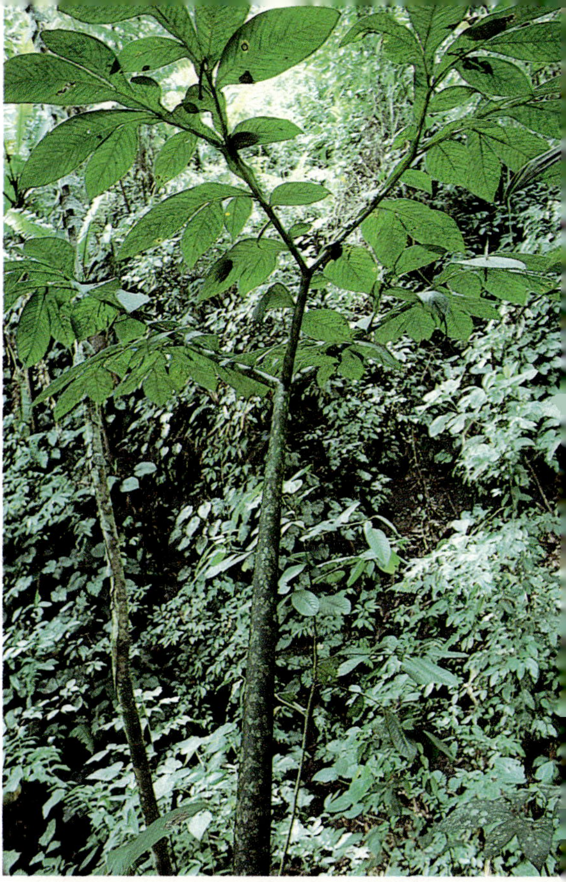
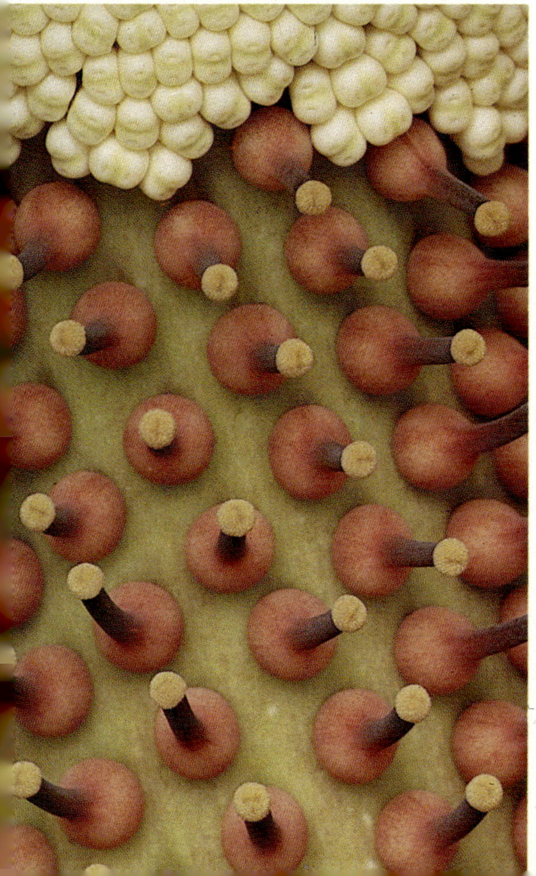
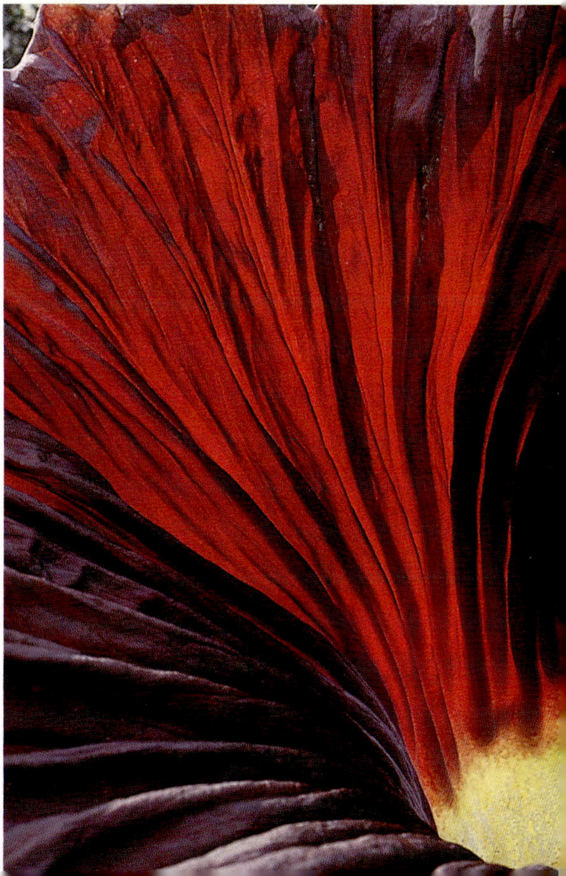

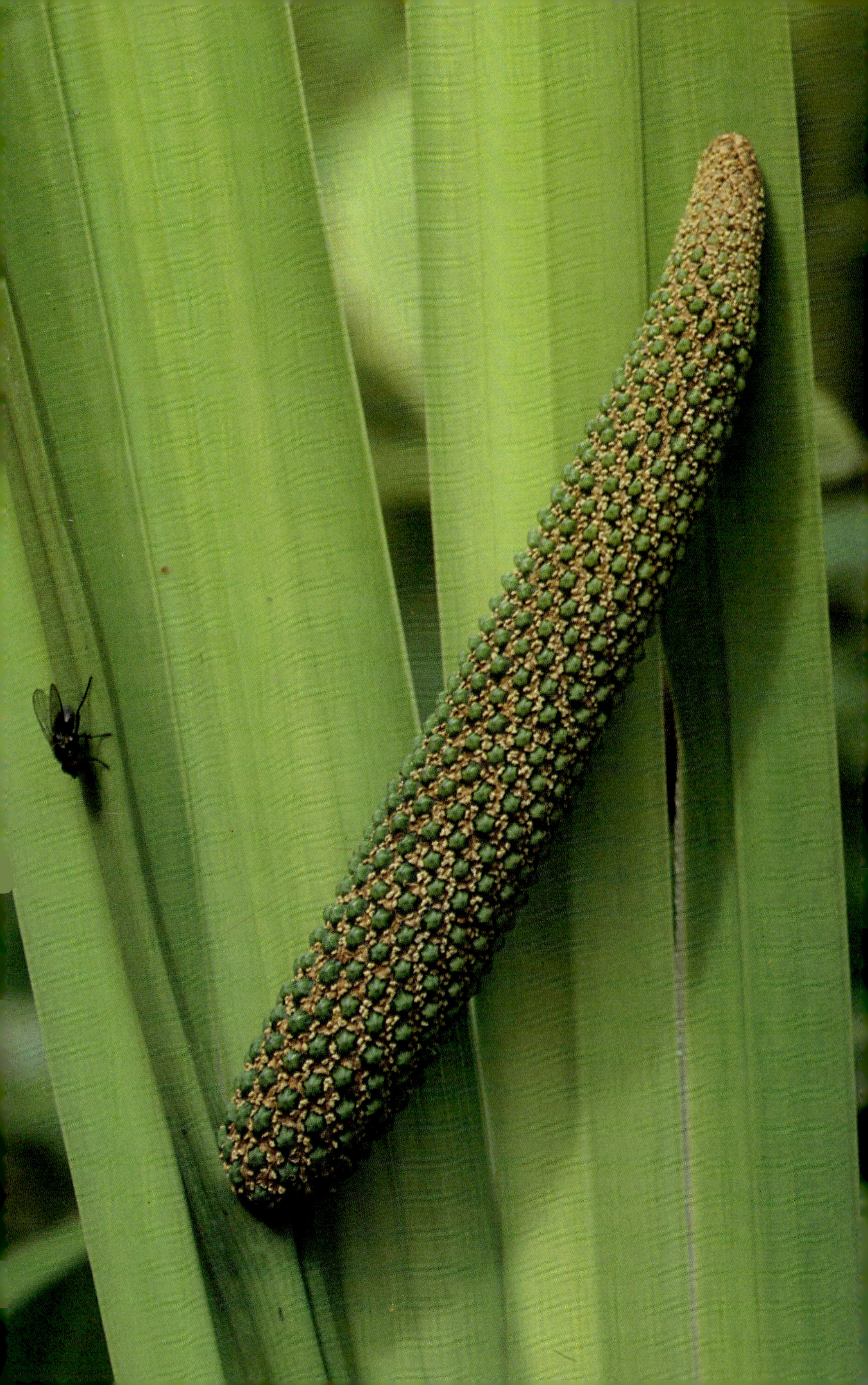

Even though most pests and diseases can be controlled chemically, the best insurance is fertile soil, sound cultivation techniques and the production of new hybrids with increased resistance. In addition, guidelines have now been drawn up by the FAO Plant Protection Committee for Southeast Asia regarding the import of both edible and ornamental aroids within the region, in an attempt to stop the spread of pathogens to disease-free areas.[12]

Yields

Yields and cropping times vary greatly according to climate and cultivation. With heavy fertilization, experimental crops in Hawaii yielded a record 50 tonnes per hectare[13] but usually a quarter of this amount or even less is obtained. Wetlands invariably give higher yields than dryland systems. The average time from planting to harvesting in the tropics is nine or ten months, but in cooler cloudier areas up to 18 months may be needed. In warm temperate regions where only six months of the year provide suitable growing conditions, cropping is still possible though yields are lower. Harvested corms will keep for about three months if kept dry, well-ventilated, and a little below 10°C. In Egypt and Samoa they are traditionally stored in underground pits but the more usual method is on raised open platforms. Where conditions are not favourable they are often left in the ground until required.

Food value

The corms, stalks, leaves, and inflorescences of taro are eaten, though it is principally grown as a root crop. In food value the corms are similar to potatoes, consisting of 13–29 per cent carbohydrate. The protein content is

OVERLEAF]
ABOVE LEFT *Amorphophallus titanum* This bud is almost 2 m tall and a day or two before opening.
ABOVE RIGHT *Amorphophallus titanum* (Sumatra) The Titan arum's solitary leaf has a fleshy trunk-like stalk over 2 m tall which divides into three at the apex and subsequently into an umbrella of leaflets up to 15 m in circumference.
BELOW LEFT *Amorphophallus titanum* The Titan arum's flowers: males at the top and colourful flask-shaped females below.
BELOW RIGHT *Amorphophallus titanum* The open inflorescence was vandalized before I could photograph it, but a portion of the vase-shaped spathe remained intact.

OPPOSITE]
Acorus calamus (Europe, temperate Asia and North America) Universally revered as an aromatic medicinal plant but not universally accepted as an aroid!

somewhat higher (1.4–3 per cent) and about twice that of sweet potato and cassava.[14] They also contain appreciable amounts of vitamins C and B complex (especially thiamine, riboflavin and niacin) and minerals. Any staple food is consumed in relatively large quantities and therefore the protein and vitamin content makes a significant contribution to the diet in areas where intake may otherwise be low.[15]

A particularly interesting feature of taro is that the starch grains are extremely small and easily digested and both the carbohydrates and proteins are non-allergenic. This has led to increasing use of processed taro in speciality foods for babies and for those with allergic and gastric disorders (notably gastric ulcers and coeliac disease).[16]

The corms may also be ground into flour for use in bakery products and desserts. They are also packaged as snacks which closely resemble potato crisps, apart from the purple veins, which are quite distinctive! The variety *Bun Long* or Chinese taro is the best for this purpose as it contains few crystals and is therefore ideal for quick frying.

The leaves cooked as a vegetable are an excellent source of vitamin C (142 mg per 100 g), vitamin A (20 385 iu per 100 g), calcium, potassium, phosphorus and iron. They are also rich in protein (4.4 mg per 100 g) which is well utilized as most of the essential amino acids are present. The stalks are considerably lower in all nutrients. Cooked taro leaves are similar to spinach and may be canned or vacuum-packed either on their own or as part of ready-to-eat dishes. In Hawaii they are often combined with coconut milk in a dish known as *luau*, which is rather like creamed spinach.

—— The staff of life ——

Although it is a fact that you can find taro growing practically anywhere in the tropics, for the most part you get the impression that it is a standby and occasional food rather than a staple. In Indonesia you can crunch your way through a packet of *kripik talas* (taro crisps) or be served the Sundanese dish *buntil* – young taro leaves cooked with a filling of freshly grated coconut – but first and last comes the rice. In the Pacific region, especially in the Hawaiian Islands, it is a different story. Here taro, or *kalo* as it is known in Hawaiian, is the staff of life and central not only to the diet but to the culture of the native islanders.

Polynesia was the last habitable area of the earth to be occupied by human beings. The pioneers originally came from Southeast Asia and within 80 to 90 generations, between 1000 BC and AD 1000, their oceanic wanderings took them from Melanesia to Micronesia and finally to far-flung Polynesia. The northernmost islands of all were the Hawaiian group and these were eventually reached from the Marquesas and Tahiti in about AD 400. Contrary to romantic notions, small volcanic and coral islands are no paradise when it comes to finding food and few of the native plants proved edible or prolific enough to sustain human life. Settlement was only possible because the people brought with them roots such as taro which could be planted and relied upon as a staple.

The early Hawaiians were completely dependent on taro and so it is not surprising that one of their legends tells that the first-born of father sky and mother earth was a taro plant and the second born was a human being. The taro plant symbolized not only life, but endless life, the same plant growing year after year, generation after generation, by the successive replanting of the cuttings, the *huli*. They named their varieties after other living things and substituted a taro plant of the same name for the animal in offerings to the gods. One of the varieties used in ritual was *Kumu*, with red stalks and a red navel, which was named after the red goat fish. Several other red taros were so highly regarded that they were considered sacred and no one but royalty was permitted to use them.[17] Times change, and one of the royal taros, *Lehua*, with a pink corm, red skin and purplish base to the stalks, is now one of the most popular and commonly grown commercial varieties.

—— Poi ——

Although taro corms are generally boiled or roasted and eaten like potatoes, in the Hawaiian Islands the favourite method of preparation is a paste known as *poi*. This is made by baking or boiling the corms, peeling them and puréeing the flesh with a little water – traditionally with a stone pounder. According to how much water is added it is known as 'one finger', 'two finger', or 'three finger' *poi*: terms which go back to the days when it was eaten by hand and the consistency was described by how many fingers were needed to scoop up a mouthful. For long journeys it was made very stiff so that it stored well and could be reconstituted or dried.

Poi can be eaten fresh but is usually left to ferment naturally through the presence of *Lactobacillus* and *Streptococcus* organisms, during which the starch is converted to dextrin, sugars and acid. The degree of sourness is indicated by the description 'one-day' or 'two-day' *poi*, and so on. The acidity preserved the *poi* in the days before refrigeration. Nowadays *poi* is packed directly into plastic bags and fermentation can be controlled by lowering the temperature. The colour of *poi* has always been important and the favourite is a dark pinkish-grey made from red-cormed varieties.

Traditionally *poi* is eaten alongside meat, fish, and other dishes. In early times the meat was likely to be that of specially reared dogs known as *poi* dogs. They were kept in kennels and fed only vegetables, mostly cooked taro, so that the flesh was tender and mildly flavoured.[18] Later pigs became more popular and were likewise fed on taro. The national dish *laulau* is a combination of pork, fish, and taro leaves accompanied by *poi*. The word *lau* is Hawaiian for leaf. The ingredients are layered and wrapped in the leaves of the *ti* plant (*Cordyline terminalis*) and baked or steamed for several hours. This is invariably served at a *luau*, a social gathering of any number of people from a few friends or extended family to a large-scale celebration. *Poi* is also popular for breakfast, eaten with milk and honey or sugar. Over the centuries Hawaiians have found that *poi* can be used for a wide range of purposes, culinary and otherwise, from thickening sauces to glueing fabric and paper, and as a poultice for skin problems.

Taro growing in Hawaii

The taro patches or *loi'i* of the Hawaiian Islands are not unlike the rice paddies of Southeast Asia. Small fields are edged with raised mud banks and flooded or drained as needed by conduits. In former times the terraces were more extensive and elaborate, with ditches and canals lined with stones which were laboriously chiselled into rectangles and meticulously fitted together. Remains of these massive structures, some 1.5–2.4 m deep, can still be seen. They not only contoured the land but shaped the social structure through a complex system of water rights.

Today the main growing areas are the Hanalei River valley on the island of Kuaui; Wailua Valley on Maui, which receives the run-off from Mount Waialeale (possibly the wettest place in the world with over 11.5 m of rain a year); the Keanae Peninsula, also on Maui and situated at the foot of the wet forested flank of Mount Haleakala, a sheer drop of more than 3000 m; and the steep-sided Waipio Valley on the 'Big Island' (Hawaii), ancient Hawaii's political and cultural centre. These areas are all perfect for wetland taro: open, sunny and amply supplied with running water from the surrounding mountains.

Upland taro accounts for little of today's commercial production. Most is grown on the rain shadow side of Hawaii around Kona, an area now more famous for its coffee plantations. Although wetland taro is preferred for *poi*, Hawaiians have always grown a certain amount under dryland cultivation. Traditionally this involved planting the cuttings in deep holes and mulching well with cut ferns and other compost material. It yields less but matures more quickly.

In 1985 the State of Hawaii had 155 taro farms which produced 6.9 million pounds of taro with a market value of $1.57 million. These figures are a record high. A significant contribution was made by Chinese taro (*Bun long*) which increased 10 per cent in value within the year.[19] It is grown both for table use and for processed foods such as deep-fried crisps or chips and for extruded products like macaroni and noodles which are made from a mixture of taro and bean flours. Although there is still relatively little acreage devoted to it (it is mainly a dryland crop), it is likely that the profitability of growing for the convenience food industry will encourage further expansion of this variety.

—— Tannia ——

After *Colocasia esculenta* the most important aroid food crop is *Xanthosoma sagittifolium*, known variously as tannia, tanier, yautia, and (new) cocoyam. It is the only New World species to be used extensively for food and was cultivated by the people of South America and the Caribbean long before Europeans arrived in the 15th century.

Between the 16th and 17th centuries, tannia reached West Africa, taken initially as supplies on board ships heading for the Slave Coast – just as taro was loaded for the return voyage. From the ports it spread further afield by traders, missionaries and other travellers and rapidly became more popular than taro. Eventually the 'new' cocoyam of the Amerindians largely replaced the 'old' cocoyam from Southeast Asia in African cultivation and diet. Confusingly, both are often referred to just as cocoyams. The main reason for tannia's popularity was that it happened to be perfect for the traditional African dish *fufu*, originally made from yams, in which cooked tubers were ground into a paste, rolled into balls and served rather like dumplings in stews and soups. Tannia now ranks third in importance to yams and cassava.

In addition to its culinary excellence, tannia proved more robust and disease resistant in adverse conditions than taro, tolerating a certain amount of drought and shade and giving heavier yields on drier soils. It can also be grown on land that is too wet for yams but not wet enough for taro. Traditionally it is cultivated in the same way as yams and cassava, on low mounds mulched with dead leaves and grass; it is widely used as a shade crop for cocoa, banana, coconut, coffee, oil palm and rubber seedlings. Because of its shade tolerance it can also be intercropped beneath mature trees. Shade does lower yields however and if too heavy will induce dormancy. In the southern United States it is grown by the field, like any other crop.

On the whole tannia makes a larger plant than taro, reaching nearer 2 m. The cordate to sagittate leaf blades are about 75 cm long and are very similar, apart from the distinct marginal vein. Unlike taro (other than *Piko* types), they are never peltate. Tannia produces a central corm surrounded by smaller cormels. They are rather different in shape from those of taro, being flask-shaped rather than oval. The two crops do however share the same reluctance to flower and set seed. Were it not for this, hybrids between them might be a more feasible proposition.

Harvesting takes place after 9–12 months, usually in the dry season. In areas with no prolonged dry spells, partial harvesting takes place, the cormels being excavated and the parent plant left to grow on. When replanting is necessary, it is done either from under-sized cormels, setts cut from corms and cormels, or from stem cuttings with 15–25 cm of stalks and a few centimetres of corm. The latter gives the highest yields.

As with taro, all parts are edible when thoroughly cooked but it is the cormels which are mostly eaten. They are either boiled, baked, parboiled and fried, or made into flour. If anything they are better than taro

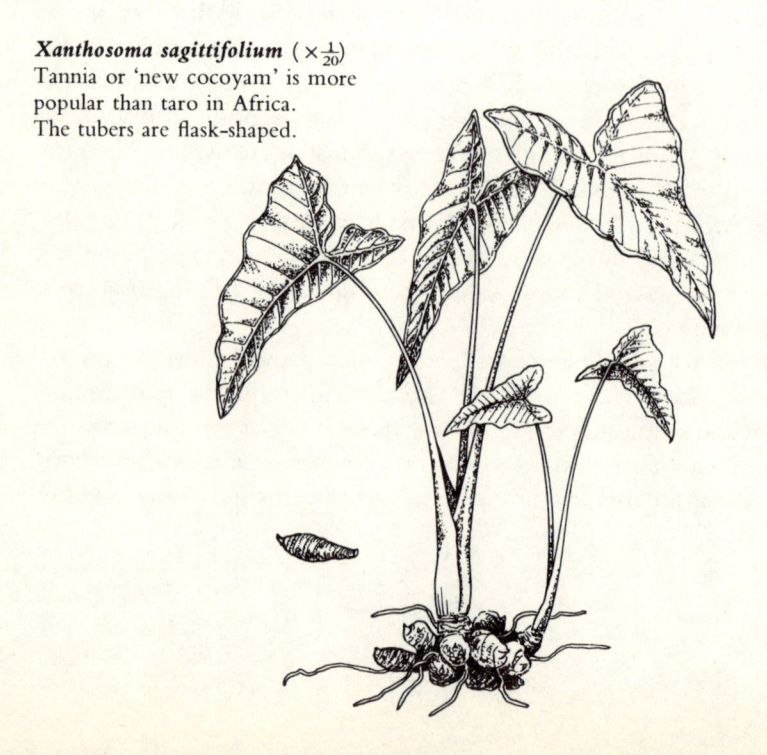

Xanthosoma sagittifolium ($\times \frac{1}{20}$)
Tannia or 'new cocoyam' is more popular than taro in Africa. The tubers are flask-shaped.

nutritionally as the mineral content is higher, but their larger starch grains make them rather more difficult to digest. Like taro, they are versatile. The national dish of Surinam, *pom*, is tannia pastry with a filling of chicken or fish and in Dominica mashed tannia is mixed with coconut, sugar, eggs and spices, and baked as a pudding. In Ghana the leaves are by far the most popular and at times the only cooked green vegetable. They and the tuber peelings are also important as feed for cattle, pigs, goats, sheep and poultry.

Around 30–40 different kinds of tannia are known and a number of these have such distinctive vegetative characteristics that they have been given species status by several authors.[20,21] It would appear however that they are indistinguishable florally and it is upon this that species can be separated. The white-fleshed malanga, so popular in Puerto Rico, has been called *X. caracu*, but now it, together with the Venezuelan *X. belophyllum*, the Mexican *X. jacquinii*, and *X. mafaffa* from northern South America, are considered variants within the polymorphic species *X. sagittifolium*.[22] There is no doubt that, as with the edible colocasias, a great deal of research on this group remains to be done.

Some edible xanthosomas are definitely not part of the *X. sagittifolium* complex though. One is *X. brasiliense*, the Carib cabbage or Tahitian spinach, which has non-acrid but tiny tubers and is grown exclusively as a leaf vegetable. It is an essential ingredient of the Trinidadian calalou soup. Another is *X. atrovirens*, the yellow-fleshed staple root crop of Dominica which is also a great favourite in Puerto Rico. Most distinctive of all is *X. violaceum*, the blue tannia, with handsome purple-veined leaves, grey-bloomed stalks and large pink tubers. It is widely grown as an ornamental but is not very popular as an edible plant. In the Philippines it is made into cakes with coconut.

Tannia is still in the course of active migration and, perhaps more than any other aroid root crop, its cultivation is increasing. The 19th and 20th centuries have seen its spread into Asia and the Pacific and into North America. Fields of malanga are now a common sight in Florida, in response to the needs of the large population of recent immigrants from Cuba, Puerto Rico and other islands of the Antilles.

Giant taro

Very similar in appearance to *X. sagittifolium* is giant taro, *Alocasia macrorrhiza* (synonymous with *A. indica*). The main distinguishing feature is that the leaf blades of giant taro are held point uppermost, whereas those of the edible xanthosomas (and colocasias) point more or less downwards. Other species of *Alocasia*, such as *A. cucullata* and *A. fornicata*, are also occasionally cultivated for food but they and giant taro are minor crops compared with taro and tannia. In Malaysia giant taro is grown more for its medicinal than culinary uses but one dish it is used for is 'curry santan'. The acridity is first removed by soaking it in water with betel nut (*Areca catechu*) chips or slaked lime. Then it is reboiled in coconut cream to which spices and dried prawns are added.[23]

Giant taro originated in India and Sri Lanka, and its cultivation spread eastward across the Pacific islands. Like taro and tannia, for optimum growth it needs high rainfall through most of the year and temperatures not falling below 10°C. It is essentially a dryland crop though, and will not stand waterlogging. Even larger than *X. sagittifolium*, it can reach 4 m, with sagittate leaf blades 90 cm long which are wavy at the margin and peltate. The cormels tend to be coarse and acrid and it is usually the stout stem, about 1 m tall and 20 cm in diameter, which is eaten. Though it is remarkably trouble free, heavy yielding and has easily digested starch, it takes 2–4 years to reach harvesting size and for this reason alone cannot rival taro and tannia as a food plant.

Swamp taro

One of the most interesting edible aroids is the swamp taro, *Cyrtosperma merkusii* (which has in the past also been known as *C. chamissonis* and *C. edule*).[24] This has the longest growing time of all, needing four years or more before it is ready to harvest. In any other root crop this would prove an unacceptable disadvantage, but as swamp taro will grow where nothing else will – in stagnant and brackish swamps – it remains unchallenged as a unique food plant.

Native to Southeast Asia, its cultivation spread eastward across the Malay and Indonesian archipelagos into the Pacific. Here, on coral atolls where food growing of any kind is a test of endurance both for the plants and the growers, it really came into its own as a crop and many islands in Micronesia adopted it as a staple. Growing it successfully on low-lying atolls means digging pits, often over 1 m deep, through the coral limestone or sand to the water table. The plant is then lowered into the wet pit in a basket filled with compost which supplies the necessary nutrients. The pits are used repeatedly and on most islands are very old. The most recent taro pit in the Marshall Islands was dug at the beginning of the century. Many pits are now disused as easier, imported, foods become available.[25]

The difficulties of cultivating crops, other than coconuts and pandanus, on 'desert' islands, has been graphically described by Arthur Grimble in his inimitable account of life in the Micronesian Gilbert and Ellice group.[26] His verdict on swamp taro, as a European accustomed to potatoes and a wide range of succulent vegetables and fruits, would never get it a mention in a good food guide but any survival guide for the region would have to rate it highly. He described it thus:

> There was also an enormous tuber of the arum family called *babai*, cultivated by the villagers in muddy pits. This could be eaten mashed with butter or in cheesy, steamed slabs and was incredibly indigestible either way.

He declares that its 'unhallowed starchiness no treatment under heaven was ever known to exorcize'. Nevertheless, he has to concede its utter reliability and, if uninspiring, *babai* – along with fish and coconuts – ensured the islanders were always well fed, if not exactly spoilt for choice.

Staple foods do not only feed a community. They influence its culture, often in the strangest ways. Varieties of taro symbolized animals in the ritual sacrifice of the Hawaiians. The Yami people, who inhabit one of the islands off Formosa, launch a new boat by loading it with taro.[27] On a more practical level, Gilbert and Ellice islanders would assess a woman's worth

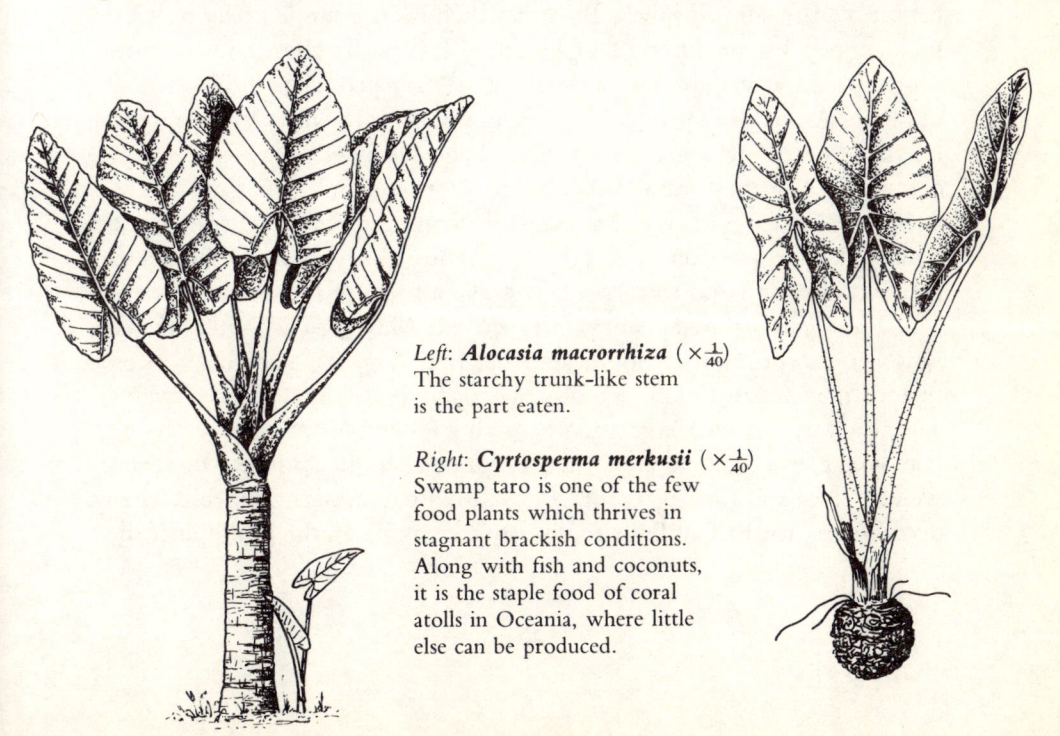

Left: **Alocasia macrorrhiza** ($\times \frac{1}{40}$)
The starchy trunk-like stem is the part eaten.

Right: **Cyrtosperma merkusii** ($\times \frac{1}{40}$)
Swamp taro is one of the few food plants which thrives in stagnant brackish conditions. Along with fish and coconuts, it is the staple food of coral atolls in Oceania, where little else can be produced.

according to her ability to produce culinary masterpieces from the somewhat intractable *babai*. Above all, her reputation and eligibility depended on the particularly difficult dish *buatoro*, a baked pudding of *babai* and coconut. Regardless of any other attributes, if she could turn this out to perfection, her desirability was ensured.[28]

Swamp taro is a commanding plant. It reaches 3–4 m and has huge upward-pointing sagittate blades with large basal lobes. The stout stalks are often spiny, though this feature and the shape and colour of the leaves is rather variable. The long pointed spathe only slightly enfolds the purplish spadix. Other than needing ample warmth, humidity and moisture, swamp taro is a tolerant plant, growing on most soil types and rarely affected by pests and diseases. The cylindrical corm weighs 60 kg or more and harvesting must pose as many problems as cooking and digesting it.

Elephant yam and konjac

Very different in appearance from all other edible aroids are the *Amorphophallus* species which are grown for food. Several species are used in times of famine but only two – *A. konjac* (commonly known as konjac) and *A. paeoniifolius* (elephant yam) – are significant as crops. Both these species have undergone name changes in recent years, the former being once known as *A. rivieri*[29] and the latter as *A. campanulatus*.

Elephant yam and konjac are similar in habit and appearance, both having a large, almost spherical corm which bears a single fleshy mottled stalk topped by an umbrella of leaflets. Periodically the corm becomes dormant and as it starts into growth again may produce a foul-smelling inflorescence before the new leaf develops. Corms usually produce offsets on their surface which detach easily. *Amorphophallus* species do best in damp rich soils and shade and are ideal for cropping under trees and shrubs. In Indonesia, corms of *A. paeoniifolius* are dug up, turned upside down and replanted after a season's growth. Apparently this stimulates the lateral buds into growth and so increases the overall size of the corm.[30]

A. paeoniifolius, the elephant yam, is native to Malaysia and Sumatra but is now grown throughout Southeast Asia, India and the Pacific. In many regions it is an important standby when the rice crop fails. Cultivated varieties (var. *hortensis*) have smoother stalks, fewer calcium oxalate crystals than wild plants (var. *sylvestris*), and less or no alkaloid content. The corms weigh 2–8 kg and vary in colour from yellowish to orange or purple. They need boiling for half an hour to destroy the toxins. In the Philippines the young shoots are considered an asparagus-like delicacy. The older leaves and corms are often cooked and fed to pigs.

The importance of edible *Amorphophallus* species has been generally underestimated in the past but there are signs now that their potential is being recognized. Indonesia's third Five-year Plan (1979–84) stressed the need for increasing production of alternatives to rice so that crop failures of the main staple would be less catastrophic. In Indonesia, elephant yam is now third in importance after rice and maize.[31] Unfortunately, because of the destruction of tropical rain forest throughout the country, most native species are becoming scarce at a time when they will be needed more than ever in the research and development of improved varieties. One of the greatest challenges to plant breeders working with *Amorphophallus* must surely be to cross *A. titanum*, which has corms weighing up to 80 kg, with the elephant yam – to produce what could only be described as the mammoth yam! So far though, no crosses between *Amorphophallus* species have been achieved.

For rather different reasons than helping to feed a large population, *Amorphophallus* species are of great interest in Japan, which perhaps more than any other country is making the most of the unusual carbohydrates they possess. In particular, it grows *A. konjac*, a hardier species than elephant yam. Though traditionally grown as food known as *konnyaku*, it is increasingly produced for specialist uses in the food technology and

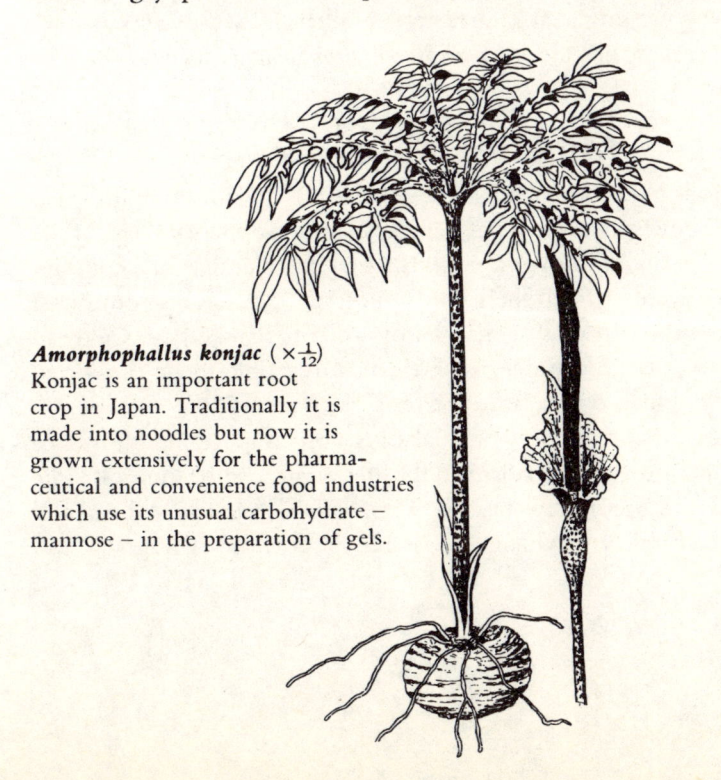

Amorphophallus konjac ($\times \frac{1}{12}$)
Konjac is an important root crop in Japan. Traditionally it is made into noodles but now it is grown extensively for the pharmaceutical and convenience food industries which use its unusual carbohydrate – mannose – in the preparation of gels.

pharmaceutical industries. The black-skinned corms are rich in the carbohydrate mannan and are used as a commercial source of mannose, a substance used in diabetic foods. The starch also has gelling properties which have many applications in convenience foods. It is also made into dietary fibre (fibre gel) supplements and slimming formulae. Judging by the markets for these products in the western world there is no doubt that konjac will continue to increase in commercial value.

For industrial processing the corms are usually harvested after three years when they weigh up to 2 kg. The domestic market prefers them younger, smaller and sweeter – about a year old and 200 g or so. Preparing the corms consists of slicing, soaking, drying, then pounding with lime and water into gelatinous grey cakes. For cooking they are rubbed with salt, rinsed, diced and either stir-fried with other vegetables or simmered in soy sauce. Alternatively the paste may be formed into vermicelli (*ito konnyaku*) or noodles (*shirataki*) and added to soups and stews. Konjac is also widely grown in southern China and Indochina for food.

The demand for mannose is such that *A. oncophyllus* (widely referred to as *A. blumei*) is collected from the wild in Java for export to Japan. In view of this, and the fact that wild populations of *Amorphophallus* species are in decline in several areas of Indonesia (particularly in Java and Bali),[32] it ought now to have potential as a cultivated crop. Apparently it is easy to grow and propagate, bearing bulbils at the base of the main leaf segments and nearly always setting seed. (It has sterile pollen and would appear to be an apomict.)[33] As is the case with most *Amorphophallus* species, the seed germinates readily.

—— Minor edible aroids ——

The edible aroids so far described are significant food plants in many parts of the world. They have been in cultivation so long – under the critical eye and discerning palate of human taste – that they have undergone constant selection and now differ from their wild counterparts in having somewhat better flavours, more attractive colours, lower acridity and higher yields. In addition to these, a large number of wild aroids are edible and are collected by local people, especially in times of food shortages.

Starchy rhizomes and tubers have always been important subsistence foods, providing most of the calories in the diet of most communities. Such different species as the yellow-fleshed *Typhonium brownii* in Australia,[34] *Montrichardia arborescens* (which has starchy roots) in Paraguay and

Argentina and the Brazilian *Dracontium asperum* (containing a gel-forming starch) are all collected for food. A number of hardy species are also recorded as edible. In some regions of North America, Indian tribes eat the rhizomes of arrow arums (*Peltandra* species, especially *P. virginica*), and both the rootstock and seeds of *Orontium aquaticum*. They also make a peppery bread from the rhizomes of *Symplocarpus foetidus*, whose young leaves are cooked as a vegetable. Further north, the Lapps grind the dormant rhizomes of *Calla palustris* to make bread.[35]

Arrowroot aroids

Several *Arisaema* species are edible after careful preparation. The best known is the North American *A. triphyllum*, the Indian turnip, whose corms produce a very fine flour similar to arrowroot. The corms are dug up in spring or late autumn when dormant. Several methods are used to prepare flour from arisaemas. They can be peeled, mashed and washed through a sieve to obtain a very fine starch solution which is then dried; boiled or roasted and then ground; or sliced and roasted before grinding, which gives a slight cocoa flavour to the flour.[36] Some *Arum* species may be put to similar uses. At one time, the Isle of Portland, a rugged limestone peninsula in southwest England, was the centre of a small industry which produced Portland arrowroot from the tubers of *A. italicum* var. *neglectum*. Both *A. italicum* and *A. maculatum* have been used in several European countries as a source of starch.[37]

Fruits

Aroid berries though often succulent and attractively coloured are usually acrid, full of calcium oxalate crystals and potentially poisonous. Many are so well endowed with these defences that even handling the crushed fruits – as horticulturists and botanists must do to extract the seeds – can cause prickling and burning sensations in the relatively tough skin of the fingers and, of course, untold misery should the tiniest amount enter the eyes or mouth. Nevertheless, in certain species this toxicity diminishes as the berries ripen and in some the fruits are surprisingly palatable. The Brazilian Indians eat those of *Philodendron bipinnatifidum* and make a jelly from the juice boiled with sugar. And in spite of the fact that *Montrichardia arborescens* is called *fruta del diablo* (fruit of the devil) in Paraguay, the massive fruiting spadices are apparently irresistible.[38]

The most delicious of all aroid fruits are those of the common houseplant *Monstera deliciosa*, as one might guess from its name. If anything could still further increase this plant's popularity, it would be if it were induced to fruit in the confines of a pot, which unfortunately seems unlikely. Known as *piñanona* in Mexico, *harpón* in Guatemala and *ceriman* in the West Indies, the ripe juicy fruiting spadix has the aroma and taste of pineapple and banana combined. It can be eaten as it is, or used in desserts and drinks. There is even a record of it being used to flavour champagne,[39] the nearest perhaps that any edible aroid has come to entering haute cuisine. This appears to be the only fruiting aroid with commercial potential.

—— Inflorescences ——

The spadices of most aroids are fleshy and tender before flowering, and many are regarded not only as edible when cooked but as delicacies. In Hawaii taro spadices have always been highly prized and in Central America the young inflorescences of several *Spathiphyllum* species are tied into bunches for sale as a vegetable in local markets.

—— The future of edible aroids ——

There is no doubt that the major edible aroids – taro, tannia, giant taro, swamp taro, elephant yam and konjac – could play a more important part in the food production and economies of developing countries if they received more attention from plant breeders and agronomists. Moves are being made in this direction and several recent publications have emphasized that the edible aroids are a much neglected source of food, especially in regions where malnutrition and food shortages are recurrent. One of the few projects so far initiated has already validated the statement made by one research paper that 'Payoffs from research into the tropical root crops have always been high'.[40] In 1984 a small breeding programme in Fiji reported that a new taro cultivar named 'Samoa Hybrid' had a 50 per cent higher yield than current varieties, with crops of 20–35 tonnes per hectare.[41] Significant progress has also been made on utilizing taro waste as silage for livestock.[42]

In recent years a number of authorities have drawn attention to the fact that in many developing countries nutritional levels are worsening and health problems increasing – often not so much from insufficiency as from a changeover to cassava and western-style convenience foods based on refined carbohydrates.[43] A return to traditional foods and diversified crops (with improved cultivars and easier growing, processing and storage methods) would not only ensure a better diet but reduce the high costs of imported commodities and make more sustainable long-term use of the land than monocultures of cash crops.

At the moment most tropical root crops, including the edible aroids, are many years behind grains and most other food plants in terms of research and development. The first step in bringing them into the 20th century must be to examine and classify existing varieties – a formidable task given the number of varieties and range of distribution, especially in taro, which has more cultivars than all the other edible aroids put together. The urgency of this is increased by the fact that, just when we need them most, these varieties and their wild ancestors are disappearing at an alarming rate. For example, in the Hawaiian Islands there were, within living memory, about 150 varieties of taro.[44] Most of these are already probably extinct. Some 70 or so still exist but only a handful are grown commercially and the islanders who know about – let alone grow – the traditional varieties are getting fewer and fewer. It is a well-established fact that monocultures with a narrow genetic base are prone to attack by pests and diseases and suffer devastating losses. So often though, yesterday's rejected cultivar has exactly the genes needed for tomorrow's resistant hybrid and therefore the importance of conserving the traditional varieties, and of course wild populations too, is far beyond mere historical or gastronomic interest.

In many ways the taro of today can be compared with the potato of 300 years ago. In the 16th century, when the potato first came into cultivation here from South America, many regarded it as scarcely fit for human consumption, if not downright poisonous (which of course all parts but the tubers are). Even in 1716, when it had gained a little ground, its eating qualities were condemned as 'inferior to skirrets and radishes'.[45] Regardless of these assertions, the potato has grown out of all recognition in yield, quality, ease of cultivation – and ensuing popularity – through breeding and cultivation trials. In taro and the other major edible aroids there is the same enormous scope for increasing yields and disease resistance, improving palatability and developing appropriate technology to take some of the back-breaking work out of the growing, harvesting, preparation and storage of these crops. At present, in spite of centuries of selection by subsistence farmers, these plants are little different from their wild ancestors – compared with most western food crops. With a hopefully shorter period of development they may well be able to make an equally great contribution to feeding a hungry world.

10
Acids and Crystals

The chemistry and toxicity of aroids; medicinal and folk uses

Be warned: aroids are acrid plants. Human contact with them over the centuries by touch, taste and smell has resulted in the discovery that the majority of them are unpleasantly caustic and can cause corrosive burns to the skin and internal tissues. The very word 'aroid' is from the Greek *aron* (and hence *Arum*), which is thought to be derived from the Arabic *ar*, meaning fire. Yet our resourcefulness has devised so many ways of using these potentially toxic plants for food and medicine and in a variety of other practical ways, that the family has reached considerable economic and cultural importance. In spite of this, the chemistry of the aroids is still bedevilled by obscurity and the key to their toxicity remains something of a mystery.

Investigating the chemical constituents of the family is important for a number of reasons. Firstly, a better understanding of the toxic principles would assist the breeding of low acidity cultivars for horticulture and agriculture (both for human consumption and animal feed). This would decrease the risks of accidental human and animal poisoning from ornamental plants and, for the edible aroids, increase palatability and reduce preparation time. Secondly, defining the active constituents is essential for assessing the effectiveness of traditional medicinal uses, a number of which are concerned with contraception and the treatment of cancer. Such a screening programme might well reveal valuable compounds for new drugs. Lastly, elucidating the chemistry sheds light on the taxonomic relationships within the family. Only recently, work on flavonoids (yellow-coloured compounds, often referred to as the vitamin P group or bioflavonoids) has shown marked differences between *Acorus* and *Gymnostachys*[1] (which have, since Engler's classification, constituted the tribe *Acoreae*) – supporting other evidence that their relationship is only leaf-shape deep.

Oxalic acid and calcium oxalate crystals

Aroids contain oxalic acid and crystals of calcium oxalate. The latter occur singly, in clusters, or as needle-sharp raphides in special cells. If the plant is damaged the raphides are released and may readily pierce skin, mucous membranes and internal tissues, causing intense irritation.

Calcium oxalate crystals occur in most families of plants, so their appearance in *Araceae* is nothing unusual. They are formed when oxalic acid combines with calcium to remove excess calcium from the tissues. What is remarkable in aroids is that they appear to do far more damage than one would expect and seem to play a specific role in the battle against herbivores. At first bite, mouth, tongue and throat are assailed by thousands of minute needles and a hasty recoil is guaranteed. Caterpillars and bugs may become immune to the attack but larger mouths are unable to find a way round the problem.

Oxalic acid and oxalates are very poisonous in excess but there is no indication that levels are particularly high in aroids – at least in those regarded as edible. Taro (*Colocasia esculenta*), for example, has about one-fifth as much as spinach and only a tenth of the amount in rhubarb.[2] Mechanically these substances cause abrasion of the alimentary tract and fine tubules of the kidneys and chemically they deplete the body fluids of calcium which is essential for the function of nerves and muscle fibres. In extreme cases, this can result in hypocalcaemia and paralysis and may be fatal. Nevertheless, the symptoms caused by aroid poisoning present a more complex picture than those of oxalic acid poisoning and the erosion of tissues by oxalates. Neither are they simply the immediate result of raphide damage. In fact, the raphides are something of a red herring. Those in other families (*Labiatae* and *Lemnaceae* for example) occur without acidity and cooked taro apparently still contains raphides although cooking has destroyed the acidity.[3]

Raphides

Raphides were first described by the Dutch microscopist Anton van Leeuwenhoek, who made significant advances in plant anatomy in the 1680s. With subsequent improvements in microscopy techniques we now know that raphides in aroids are intricately engineered. Some are grooved with an H-shaped cross-section and one end more pointed than the other. They may also be barbed, as has been observed in *Xanthosoma sagittifolium*.[4]

Comparison with the stings and fangs of venomous insects and reptiles is unavoidable. They too have grooves which facilitate the entry of toxins and cause leakage of cell contents, and barbs to work their way into the tissues. Though damaging, they themselves are not lethal but merely instruments for administering a dose of poison.

—— Deadly dumb cane ——

As far as is known, *Dieffenbachia* – commonly known as dumb cane – is the most poisonous genus and certainly the one which has received most attention. Most investigations have involved *D. seguine*, the most widely cultivated species. Although leaves are less toxic than stems, and species and cultivars may vary considerably in acridity, most dieffenbachias cause a temporary loss of speech, and worse, if even a small amount is eaten.

Though microscopically small (150–300 μ in length and 8–15 μ in diameter) the raphides in *Dieffenbachia* are particularly alarming. They occur in special cells (idioblasts) that differ markedly from those around them. These cells are like capsules with nipple-shaped ends and lie at various orientations in the tissues. Each contains a bundle of precisely parallel raphides surrounded by a jelly-like substance which, it is thought, creates a high degree of turgidity by the absorption of water from adjoining tissues.[5] The contents are thus under pressure and when the cell tip is damaged, they are discharged with considerable force. Biting into a dieffenbachia plant therefore triggers the ejection of thousands of raphides, along with various

Dieffenbachia seguine ($\times \frac{1}{12}$)
The popular ornamental dieffenbachias, known as dumb cane, are poisonous and if chewed will cause sufficient swelling of the mouth and tongue to make speech impossible.
Far right, above: Needle-sharp raphides (150–300 μ in length) in special ejector cells are partly responsible for the symptoms.
Far right, below: The raphides are released with considerable force when the plant is damaged, piercing skin and mucous membranes and facilitating the entry of toxins.

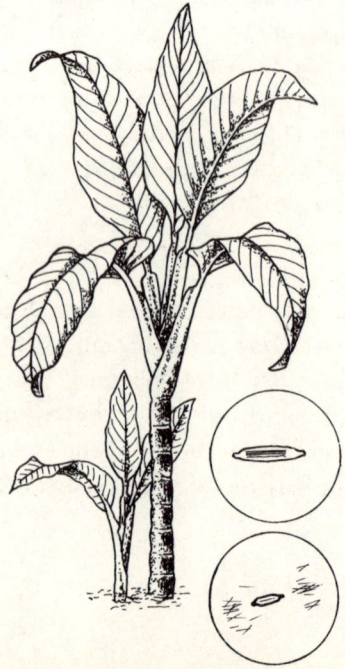

poisons, into the mucous membranes causing a burning sensation, increased salivation and swelling of the tongue, mouth and throat. In serious cases the burning pains and swelling are so severe that both speech and breathing are impaired and the oesophagus, larynx and alimentary tract become corroded. Internal bleeding, cramps, diarrhoea and vomiting, respiratory failure and death may follow.

The initial discomfort of biting the plant is usually sufficient to deter excessive consumption but some reports of dieffenbachia poisoning have come from deliberate overdoses. Among black Americans in the slave era, it was a chosen method of suicide, though not always successful. One record describes how 'a negro woman, who had long been ailing, in a fit of despair ate a good deal of dumb cane, with a view to destroy herself. It excoriated her mouth and throat much and she voided many worms, but recovered her health soon after'.[6] It was also employed as a punishment for slaves on West Indian plantations where it was grown for use in the granulation of particularly viscid sugar. They were made to bite the stem – the most virulent part of the plant.[7] It was even used to silence a witness in a court case.[8]

A possible side effect of dieffenbachia poisoning is temporary sterility. The fresh juice fed to laboratory rats was reported to cause atrophy of the reproductive organs and sterility in males for 40–90 days and in females for 30–50 days. This experiment was carried out by two doctors in Germany during the Third Reich (1933–45). They used *D. seguine* (*Caladium seguinum* as they knew it) and passed on the results to Heinrich Himmler who initiated a programme to sterilize three million Soviet concentration camp prisoners. The plan foundered when it proved impracticable to grow enough plants for the purpose.[9]

Later experiments have, however, cast doubt on these early findings, with rats suffering severe weight loss and symptoms of poisoning, but no changes in the sexual organs.[10] There is, nevertheless, a long history of *Dieffenbachia* being used to bring about sterility. Brazilian Indians fed it to their enemies for this purpose[11] and in the Caribbean it is a traditional means of contraception and was often planted under bedroom windows. According to users of this method, one becomes immune or accustomed to the discomfort when eating it regularly.[12]

In spite of its toxicity, *Dieffenbachia* has, for many years, been a very popular ornamental plant. Some producers now print a warning on the label but each year in homes throughout the world, and in the horticultural industry, there are many cases of poisoning. In the Netherlands, the Poison Information Centre at Utrecht reported 206 cases over a 13-year period (181 children, 11 adults and 14 animals). In Great Britain there were 183 cases of poisoning in people (the majority in children under five years of age, but mostly with minimal symptoms) and an average of 1–4 cases per annum in animals (mostly in dogs) from 1983 to May 1987, according to statistics compiled by the National Poisons Information Service at Guy's Hospital, London. Propagating or trimming plants is hazardous and many accidents have resulted from the most trivial contact with the sap, such as rubbing the eyes or holding a piece of tying material in the mouth. A minute amount of sap in the eye causes swelling, pain, watering and spasms, impairing vision for several weeks.[13] Just 0.2 cc has caused permanent blindness in laboratory animals.[14] The raphides are also capable of entering the relatively tough skin of the fingers, however carefully plant material is handled. One Miami horticulturist avoided touching the cut surfaces but still had blistered fingers which burned 'like fire'.[15]

Poisonous plants are almost always used medicinally, their toxic principles often being therapeutic when administered in appropriate doses and according to certain techniques. *Dieffenbachia* is no exception. In the Caribbean the wilted leaves have been used as a poultice for oedematous legs and the acrid juice is considered effective in lotions for swellings and pruritis. It may work by counter-irritation: a form of treatment which induces skin inflammation so that the increased flow of blood and lymph to and from the area speeds the elimination of toxins from deeper tissues. In Cuba the juice is applied directly to the genitals as an aphrodisiac. This almost certainly works by irritation, but supposedly only for women.[16]

Dieffenbachia has not entered the mainstream of western medicine but was introduced (as *Caladium seguinum*) to homoeopathy in 1832 as a cure for frigidity and impotence.[17]

In spite of the fact that the economic importance of *Dieffenbachia* and its history of medicinal uses (and abuses) has motivated a good deal of research, its chemistry is still not fully understood. Consumption of the plant in sufficient quantity can cause severe corrosive burns, symptoms of oxalate poisoning, lowering of blood pressure, vasodilation, muscle twitching and respiratory depression, as well as intense itching and pain. Just what is responsible for such severe symptoms? The list of possible culprits includes: concentrated oxalic acid;[18] hydrogen cyanide;[19] alkaloids;[20] glycosides;[21] saponins (which are thought partly responsible for acridity in *Arum* and *Pinellia*); and an enzyme named dumbcain which breaks down protein.[22] This last substance is of particular interest. A protease triggers the release of histamine and/or kinins. Histamines are well known for causing swelling and itching. One effect of kinins is to make smooth muscle contract — which has been observed in laboratory tests with dieffenbachia juice.[23] Is it possible that this could constrict the ducts of the reproductive organs, causing temporary sterility? Proteolytic enzymes are commonly found in snake, scorpion and spider venoms. They are responsible for the necrosis which develops round the site of injury — another feature of dieffenbachia damage — and also produce intense itching to surface tissues and pain at deeper levels. The existence of a protease in *Dieffenbachia* was disputed by a later study[24] but reconfirmed by another[25] which did, however, stress that the picture is still incomplete. On top of everything so far mentioned is the possibility of both heat-stable and unstable compounds which so far defy analysis. Researchers working on the toxicity of taro (*Colocasia esculenta*) faced a similar problem, finding over 100 volatile compounds alone.[26]

Sweet flag

Although many aroids are used medicinally, nearly all are gathered from the wild. Only one is cultivated to any extent for its chemical constituents and that is *Acorus calamus*: sweet flag, sweet rush, or calamus. It has a well-documented history of use in India, China, Russia and North America. Highly valued in ancient times, it was found in the tomb of Tutankhamen[27] and is celebrated in the Old Testament Song of Solomon: 'Thy plants are an orchard of pomegranates, with pleasant fruits; camphire with spikenard, Spikenard and saffron; calamus and cinnamon, with all trees of frankincense.'

In England its cultivation, both for medicinal uses and as a strewing herb, was centred on the waterways of the Norfolk Broads where it was known as gladdon and gathered by boat at the annual gladdon harvest. One of the charges levelled against Cardinal Wolsey, as he fell from Henry VIII's favour, was the exorbitant amount he spent on Norfolk sweet flag for the floors of his London residence.[28]

Sweet flag is warm, bitter and aromatic to the taste and fragrant in all its parts. The thick iris-like rhizome is used most and trade in this commodity, together with a talent for thriving in cultivation and escaping from it, has resulted in worldwide distribution. The species includes polyploids and there is some correlation between distribution and genetic conformation, with diploids (var. *americanus*) predominating in North America and Siberia, triploids (var. *calamus* or var. *vulgaris*) – which never set seed – in Europe and temperate India and tetraploids (var. *angustatus*) in temperate east and tropical south Asia.[29] Tetraploids have the most pronounced midrib, the thinnest texture of leaf blade and the greatest number of airspaces in the rhizome.[30]

The genetic groups differ in chemistry as well. The principal ingredient is a complex yellow volatile oil (*Ol. calami*) which makes up 1–4 per cent of the rhizome. Among its constituents is asarone, a tranquillizing and antibiotic substance. The triploid form of this is potentially toxic and carcinogenic[31] but diploid oils are asarone-free.[32] In addition, there are eight coumarins (see page 224) effective against tuberculosis.[33]

Acorus calamus was one of 4000 species screened for antimicrobial activity in a recent Soviet study. They were examined for effectiveness against various organisms, including *Staphylococcus aureus*, which causes abscesses, septicaemia and impetigo, and several *Shigella* species responsible for dysentery. It proved effective against all the strains used.[34] The main sphere of action is on the digestive system and it has a long history of use in many countries for colic, gastritis, dysentery and anorexia. The Chinese herbal pharmacopoeia also prescribes it for rheumatoid arthritis, epilepsy and

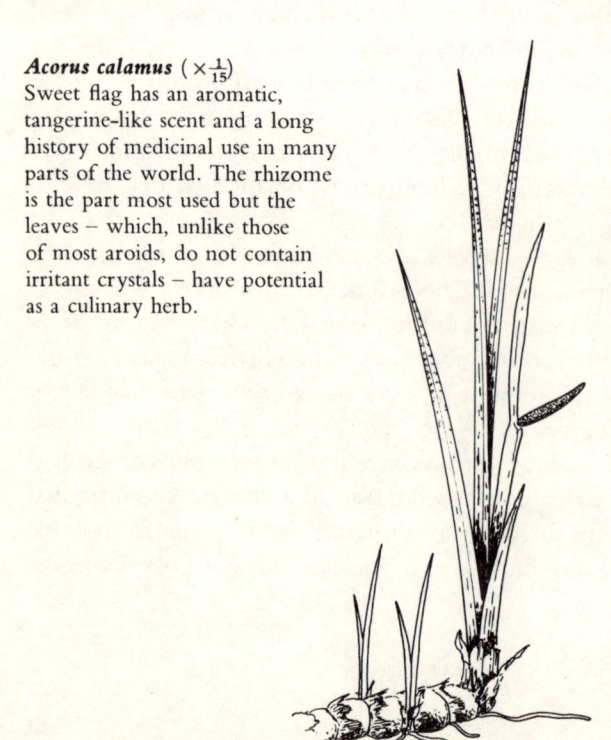

Acorus calamus ($\times \frac{1}{15}$)
Sweet flag has an aromatic, tangerine-like scent and a long history of medicinal use in many parts of the world. The rhizome is the part most used but the leaves – which, unlike those of most aroids, do not contain irritant crystals – have potential as a culinary herb.

strokes. A wide variety of uses are recorded among North American Indian tribes: a decoction to induce abortion; boiled and mashed as a dressing for burns; powdered as snuff for nasal congestion; and chewed raw for colds and toothache (a fraction of the oil, termed eugenol, is also found in oil of cloves (*Eugenia caryophyllata*), a well-known dental antiseptic and analgesic). The powdered rhizome also deters or kills insects and is an old Chinese way of getting rid of bed bugs.[35]

At one time, sweet flag was an official drug in the pharmacopoeias of 24 countries. In 1710 the herbalist William Salmon listed 16 different ways of preparing calamus for medicinal use, from a decoction in wine and a preserve to a 'liquid juice' which would 'prevail against the bitings of mad dogs and other venomous creatures'. He went on to say that it is 'a peculiar thing against poison, the plague, and all contagious diseases'.[36]

The mention of plague is significant. Plague is a highly infectious disease carried by rats and their fleas. Known as the Black Death, it killed a third of the population of Europe in the 14th century and continued sporadically until the 17th century when in London alone 68,000 people – a sixth of the population – died within a year (1664–65). Contact with the dead and dying was greatly feared: doors were marked with crosses and bells warned of the approach of the hearse. But during its height, a notorious band of thieves went about robbing corpses with impunity – and immunity. When finally brought to trial, they were promised acquittal if they would reveal how they escaped catching the disease. The secret was a herbal potion which they drank and doused themselves with before going about their macabre business. Known as 'four thieves' vinegar', it rapidly became an essential item in every medicine chest. One of its ingredients was of course calamus.

This interesting aroid has uses other than medicinal. The oil, with its long-lasting aroma, is a valuable addition to toiletries and perfumes, especially those with woody and leather overtones. It is cultivated all over the world for oil production, from North America to India, Burma, the Soviet Union and Japan. Eastern Europe is a significant producer and oils from Poland and Yugoslavia are of very fine quality.[37] It is also a flavouring agent and has been widely used in brewing and distilling, especially in liqueurs. The leaves are pleasant in cooked dishes: Culpeper recommended them in a sauce for fish.[38] They do not contain raphides or caustic substances.

Healing aroids

An estimated one in seven of all plant species has curative properties,[39] so it is no surprise that many aroids have been recorded by ethnologists and travelling botanists as being used medicinally. Unfortunately, most accounts are little more than lists of species and complaints, and rarely is there any information on effective doses or synergistic combinations (a blend of several plant extracts which works more effectively than an individual herb on its own).

One of the largest categories is external healing, for which poultices of crushed leaves or roots, wilted heated leaves, or lotions made by infusing or decocting the plant are used. Herbs chosen to treat injuries usually cauterize and disinfect the tissues and stop bleeding. A valuable ingredient in wound healing is tannin, an astringent which precipitates proteins and thus coagulates and seals damaged flesh. Tannins are known to be present in a number of aroids[40] but unfortunately most tests have been carried out on leaves, and tannin is usually concentrated in the roots (or bark if present). *Pistia* contains tannin and the dried powdered plant is used in West Africa for healing wounds.[41] Skunk cabbage (*Symplocarpus foetidus*) probably does too, as North American Indians applied the dried powdered root, or root hairs alone to bleeding wounds.[42] Whatever the active constituents, the juice from the stalks of taro (*Colocasia esculenta*) is reputed to stop bleeding, even in the case of arterial haemorrhage,[43] and in Brazil the sap of *Xanthosoma auriculatum* is considered sufficiently effective to be used on serious wounds.[44]

Perhaps partly for psychological reasons, certain aroids are respected remedies for bites and stings, their mottled stalks symbolizing the reptilian and venomous. Several South American *Dracontium* species are used both internally and externally for snake and spider bites and stingray wounds. Their local names, such as *tája de cobra*, are mostly taken from those of snakes.[45] In Indonesia the milky sap from the stalks of *Colocasia esculenta* is used against snake bites.[46] In Chinese medicine *Amorphophallus konjac* corms are crushed and applied to snake and rodent bites and *Aglaonema modestum* is mashed into a poultice for dog bites.[47] By comparison with these potentially fatal injuries, nettle stings seem trivial, but not so in Australia and Polynesia where some of the most virulent species of Urticaceae grow. The monster of the family is *Laportea gigas*, a tree 12 m tall which dispenses stings as bad as those from bees and whose effects may be felt for several months. Fortunately, the juice from giant taro (*Alocasia macrorrhiza*) gives instant relief.[48]

Poultices are applied to draw out toxins and are often made from plants that contain mucilages. These substances are produced by plants to store water and food reserves and are often protected by antibiotic polysaccharides. They retain heat and make soothing, healing compresses. Genera known to contain mucilages include *Pistia, Calla, Xanthosoma, Monstera, Alocasia* and *Colocasia*, all of which have species used for external healing. The popular houseplant *Caladium bicolor* has medicinal uses in tropical America, as well as being eaten as a vegetable. Its dried powdered leaves make a dressing for wounds and the heated rootstock draws splinters.

Some of the recipes in folk medicine contain an odd assortment of ingredients. One concoction made by German settlers in Pennsylvania to cure whitlows and felons (septic finger infections) called for *Arisaema triphyllum* blended with rotten cheese, bread crusts, butter, whisky and various herbs.[49] It would be easy to dismiss such a remedy were it not for the fact that *A. triphyllum* – commonly known as Indian turnip – was listed in the *United States Pharmacopoiea* from 1820 to 1893 and although dangerously acrid when fresh, the tubers were used in a variety of ways, notably for healing rattlesnake bites and ringworm.[50]

In most cultures skin complaints are treated primarily by external applications. The range of aroids used is as wide as the variety of ailments: *Zantedeschia aethiopica* for burns; *Spathiphyllum cochlearispathum* for warts; *Urospatha caudata* for herpes;[51] *Lysichiton americanus* for ringworm and scrofula (skin eruptions associated with certain forms of tuberculosis);[52] *Anchomanes welwitschii* for leprosy;[53] and *Gonatanthus pumilus* for allergic urticaria (nettle rash). The effectiveness of the last species for allergic conditions, including asthma, is based on its ability to reduce antibody activity.[54] It is used in Chinese herbal medicine.

—— Stimulating aroids ——

Arthritic complaints are another major category to be treated externally. Poultices and embrocations are favourite remedies, many of which act as rubefacients. These cause a temporary controlled inflammation – seen as a reddening of the area – which stimulates the circulation and lymph flow and thus speeds the excretion of toxins (such as uric acid in the case of gout). Aroids seem well suited to this kind of therapy, primarily because of the raphides of calcium oxalate and other irritating substances. The tubers of

Remusatia vivipara, roots of *Homalomena occulta*,[55] and leaves of several philodendrons[56] are just a few examples of aroids used for this purpose. In the light of recent research, however, it would seem inadvisable to rub *Philodendron* into the skin, as allergenic resorcinols have been isolated from several species.[57] Contact dermatitis from regular handling of plants is a recognized hazard in horticulture.

——— Expectorant aroids ———

The expectorant and decongestant properties of herbs used for bronchial complaints are largely dependent on saponins, essential oils or alkaloids. Many aroids contain saponins, some possess volatile oils and a few have been reported as having alkaloids. *Arum maculatum* and *A. italicum*, for example, are reported to contain arin (a saponin) and a volatile foul-smelling alkaloid similar to coniine in hemlock (*Conium maculatum*), though the existence of the latter has been disputed.[58] Both species are used as stimulating expectorants in tuberculosis, asthma, catarrh and bronchitis.

Saponins are soap-like glycosides which increase the permeability of membranes and in therapeutic amounts assist the absorption of minerals and other substances derived from food and medicines. They also irritate mucous membranes and bring about more productive coughing. Volatile oils are generally anti-inflammatory, antiseptic and antispasmodic. Skunk cabbage (*Symplocarpus foetidus*) has both and its powdered rhizome and roots are used for bronchitis, asthma and whooping cough. It is listed in the *British Herbal Pharmacopoeia* of 1976, which recommends it combined with gumplant, *Grindelia camporum* (*Compositae*) and pill-bearing spurge, *Euphorbia hirta* (*Euphorbiaceae*). Confusingly, in herbal medicine it is sometimes referred to as dracontium, but has no connection with the South American genus of that name. Another skunk cabbage, *Lysichiton americanus*, is also used for chest infections. It was the main ingredient of a North American patent medicine called 'skookum' which relieved coughs and bronchial congestion.

An east African remedy for coughs is to lick the ash of dried burnt *Pistia* plants,[59] though anything more likely to cause coughing is hard to imagine. A South American version sounds slightly more palatable – the leaves mixed with sugar and rosewater – were it not for the fact that 'they at first taste grassy and slimy, followed by a burning sensation, then a numbing of the mouth and lips'.[60]

Contraceptive aroids

Three unrelated species in South America have been recorded as oral contraceptives: *Anthurium tessmannii*, an epiphyte from eastern Peru; *Philodendron dyscarpium*, a native of quarzite rocks in the savannahs of eastern Colombia; and *Urospatha antisylleptica*, a swamp-dwelling species of the northern Amazon with large spirally twisted spathes. In each case, the dried inflorescence is pulverized and eaten by the woman – the mature spadix, whole inflorescence, and immature spadix respectively. There is no information on their effectiveness or active ingredients. An American pharmaceutical company analysed some of the *Philodendron* inflorescences but found nothing of interest. This might be because the specimens were oven-dried at 60°C.[61] Medicinal herbs should be dried below 42°C to preserve their active ingredients, or 37°C if they have volatile constituents.

Obviously, the use of inflorescences suggests phallic symbolism rather than pharmacology may be at work here but the effects of several other aroids on the reproductive system may not be wishful thinking. As has already been described, *Dieffenbachia* is said to cause temporary sterility. So apparently does *Arisaema triphyllum*. One teaspoonful of the powdered corm was decocted in half a cup of cold water, strained, and drunk by Hopi Indians. This much, they believed, would prevent conception for a week, and twice the amount, taken hot, would cause permanent sterility.[62]

Several different aroids are reputed to contain oxytocic substances which cause uterine contractions and are therefore effective for menstrual problems, childbirth or abortion. Exactly what they contain has not been established, but it is known that various volatile oils, alkaloids, hormonal and bitter compounds can stimulate reflexes and glandular responses. Interestingly, the rhizome of *Lysichiton americanus* is used raw by one North American Indian tribe (the Makah) to induce abortion and boiled by another (the Quileute) to ease labour – which suggests that the active constituent is reduced by heating but is still present in sufficient quantity to strengthen contractions.[63] In tropical Africa, *Culcasia angolensis* and *Anchomanes difformis* are known as oxytocics.

Insecticidal aroids

Herbs used to deter and kill insect pests and intestinal parasites are usually bitter and contain poisonous alkaloids or essential oils. The woman who took an overdose of dieffenbachia with the intention of committing

Culcasia
This widespread tropical African genus has several species which are used by local people.
Right: ($\times\frac{1}{6}$) ***Culcasia striolata*** is aromatic and valued in body oils and ornaments.
Far right: ($\times\frac{1}{4}$) ***Culcasia parviflora*** has insecticidal properties.

suicide because of chronic ill health, not only survived but ironically recovered her former fitness when the poisons in the plant destroyed the infestation of worms. Exactly which constituents act as a vermicide in *Dieffenbachia* – and other less drastic aroids used for this purpose – has not been determined. In Thailand the pounded roots of *Rhaphidophora falcata* kill threadworms[64] (a common problem in children) and in Brazil the juice of *Caladium bicolor* is used against roundworms. Like *Dracontium asperum*, the latter will also destroy maggots. In Cuba, the exceedingly caustic *Philodendron krebsii* is decocted to expel worms and is also used to induce abortion. The seeds of *P. bipinnatifidum* (formerly known as *P. selloum*) are vermicidal too.[65]

Other aroids poisonous to insects are *Acorus*, several *Arisaema* species, the aptly named *Synandrospadix vermitoxicus* from the Argentinian Andes, which is reported to be very toxic indeed,[66] and *Culcasia parviflora*, a west African species used in Sierra Leone for wrapping cola nuts to prevent weevil damage.[67]

Anti-cancer aroids

The search for plants to cure cancer or delay its progress has preoccupied medicine for decades and already produced some remarkable results, most **notably with childhood leukaemia and Hodgkin's disease which respond** well to alkaloids in the Madagascan periwinkle (*Catharanthus roseus*). A drug such as this is directly destructive of cancer cells and is inevitably extremely poisonous to normal cells too, with unpleasant side-effects. There are other types of herbs used in cancer treatment though, which may not destroy the cancer cells themselves but form a vital role in mobilizing the body's defences, improving detoxification and increasing the level of tolerance to chemotherapy and radiation therapy.

The aroids used against cancer come under both categories: the drastic and the supportive. The crushed raw tubers of *Dracunculus vulgaris*, *Arum maculatum* and the closely related *A. italicum*,[68] and several *Arisaema* species[69] are applied directly to cancerous growths, sometimes mixed with honey or in arsenic paste. It is probably their corrosive properties which achieve results.

A North American Indian remedy involves taking a mixture made from the rhizome of skunk cabbage (*Symplocarpus foetidus*) and leaves of yellow dock (*Rumex crispus*). Yellow dock is a purifying herb which strengthens liver function and promotes the excretion of toxins. Skunk cabbage is an expectorant and mild sedative which relieves and counteracts muscular spasms. Their action together in the background treatment of cancer might merit investigation.

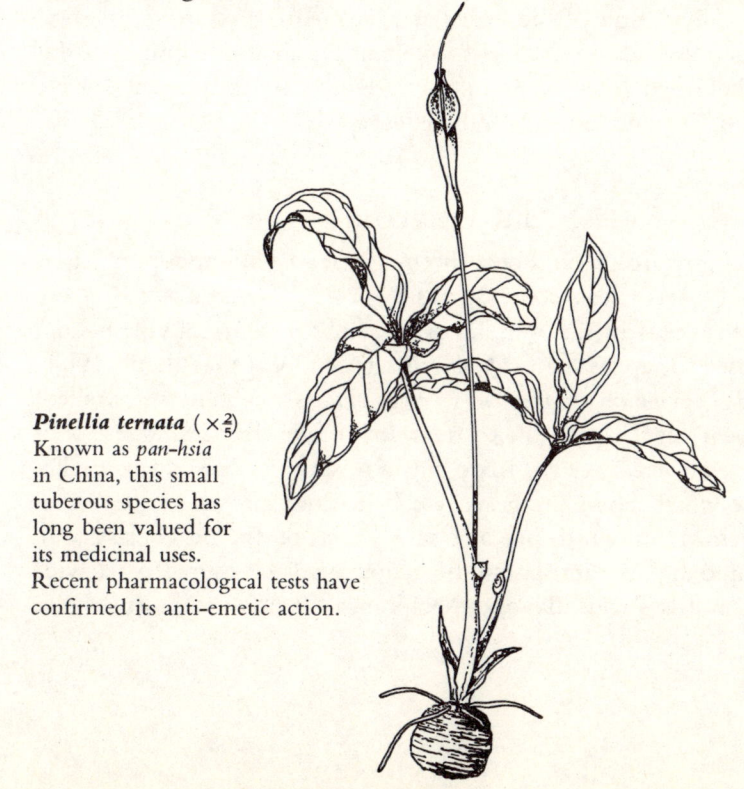

Pinellia ternata ($\times \frac{2}{5}$)
Known as *pan-hsia* in China, this small tuberous species has long been valued for its medicinal uses. Recent pharmacological tests have confirmed its anti-emetic action.

Modern Chinese medicine, which aims to make the best use of both traditional and western methods, has reported encouraging results from several species of *Pinellia*, an eastern Asian genus. *P. ternata* (*pan-hsia*) was used as long ago as the first century BC for breast and stomach cancer and leukaemia but its chemistry was only analysed in recent years. Besides starch, the tubers contain β-sitosterol glucoside, choline, volatile oils, sterols, saponin, fatty acids, and the glycoside 3,4-dihydroxybenzaldehyde. They are the source of a drug known as *Tubera Ari*. Traditionally, it was used as a decongestant for bronchial diseases, as a carminative in flatulent colic and for the control of vomiting in pregnancy. Tests have confirmed that it does indeed relieve nausea. This in itself could contribute to cancer therapy in which severe nausea is an unpleasant side-effect of treatment based on chemotherapy and radiation. Perhaps this aspect also accounts for its use in the successful Nan-Kai Hospital prescription for removing gallstones without surgery, in which it is one of five herbs combined in equal quantities to stimulate contractions of the gallbladder – a process which causes considerable pain and nausea.[70]

P. pedatisecta has larger tubers than *P. ternata* and is used in similar ways and *P. tripartita* is another anti-emetic that also relieves coughs and improves cardiac function. *P. tuberifera* is also used for several forms of cancer, especially cancer of the breast and leukaemia. Again, it has anti-emetic properties and is reported to contain an alkaloid similar to the coniine of hemlock *(Conium maculatum)*, which could account for its additional sedative and antispasmodic effects.

Sedative aroids

The idea that any aroid could be a soothing remedy might appear rather far-fetched. Nevertheless, some aroids – properly administered – are actually taken as sedatives. A number of arisaemas are anti-convulsants and have been used for epilepsy. *A. thunbergii* acts as a local anaesthetic. *Pothos scandens* and *Typhonium giraldii* (or *T. giganteum* as it is sometimes called) have also been used in the east to treat convulsive illnesses.[71] The west African *Cercestis afzelii* is very toxic but in measured doses is a painkiller and sedative which slows the heart rate.[72] It is not known exactly what these aroids may contain to produce such effects on the nervous system. Sedative, anodyne and antispasmodic actions may be caused by a wide range of chemicals – acids, flavonoids, saponins, alkaloids, or essential oils.

Hallucinogenic aroids

A few aroids have been recorded as possible hallucinogens. The essential oil in the rhizome of *Acorus calamus* includes the fractions α-asarone and β-asarone, which are alleged to have hallucinogenic properties,[73] and the rhizome of *Symplocarpus foetidus* yields an antispasmodic, slightly narcotic drug which contains 5-hydroxytryptamine.[74] Tryptamines are alkaloids with an indole nucleus. They form the base of many well-known hallucinogens, such as *Psilocybe* fungi. However, in large amounts *Symplocarpus* causes vomiting and headaches.

Homalomena is a large genus of about 110 species, mostly Asian but with around ten in tropical America. Little is known about the chemistry of the genus but several are highly aromatic. In India, the roots of *H. aromatica* are distilled to obtain a yellow essential oil that is used medicinally and added to tobacco. When bruised, *H. lindenii* smells like aniseed, and *H. rubescens* resembles coriander. The latter is used as an arrow poison in Malaysia, and medicinally in India.[75] Homalomenas are used for a variety of conditions, from bites, colic and diarrhoea to rheumatism, venereal disease and childbirth. Some are hallucinogenic and narcotic, though the substances involved have not been isolated. *H. occulta*, an antirheumatic herb, may be one, as its name suggests.[76] In Papua New Guinea an unidentified species is decocted with the leaves and bark of a tree called *agara* (*Galbulimima belgraveana*: Himantandraceae), a potion which causes extreme intoxication followed by visions, deep sleep and vivid dreams. The *Homalomena* in question is known locally as *ereriba*.[77] The *agara* tree is the main hallucinogen, but adding an extract from the *ereriba* plant may increase, lengthen or change the nature of the experiences.

Hunting potions

The first sample of arrow poison from tropical America was brought back to England by Sir Walter Raleigh in 1595 but its ingredients remained a mystery until the 19th century. One of them, curare, was to become of major importance as the source of muscle relaxants used in surgery. Curare is a black resin obtained from several trees of the genera *Chondrodendron* and *Strychnos*. The preparation of arrow poison is a complex and usually secret process involving the making of the curare to which other extracts are added. According to an early account, one ingredient used by the Tecuna Indians is an extract of the stems and leaves of a *Dieffenbachia* species.[78]

A well-documented method of fishing by some South American Indians is the adding of a poison to the water which stupefies the fish. Those needed for eating are then collected from the surface and the remainder recover as the toxins are broken down. *Philodendron craspedodromum* is used for this purpose by a tribe in the Colombian Amazon.[79] They cut the leaves and put them in bundles for a few days to ferment. Then they are crushed and thrown into the river to stun the fish. There is no information on the chemistry of this aroid but it is known that certain coumarins are toxic to fish. Coumarins are formed in certain plants by chemical changes during drying and fermentation. Characteristically, a smell of new-mown hay is given off which is not apparent in the living plant. Coumarins are certainly present in some aroids – possibly in *Anthurium oxycarpum*, which is said to be odourless when fresh but develops a musky vanilla scent when dried.

Magic and ritual

Some form of magic or ritual is performed in every society with the hope of influencing the outcome of events. Aroids have had an important part to play in several different cultures. The Quileute tribe of western Washington put the large oval leaves of *Lysichiton americanus* beneath the bow of a canoe before setting off to hunt seals. The sealers believe that, as the leaves lie flat and ummoving in the water, so the seals will keep still and be caught easily.[80] And in Micronesia there is a belief that anyone who has been porpoise hunting must not enter a field where taro (*Colocasia esculenta*) is growing, as this would ruin the crop.[81]

A peculiar ritual was performed by the Menomini Indians of North America. When someone recovered from an illness, they were tattoed wherever it had caused pain. This was done with the teeth of a gar pike dipped into a decoction of the roots of *Symplocarpus foetidus*. They believed the procedure would ensure that the illness did not recur.[82]

Dyes and adornment

Personal adornment can take many forms. In the Philippines, two species of *Rhaphidophora* are chewed to blacken the teeth: the aerial roots of *R. merillii* and the inner bark of *R. korthalsii* (which also turns the saliva red).[83] Less alarming is the use of leaves and fragrant oils to enhance the appearance. The tropical African *Culcasia striolata* is an aromatic plant which retains its scent when dried. In Gabon the leaves are worn round the neck and in the hair, and the crushed dried roots are added to body oils.[84] Many societies associate the colour red with royalty and power. A Colombian tribe, the Kubeo, have a high regard for the spathes of *Philodendron haematinum* (named after the Greek *haemato*, indicating blood). Before treating a sick person, the witch doctor takes hold of the spathes and dyes his hands red with their juice.[85] Red pigments are due to the presence of pelargonidin, a precursor of anthocyanins which are widely distributed in *Araceae*.[86]

Any number of plants will produce green or yellow shades for dyeing fabrics. In Lesotho, both *Zantedeschia aethiopica* and *Z. albomaculata* are gathered for this purpose and in Togo a yellow dye is extracted from *Stylochaeton lancifolius*.[87] The juice of *Dieffenbachia* leaves can be used as a stain, but the fact that it produces an indelible brown dye, ideal for marking laundry, suggests that in this case the substance may be tannin.

Starch

All tuberous aroids contain starch and many are edible after careful preparation, but there are numerous other uses for this material. One of the common names for *Arum maculatum* is starchwort and this goes back to Elizabethan times when it was considered the finest laundry starch. In the 16th and 17th centuries starch played a crucial role in fashion for stiffening ruffs – pleated or fluted collars of lawn or muslin – which were worn at the neck and wrists of garments by men, women and children alike. Starching and ironing these items was a painstaking process in more ways than one. John Gerard wrote in his herbal that 'the most pure and white starch is made from the roots of the cuckoo-pint, but most hurtful for the hands of the laundress that have the handling of it; for it chappeth, blistereth, and maketh the hands rough and rugged and withall smarting'.[88] In spite of – or possibly because of – its abrasive effects on skin, it was also used as a face powder for removing freckles. The saponin content of the plants was utilized too. During flowering, the stalks were cut and soaked in water for three weeks and the residue dried as a substitute for soap.[89]

Hardware

Many tropical aroids have tough flexible aerial roots and stems which provide local people with useful materials. Most are turned into string or rope for binding, or are woven into household items. Lengths of strong cordage are obtained from *Anthurium flexuosum* and *A. palmatum* and that from *A. nymphaeifolium* is fine enough for guitar strings. The roots of *A. scandens* are especially tough and are used not only for basketwork but as wire and nails in the construction of houses. *Heteropsis jenmani* is similarly used. *Monstera* roots can be easily stripped and absorb natural dyes well, so are good for making hats and decorative articles. *Philodendron bipinnatifidum* has thick robust roots, ideal for heavy duty purposes in the construction of buildings, boats and tools. Those of *P. imbe* are so long and strong that even the bark stripped from them can be made into baskets. In the West Indies, the hard stems of *P. scandens* (also known as *P. oxycardium*) serve as basket handles. The 4 m tall *Montrichardia arborescens* has fibrous stems suitable for string, rope, and paper-making.[90]

In Australia the endemic *Gymnostachys anceps* is known as settlers' twine (or flax), sword sedge and travellers' grass. This strange primitive aroid grows mostly in forests in New South Wales and Queensland, sometimes with its tuberous roots wedged into rock crevices. It has tough sword-shaped leaves up to 2 m long and clusters of stringy inflorescences which are virtually spatheless and produced on a woody flowering stem. In the early days, the leaves were seared in hot ashes and used for a variety of farmyard tasks needing particularly strong cords, such as sowing sacks or tying the legs of pigs for market.[91]

The African *Cercestis afzelii* has flexible stems about 1 cm thick which are woven into baskets and mats for drying the cocoa harvest. They are even tough enough to serve as nooses for catching antelope.[92]

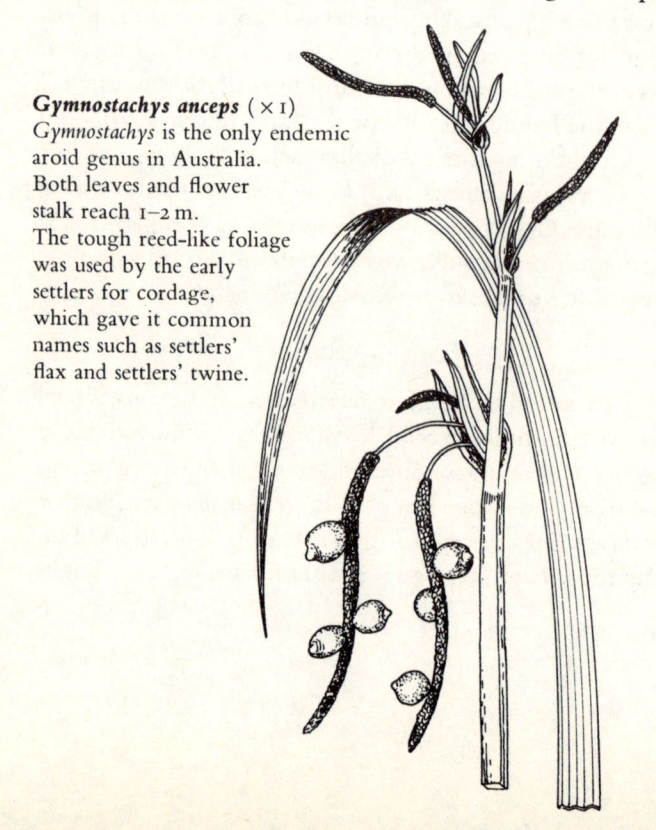

Gymnostachys anceps (× 1)
Gymnostachys is the only endemic aroid genus in Australia. Both leaves and flower stalk reach 1–2 m. The tough reed-like foliage was used by the early settlers for cordage, which gave it common names such as settlers' flax and settlers' twine.

The outstanding size of some aroid leaves enables them to be turned into wrappings and coverings, usually for items of food but on occasion for much larger objects, such as a person caught in a sudden downpour. The massive leaves of *Xanthosoma robustum* and *X. roseum* provide cape-like protection, though more modestly-sized species make quite adequate umbrellas, complete with handle. Such improvisations are a common sight in tropical countries.

—— Future uses ——

As this and the last chapter have tried to show, there is ample evidence that aroids have long been used for a wide variety of purposes wherever they grow. Many of these are of course diminishing as lifestyles change but this is not to say that aroids have no potential in the modern world. Some are already making the transition from ethnobotany to technology: the wastes from taro (*Colocasia esculenta*) can be turned into fuel alcohol; taro starch – which has particles one-tenth those of maize – may be used for pharmaceutical grade powders and added to biodegradable plastics to improve breakdown;[93] konjac (*Amorphophallus konjac*) and related species provide gels for the food and pharmaceutical industries; *Pistia* has possibilities in sewage treatment and fertilizer manufacture.

Meanwhile, the medicinal uses remain firmly rooted in the past, with only a handful of aroids in modern Chinese and European medicine. Though in many cases documented as having medicinal uses, hundreds of tropical species are unexamined – and, worse still, may continue that way as knowledge of them, and the rain forest, is lost. We need to find out much more about their chemistry and their traditional uses, especially in countries where tropical forest is intact and where people still know the plants and how to use them. As Timothy Plowman wrote about the family in his fascinating survey of the folk uses of New World aroids, 'there is every expectation that it will prove to be chemotaxonomically unusually interesting.'

In practical terms we have, at one extreme, drugs and poisons; at the other, foods; and in between, serviceable materials – not to mention hundreds of beautiful horticultural species and cultivars. Learning to utilize such well-defended plants has already taken a good deal of human ingenuity. Revealing the secrets of their defences and extending their applications needs still more. Appreciating and enjoying them in all their diversity and complexity should provide the inspiration.

Check List of Aroid Genera

The *Araceae* is undergoing extensive revision but is at present generally considered to consist of approximately 110 genera (arranged into nine subfamilies in order of evolutionary advancement) and over 2500 species.

SUBFAMILIES

Acoroideae: *Acorus, Gymnostachys.*
Pothoideae: *Pothos, Pedicellarum, Pothoidium.*
Monsteroideae: *Anadendrum, Amydrium, Rhaphidophora, Epipremnum, Scindapsus, Alloschemone, Stenospermation, Rhodospatha, Monstera, Heteropsis, Spathiphyllum, Holochlamys.*
Calloideae: *Calla.*
Lasioideae: *Lysichiton, Symplocarpus, Orontium, Anthurium, Cyrtosperma, Lasia, Anaphyllum, Anaphyllopsis, Podolasia, Urospatha, Dracontioides, Dracontium, Pycnospatha, Zamioculcas, Gonatopus, Plesmonium, Amorphophallus, Pseudodracontium, Callopsis, Pseudohydrosme, Anchomanes, Nephthytis, Cercestis, Culcasia, Montrichardia.*
Philodendroideae: *Furtadoa, Homalomena, Schismatoglottis, Piptospatha, Hottarum, Bucephalandra, Phymatarum, Aridarum, Heteroaridarum, Philodendron, Anubias, Bognera, Aglaonema, Aglaodorum, Dieffenbachia, Zantedeschia, Typhonodorum, Peltandra.*
Colocasioideae: *Xanthosoma, Caladium, Scaphispatha, Aphyllarum, Chlorospatha, Jasarum, Steudnera, Remusatia, Gonatanthus, Hapaline, Protarum, Colocasia, Alocasia, Xenophya, Syngonium, Ariopsis.*
Aroideae: *Stylochaeton, Carlephyton, Colletogyne, Arophyton, Mangonia, Taccarum, Asterostigma, Gorgonidium, Synandrospadix, Gearum, Spathantheum, Spathicarpa, Zomicarpa, Zomicarpella, Filarum, Ulearum, Arum, Dracunculus, Heliodiceros, Theriophonum, Typhonium, Sauromatum, Eminium, Biarum, Arisarum, Arisaema, Pinellia, Ambrosina, Lagenandra, Cryptocoryne.*
Pistioideae: *Pistia.*

GENERA

KEY
- Aq aquatic (growing in water – submerged or emersed)
- Arb arborescent (tree-like)
- Cl climber
- Ep epiphytic (growing on a tree rather than the ground)
- Es erect-stemmed
- H hardy
- Hh half-hardy
- Lith lithophytic (growing on rock)
- Rh rhizomatous
- Rheo rheophytic (growing in fast-flowing water)
- Sc scandent (climbing or scrambling)
- Semi-aq semi-aquatic (growing in swamps, marshes, etc.)
- Subtr subtropical
- Terr terrestrial
- Tr tropical
- Tu tuberous

CHECK LIST OF AROID GENERA

GENUS	SPECIES	DISTRIBUTION	HABIT	ECOLOGY
Acorus	2	N Temp	Rh	H/Aq/Semi-aq
Aglaodorum	1	SE Asia	Rh	Tr/Semi-aq
Aglaonema	21	SE Asia	Es/Rh	Tr/Terr
Alloschemone	1	S Am	Sc	Tr/Cl
Alocasia	70	SE Asia/Austr/Pacific	Rh/Tu/Arb/Es	Tr/Terr/Lith
Ambrosina	1	Medit	Tu	Hh/Terr
Amorphophallus	100	Asia/Afr/Austr/ Pacific/Madagascar	Tu	Tr/Terr
Amydrium	6	SE Asia	Sc	Tr/Cl
Anadendrum	7	SE Asia	Sc	Tr/Cl
Anaphyllopsis	3	S Am	Rh	Tr/Semi-aq
Anaphyllum	2	India	Rh	Tr/Terr/Semi-aq
Anchomanes	5	Africa	Rh/Tu	Tr/Terr
Anthurium	700+	C and S Am	Ep/Es/Sc	Tr/Terr/Cl/Lith/Semi-aq
Anubias	8	W and C Africa	Rh	Tr/Semi-aq/Rheo
*Aphyllarum	1	S Am	Tu	Tr/Terr
Aridarum	7	Borneo	Es	Tr/Rheo
Ariopsis	1	India/Burma	Tu	Tr/Terr
Arisaema	177	E and NE Afr/ S Arabia/ S and E Asia/N Am	Tu/Rh	H/Hh/Tr/Terr/Ep
Arisarum	3	Medit	Tu/Rh	H/Hh/Terr
Arophyton	7	Madagascar	Tu/Rh	Tr/Terr/Ep
Arum	±20	Eur/W Asia	Tu	H/Hh/Terr
Asterostigma	7	S Am	Tu	Tr/Terr
Biarum	16	Medit	Tu	H/Hh/Terr
Bognera	1	S Am	Rh	Tr (habitat unknown)
Bucephalandra	1	Borneo	Rh	Tr/Rheo
Caladiopsis (now Chlorospatha)				
Caladium	8	C and S Am/W Indies	Tu	Tr/Terr/Semi-aq
Calla	1	N Temp	Rh	H/Aq/Semi-aq
Callopsis	1	E Afr	Rh	Tr/Terr
Carlephyton	3	Madagascar	Tu	Tr/Terr
Cercestis	10	W and C Africa	Sc	Tr/Cl
Chlorospatha	15	C and S Am	Tu	Tr/Terr
Colletogyne	1	Madagascar	Tu	Tr/Terr
Colocasia	8	SE Asia (now pantropical)	Rh/Tu	Tr/Terr/Semi-aq
Cryptocoryne	56	India/SE Asia	Rh	Tr/Aq/Semi-aq
Culcasia	27	Trop Afr	Es/Sc	Tr/Cl/Terr

CHECK LIST OF AROID GENERA

GENUS	SPECIES	DISTRIBUTION	HABIT	ECOLOGY
Cyrtosperma	11	India/SE Asia/Pacific	Rh/Tu	Tr/Semi-aq
Diandriella (now *Homalomena*)				
Dieffenbachia	25–30	C and S America/ W Indies	Es	Tr/Terr
Dracontioides	1	Eastern S Am	Rh	Tr/Semi-aq
Dracontium	±15	S and C Am	Tu	Tr/Terr
Dracunculus	2	Medit	Tu	Hh/Terr
Echidnium (now *Dracontium*)				
Eminium	6	W and C Asia	Tu	Hh/Terr
Epipremnum	20	SE Asia/Pacific/Austr	Sc	Tr/Subtr/Cl
Filarum	1	S Am	Tu	Tr/Terr
Furtadoa	2	SE Asia	Es	Tr/Terr
Gearum	1	S Am	Tu	Tr/Terr
Gonatanthus	2	SE Asia	Tu	Subtr/Terr/Ep
Gonatopus	7	E Africa	Tu/Rh	Tr/Terr
Gorgonidium	2	S Am	Tu	Tr/Terr
Gymnostachys	1	Austr	Rh	Subtr/Terr
Hapaline	6	SE Asia	Tu	Tr/Terr
Helicodiceros	1	Corsica/Sardinia	Tu	Hh/Terr
Heteroaridarum	1	Borneo	Rh	Tr/Rheo
Heteropsis	13	S Am	Sc	Tr/Cl
Holochlamys	1	New Guinea	Es	Tr/Terr
Homalomena	108	S Asia/C and S Am/ SW Pacific	Es/Rh	Tr/Terr/Rheo
Hottarum	4	Borneo	Es/Rh	Tr/Rheo
Jasarum	1	S Am	Es	Tr/Aq
Lagenandra	13	India/Sri Lanka	Rh	Tr/Aq/Semi-aq
Lasia	1	India/SE Asia/ Melanesia	Rh	Tr/Semi-aq
Lasiomorpha	1	Trop W and C Afr	Rh	Tr/Semi-aq
Lysichiton	2	N E Asia/WN Am	Rh	H/Semi-sq
Mangonia	2	S Am	Tu	Tr/Terr
Monstera	24	C and S Am/W Indies	Sc	Tr/Cl
Montrichardia	2	C and S Am/W Indies	Arb	Tr/Aq/Semi-aq
Nephthytis	8	W Afr	Rh	Tr/Terr
Orontium	1	Eastern N Am	Rh	H/Aq
Pedicellarum	1	Borneo	Sc	Tr/Cl
Peltandra	3	Eastern N Am	Rh	H/Semi-aq
Philodendron	350+	C and S Am/W Indies	Sc/Ep/Arb	Tr/Terr/Cl/Lith/ Semi-aq
Phymatarum	2	Borneo	Rh	Tr/Rheo
Pinellia	6	E Asia	Tu	H/Terr
Piptospatha	9	SE Asia	Es/Rh	Tr/Rheo
Pistia	1	Pantrop	Aq	Tr/floating Aq
*Plesmonium	3	India/SE Asia	Tu	Tr/Terr

CHECK LIST OF AROID GENERA

GENUS	SPECIES	DISTRIBUTION	HABIT	ECOLOGY
Podolasia	1	SE Asia	Rh	Tr/Semi-aq
Porphyrospatha (now *Syngonium*)				
Pothoidium	1	SE Asia	Sc	Tr/Cl
Pothos	±50	India/SE Asia/Madagascar	Sc	Tr/Cl
Protarum	1	Seychelles	Tu	Tr/Terr
Pseudodracontium	6	Indochina/Thailand	Tu	Tr/Terr
Pseudohydrosme	2	W Afr	Rh	Tr/Terr
Pycnospatha	2	Laos/Thailand	Tu	Tr/Terr
Remusatia	2	S Asia/Afr	Tu	Subtr/Terr/Ep
Rhaphidophora	122	India/SE Asia/Pacific/Afr	Sc	Tr/Subtr/Cl/Rheo
Rhektophyllum (now *Cercestis*)				
Rhodospatha	23	C and S Am	Sc	Tr/Cl
Sauromatum	2	E, W and C Afr/India/Arabian Peninsula/S Asia	Tu	Hh/Subtr/Terr
Scaphispatha	1	S Am	Tu	Tr/Terr
Schismatoglottis	123	SE Asia/S Am/Melanesia/	Es/Rh	Tr/Terr/Rheo
Schizocasia (now *Alocasia*)				
Scindapsus	36	Asia/Pacific	Sc	Tr/Cl
Spathantheum	2	S Am	Tu	Tr/Terr
Spathicarpa	2	S Am	Tu	Tr/Terr
Spathiphyllum	41	C and S Am/SE Asia	Es	Tr/Terr
Stenospermation	20	C and S Am	Es	Tr/Ep/Terr
Steudnera	8	SE Asia	Tu	Tr/Terr
Stylochaeton	15	Afr	Tu	Tr/Subtr/Terr
Symplocarpus	1	E Asia/N Am	Rh	H/Semi-aq
Synandrospadix	1	Argentina/Bolivia	Tu	Subtr/Terr
Syngonium	33	C and S Am/W Indies	Sc	Tr/Cl
Taccarum	5	S Am	Tu	Tr/Subtr/Terr
Theriophonum	5	S India	Tu	Tr/Subtr/Terr
Thomsonia (now *Amorphophallus*)				
Typhonium	25	India/SE Asia/Austr/China/Melanesia (now naturalized in Afr/Madagascar/Trop Am)	Tu	Tr/Subtr/Terr
Typhonodorum	1	Madagascar	Rh	Tr/Aq
Ulearum	1	S Am	Rh	Tr/Terr
Urospatha	12	C and S Am	Rh	Tr/Semi-aq
Xanthosoma	57	Trop Am	Tu/Rh/Arb	Tr/Terr
Xenophya	2	New Guinea	Es	Tr/Terr
Zamioculcas	1	E Afr	Rh	Tr/Terr

CHECK LIST OF AROID GENERA

GENUS	SPECIES	DISTRIBUTION	HABIT	ECOLOGY
Zantedeschia	6	E and S Afr	Rh/Tu	Hh/Semi-aq/Terr
Zomicarpa	3	Eastern S Am	Tu	Tr/Terr
Zomicarpella	1	S Am	Tu	Tr/Terr

Please note: C Am (Central America) includes Mexico.
Aphyllarum: considered by some botanists as indistinguishable from *Xanthosoma*.
Plesmonium: another doubtful genus. It may be transferred to *Amorphophallus*.
Xenophya: may be transferred to *Alocasia*.

The above list has been prepared with reference to the following works:
Bogner, J. 1979. 'A critical list of the aroid genera'. *Aroideana* 1:3, 63–73.
Bogner, J. and D. H. Nicolson (in press). 'A revised classification of *Araceae* with dichotomous keys'. In J. Arditti and L. Nyman (eds.), *The Biology of Araceae.*
Hay, A. 1986. 'Cyrtosperma Griffith and the origin of the aroids'. Ph.D. thesis: Univ. of Oxford.

In addition, I am indebted to information and advice received from
Dr Simon Mayo (Royal Botanic Gardens, Kew) and
Josef Bogner (Munich Botanic Gardens).

Selected Bibliography

Aroideana (Journal of the *International Aroid Society) vols. 1–9.
Bailey, L. H. and E. Z. Bailey 1976. *Hortus Third*. New York: Macmillan. London: Collier Macmillan. (Horticultural aroids listed in alphabetical order according to genus.)
Birdsey, M. 1951. *The Cultivated Aroids*. Berkeley, California: Gillick Press.
Bogner, J. 1979. 'A critical list of the aroid genera'. *Aroideana* 1: 3, 63–73.
Bogner, J. and D. H. Nicolson (in press). 'A revised classification of *Araceae* with dichotomous keys'. In J. Arditti and L. Nyman (eds.), *The Biology of Araceae*.
Bunting, G. S. 'A revision of *Spathiphyllum (Araceae)*'. *Mem. New York Bot. Gard.* 10: 1–53.
Burnett, D. 1984. 'The cultivated *Alocasia*'. *Aroideana* 7: 3 and 4.

Chandra, S. (ed.) 1984. *Edible Aroids*. Oxford: Clarendon Press.
Croat, T. B. 1979. 'The distribution of *Araceae*'. In K. Larsen and L. B. Holm-Nielsen (eds.), *Tropical Botany*, pp. 291–308. London: Academic Press.
 1981. 'A revision of *Syngonium*'. *Ann. Missouri Bot. Gard.* 68: 565–651.
 1983. 'A revision of the genus *Anthurium (Araceae)* of Mexico and Central America'. Part 1: Mexico and Middle America. *Ann. Missouri Bot. Gard.* 70: 211–420.
 1986. 'A revision of the genus *Anthurium (Araceae)* of Mexico and Central America'. Part 2: Panama. *Monographs in Systematic Botany* 14 (Missouri Botanical Garden).
Croat, T. B. and R. B. Sheffer 1983. 'The sectional groupings of *Anthurium (Araceae)*'. *Aroideana* 5: 2, 85–123.
Crusio, W. 1979. 'A revision of *Anubias* Schott *(Araceae)*'. *Meded. Landbouwhogeschool Wageninen* 79–14: 1–48.

Engler, A. and K. Krause 1905–20. '*Araceae*'. In A. Engler (ed.), *Das Pflanzenreich* 4: 23 A–F, Heft 21, 37, 48, 55, 60, 64, 71, 73, 74. Leipzig: Engelmann. (In Latin.)
Everett, T. H. 1980. *The New York Botanical Garden Illustrated Encyclopedia of Horticulture* (10 vols.). New York and London: Garland Publishing. (Horticultural aroids in alphabetical order according to genus.)

Grayum, M. H. 1984. *Palynology and Phylogeny of the* Araceae. Ph.D. thesis. Ann Arbor, Michigan: Univ. Microfilms International.

Hooker, J. D. 1883. '*Aroideae*'. In G. Bentham and J. D. Hooker, *Genera Plantarum* 3: 955–1000. London: L. Reeve.
Hotta, M. 1971. 'Study of the family *Araceae* – general remarks'. *Jap. J. Bot.* 20: 4, 269–310.
Hutchinson, J. 1959. '*Araceae*'. In *The Families of Flowering Plants*, 3rd ed., pp. 774–85. Oxford: Clarendon Press.

Jacobsen, N. 1976. 'Notes on *Cryptocoryne* of Sri Lanka (Ceylon)'. *Bot. Not.* 129: 179–90.
 1979. *Cryptocoryner*. Copenhagen: Clausen Boger. (In Danish, also published in German by Kernen Verlag in 1982.)
 1979. *Aquarium Plants*. Poole, Dorset: Blandford Press.

SELECTED BIBLIOGRAPHY

Jervis, R. N. 1978 (revised 1980). *Aglaonema Grower's Notebook.* Clearwater, Florida: Roy N. Jervis.

Knecht, M. 1983. *Aracées de la Côte d'Ivoire.* Vaduz: J. Cramer. (In French.)

Letty, C. 1973. 'The genus *Zantedeschia*'. *Bothalia* 11: 5–26.

Li, H. 1979. '*Arisaema*'. In C. Y. Wu and H. Li (eds.), *Flora Reipublicae Popularis Sinicae* Vol. 13:2. China: Institutum Botanicum Kunmingense Academiae Sinicae. (In Chinese.)

Madison, M. 1981. 'Notes on *Caladium (Araceae)* and its allies'. *Selbyana* 5: 342–77.
  1977. 'A revision of *Monstera (Araceae)*'. *Contr. Gray Herb. Harvard Univ.* 207: 3–100.

Mayo, S. J. (with notes on cultivation by T. Hall) 1984. 'Some choice cultivated arisaemas'. *Kew Mag.* Vol. 1, part 2.

Murata, J. 1984. 'An attempt at infrageneric classification of the genus *Arisaema*'. *J. Fac. Sci. Univ. Tokyo, Sect 3, Bot.* 13: 431–82.

Nicolson, D. H. 1960. 'A brief review of classifications in the *Araceae*'. *Baileya* 8: 62–7.
  1969. 'A revision of the genus *Aglaonema (Araceae)*'. *Smithsonian Contr. Bot.* 1: 1–69.
  1982. 'Translation of Engler's classification of *Araceae* with updating'. *Aroideana* 5: 3, 67–88.
  1987. 'Derivation of aroid generic names'. *Aroideana* 10.

Ohashi, H. and J. Murata 1980. 'The taxonomy of the Japanese *Arisaema (Araceae)*'. *Jap. J. Sci. Univ. Tokyo, Sect. 3, Bot.* 12: 281–336.

Plowman, T. 1969. 'Folk uses of New World aroids'. *Econ. Bot.* 23: 2, 97–122.

Prime, C. T. 1960. *Lords and Ladies.* London: Collins.

Schott, H. W. 1856. *Synopsis aroidearum.* Vienna. (In Latin.)
  1860. *Prodromus systematis aroidearum.* Vienna: Congregationis Mechitharisticae. (In Latin.)

Walters, S. M. *et al.* (eds.) 1984. '*Araceae*'. In *The European Garden Flora* Vol. 2, *Monocotyledons* pp. 75–112. Cambridge: Cambridge University Press.

Wang, J.-K. (ed.) *Taro: a review of* Colocasia esculenta *and its potentials.* Honolulu: University of Hawaii Press.

Wit, H. C. D. de 1978. 'Revisie van het genus *Lagenandra* Dalzell *(Aracea)*'. *Meded. Landbouwhogeschool Wageninen* 78–13: 5–45.
  1982. *Aquariumplanten.* Baarn: Uitgeverij Hollandia. (In Dutch.)

*THE INTERNATIONAL AROID SOCIETY
is a non-profit corporation engaged primarily in the study of the aroid family
in all its aspects throughout the world.
Membership is open to all those interested in the family.
Aroideana, the journal of the International Aroid Society,
is sent to all members.
Enquiries should be addressed to:
The International Aroid Society
P.O. Box 43-1853
South Miami
Florida 33143.

Glossary

ADVENTITIOUS—describing roots which grow from the stem or a leaf.
ALKALOID—a nitrogen-containing compound which is usually poisonous to herbivores.
ANGIOSPERM—a seed-bearing plant of the division *Angiospermae*, in which the ovules are enclosed in an ovary which develops into the fruit after fertilization; any flowering plant.
ANTHER—the terminal part of a stamen, in which pollen is produced.
ANTHOCYANINS—a group of red, blue or violet pigments found in the sap of flowers, fruits, leaves and stems.
APOMICT—a species which produces seeds without fertilization.
APPENDIX—the terminal portion of the spadix in some unisexual species, composed of sterile florets which produce odour and heat.
ARBORESCENT—resembling a tree in shape.
AXIL—the point of the angle where the upper side of a leaf stalk joins the stem.

BISEXUAL—having male and female reproductive parts in the same flower.
BRACT—a leaf-like structure immediately below a flower or an inflorescence.
BULBIL—a small bulb-like organ of vegetative reproduction growing on a stalk or in a leaf axil.

CALYX—the outer whorl of the perianth, composed of sepals.
CARPEL—the female reproductive part of a flower, consisting of an ovary, style and stigma.
CATAPHYLL—a simplified leaf which protects the shoot.
CHEMOTAXONOMY—the use of chemical analysis of plant compounds (especially secondary metabolites) in the classification of plants.

CLASS—a taxonomic group consisting of orders.
CLONE—a population of plants which are genetically identical, having arisen by vegetative reproduction.
CONVERGENT EVOLUTION—the development of similar features and adaptations in unrelated groups of organisms as a result of similar environmental pressures (e.g. the same kind of habitat or pollinator).
CONVOLUTE VERNATION—a manner of folding of the leaf in bud in which the margin of one side of the blade is wrapped round the other.
CORM—a globular stem base, swollen with food reserves and surrounded by papery or fibrous scale leaves, which serves as an organ of vegetative reproduction and enables the plant to resume growth after dormancy.
CORMEL—a new small corm arising on the surface of one fully grown.
COROLLA—the inner whorl of the perianth, composed of petals.
COTYLEDON—the first or seed leaf of the embryo of seed-bearing plants.
COUMARINS—toxic acidic crystalline compounds which have a characteristic smell of vanilla or new-mown hay in the dried plant.
CULTIVAR—a cultivated variety of a species which may first have been found in a wild population but is maintained and propagated in cultivation.

DECOCTION—the extraction of non-volatile water-soluble constituents in medicinal plants by boiling in water.
DICOT (DICOTYLEDON)—a flowering plant of the subclass *Dicotyledonae*, the division of flowering plants characterized by the embryo having two cotyledons.

GLOSSARY

DIOECIOUS—having male and female flowers on different plants.
DIPLOID—having two sets of homologous chromosomes.

ELAIOSOME—an oil-filled protrusion on a seed which serves to attract ants as dispersal agents.
EPIDERMIS—the protective outer layer of cells.
EPIPHYTE—a plant that lives above the ground, supported by another plant or object and obtaining its nutrients from rain water, atmospheric moisture and the accumulation of organic debris around its base.
EXTRA-FLORAL—situated outside the flower.

FAMILY—a taxonomic group consisting of related genera.
FENESTRATION—the occurrence of natural perforations in the leaf.
FILAMENT—the stalk of a stamen; any slender structure, such as the thread-like extension of a staminode.
FLAGELLUM—a whip-like leafless shoot produced by climbers in low light conditions or after flowering.
FLAVONOID—a group of phenolic compounds which includes many plant pigments and vitamin P.
FLORET—a small flower, especially one of many making up an inflorescence.
FORMA (FORM)—a minor but distinctive variant within a subspecies or variety (and occasionally within a species).

GENICULUM—another word for PULVINUS.
GENUS (*pl* GENERA)—a taxonomic group of related species. The first Latin name of a species is that of its genus.
GLYCOSIDE—a compound formed by a sugar molecule bonded to a non-sugar molecule (a hydrocarbon) which may be an alcohol, a phenol or a constituent containing nitrogen or sulphur.

HERMAPHRODITE—having male and female reproductuve parts in the same flower.
HYBRID—a plant which results from the cross-fertilization of genetically unlike individuals (i.e. two different species, subspecies, varieties etc.).

INFLORESCENCE—a flowering shoot with more than one flower.
INFRUCTESCENCE—a cluster of fruits derived from an inflorescence.
INFUSION—the extraction of volatile and other constituents in medicinal plants by soaking in near-boiling water.
INTERNODE—the section of stem between two nodes.
INVOLUTE VERNATION—a manner of folding of the leaf in bud in which the margins are rolled inwards towards the midrib on the upper side of the leaf.

LITHOPHYTE—a plant which grows on rock.

MIDRIB—the central largest vein of a leaf.
MONOCOT/MONOCOTYLEDON—a flowering plant of the subclass *Monocotyledonae*, the division of flowering plants characterized by the embryo having only one cotyledon.
MONOPODIAL—describing a type of growth in which the main stem may put out side branches but continues to grow from the tip.
MONOTYPIC—a family, subfamily or genus with only one species.
MUCILAGE—a complex glutinous substance which swells in water to form a slimy solution, consisting mainly of sugars, protein and cellulose.
MYRMECOPHILE—living in association with ants.

NECTARY—a gland which secretes a sugary solution.
NODE—a point on the stem from which a leaf grows.

GLOSSARY

ORDER—a taxonomic group consisting of families.

OSMOPHORE—a scent-producing organ.

OVARY—the part of the female reproductive organ containing the ovules.

OVULE—the structure in seed-bearing plants which at maturity contains the egg cell and develops into the seed after fertilization.

PEDICEL (*adj.* PEDICELLATE)—the stalk of an individual flower in an inflorescence.

PEDUNCLE—the stalk of an inflorescence.

PERIANTH—the outer whorls of the flower, usually consisting of a calyx and corolla. In aroids it consists of minute tepals or is missing altogether.

PHOTOSYNTHESIS—the process by which plants use the energy from sunlight to produce complex organic compounds (such as carbohydrates) from carbon dioxide and water, releasing oxygen as a by-product.

PHYLLODE—a flattened stalk which resembles and performs the functions of a leaf.

PINNATE—a compound leaf with leaflets arranged in rows on either side of the midrib.

PINNATIFID—pinnately divided into lobes reaching more than halfway to the midrib.

PINNATISECT—pinnately lobed almost to the midrib (but not into separate leaflets).

PISTIL—the female reproductive part of a flower.

POLYMORPHIC—having more than one form.

POLYPLOID—an organism with more than twice the basic number of chromosomes.

POLYSACCHARIDES—a group of carbohydrates which includes starch, cellulose and mannose.

PROTOGYNY (*adj.* PROTOGYNOUS)—the functioning of female parts before male.

PULVINUS—a swelling in a leaf stalk consisting of special cells which respond to changes in conditions by moving water in or out of their vacuoles, thus altering the position of the leaf or leaflet.

RAPHIDE—a needle-like crystal of calcium oxalate.

RESORCINOL—a sweet-tasting colourless crystalline phenol.

RETICULATE—a type of venation in which the veins are arranged in a network rather than being parallel.

RHEOPHYTE—a plant that is adapted to living partly or wholly submerged along watercourses which are subject to regular flash floods.

RHIZOME—a stem which grows under the ground.

ROSULATE—having a vase-shaped rosette of leaves.

RUNNER—a stolon which produces a new plant at its apex, after which it decays.

SAPONINS—toxic soap-like compounds with a steroid structure.

SCANDENT—climbing.

SCLEROPHYLL—a plant with thick leathery leaves.

SECTION—a subdivision of a genus, consisting of species which share certain characteristics.

SELF-INCOMPATIBLE—incapable of self-fertilization.

SESSILE—unstalked.

SHEATH—a leaf base which encases the stem and/or a developing leaf.

SINUS—the notch or space between the basal lobes of a leaf.

SKOTOTROPISM—movement towards dark objects.

SPADIX—an inflorescence with a fleshy axis bearing stalkless flowers.

SPATHE—a specialized bract enclosing a spadix.

SPECIES—a group of individuals which are alike and can interbreed. In the Latin name of a species, the first word denotes its genus, the second its species.

GLOSSARY

SPIKE—an inflorescence with a central axis and stalkless flowers.

STAMEN—the male reproductive part of a flower.

STAMINODE—a sterile stamen.

STIPE—a section of peduncle between the point of attachment of the spathe and the spadix (absent in many species).

STOLON—a stem which grows along the surface of the substrate, producing new plants at the nodes.

STYLE—the slender extension of the ovary, bearing the stigma.

SUBFAMILY—a subdivision of a family, consisting of a genus or genera with certain distinctive characteristics.

SUBSPECIES—a subdivision of a species, consisting of individuals which differ in small but consistent ways from the species.

SUBTRIBE—a subdivision of a tribe, consisting of genera which share certain characteristics.

SYMPODIAL—describing a type of growth in which side branches develop successively as the main shoot.

SYNCARP—an ovary in which two or more carpels are joined together.

TANNINS—bitter-tasting astringent yellowish or brownish substances found in the outer tissues of many plants.

TAXON (*pl.* TAXA)—any named taxonomic group.

TAXONOMY—the study and description of variation in living things; classification.

TEPAL—a subdivision of the perianth which is not differentiated into a calyx and corolla.

TETRAD—a cluster of four cells, e.g. pollen grains.

TRANSPIRATION—loss of water by evaporation from the plant's aerial surface, which causes water to be drawn from the substrate surrounding the roots.

TRIBE—a group of closely related genera within a family.

TRICHOSCLEREIDS—tough needle-like cells which protect the plant against herbivores.

TRIPLOID—an organism with three times the basic number of chromosomes.

TUBER—an underground stem or root swollen with food reserves which serves as an organ of vegetative reproduction and enables the plant to begin growth after dormancy. (In aroids it is the stem, rarely the roots, which become tuberous.)

UNDERSTOREY—the part of the woodland or forest underneath the canopy, consisting of shrubs, saplings and herbaceous plants.

UNISEXUAL—having male and female reproductive parts in separate flowers on the same plant.

VARIETY (*abbr.* VAR.)—a population within a species which varies in certain respects.

VENATION—the pattern of veins on the surface of the leaf.

VERRUCOSE—having a surface covered in small wart-like protrusions.

VIVIPARY (*adj.* VIVIPAROUS)—the germination of seeds before release from the mother plant.

XEROPHYTE—a plant adapted to living in dry conditions.

TYPICAL AROID LEAF AND INFLORESCENCE SHAPES

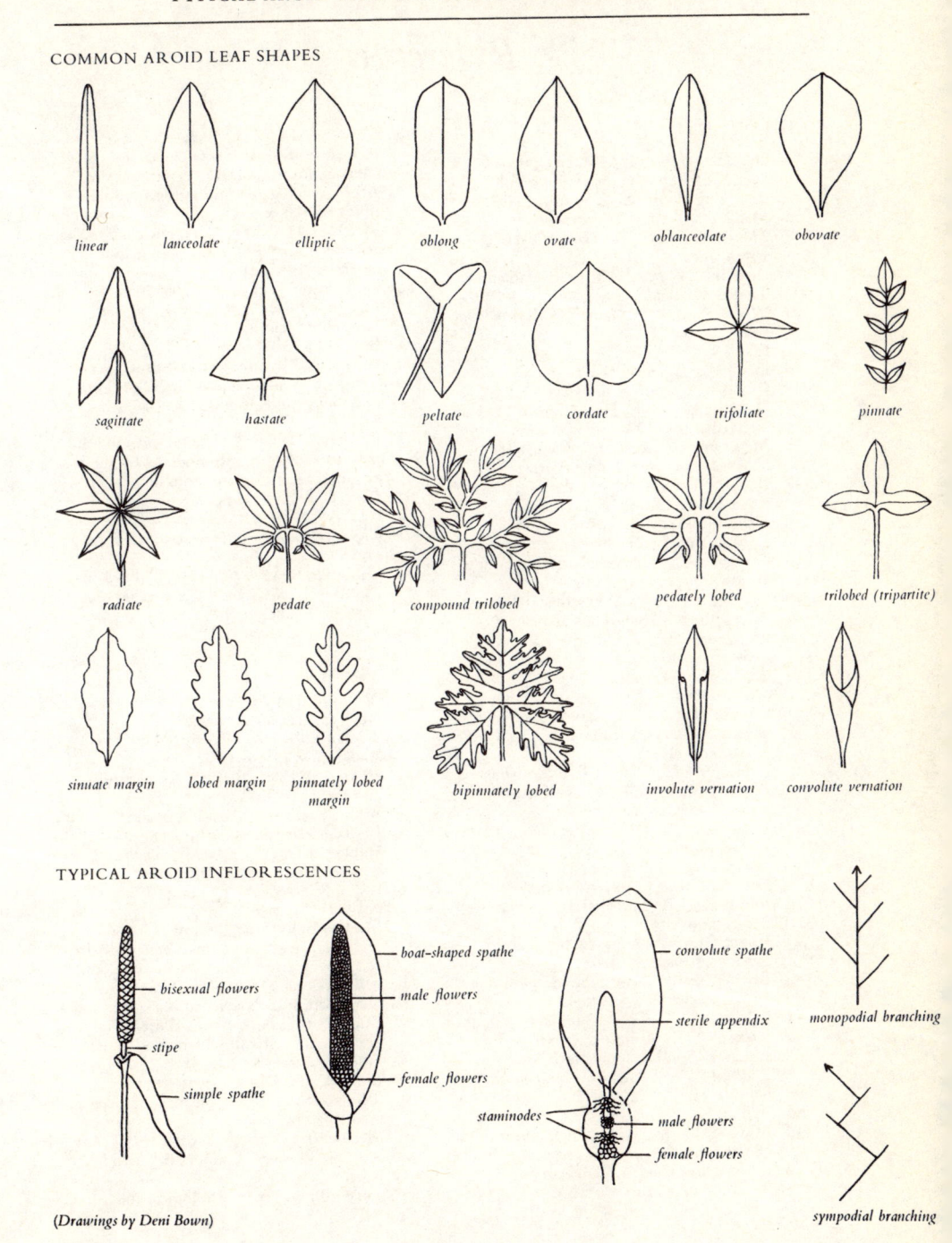

COMMON AROID LEAF SHAPES

linear lanceolate elliptic oblong ovate oblanceolate obovate

sagittate hastate peltate cordate trifoliate pinnate

radiate pedate compound trilobed pedately lobed trilobed (tripartite)

sinuate margin lobed margin pinnately lobed margin bipinnately lobed involute vernation convolute vernation

TYPICAL AROID INFLORESCENCES

bisexual flowers — stipe — simple spathe

boat-shaped spathe — male flowers — female flowers

convolute spathe — sterile appendix — staminodes — male flowers — female flowers

monopodial branching

sympodial branching

(Drawings by Deni Bown)

References

CHAPTER 1

1. Gottsberger, G. 1977. 'Some aspects of beetle pollination in the evolution of flowering plants'. *Evol. Suppl.* 1, 211–26.
2. Madison, M. 1977. 'A Revision of *Monstera*'. *Contr. Gray Herb. Harvard Univ.* 207: 3–100.
3. Brown, R.F. 1984. 'The new aglaonemas of Thailand'. *Aroideana* 7: 2, 42–52.
4. White, R. 1982. 'Panama West'. *Aroideana* 5: 4, 116–21.
5. Ibid.
6. Idris, S. 1974. '*Amorphophallus titanum* Becc. (Bunga bangkai)'. *Bull. Kebun Raya* 1, 7–10.
7. Dahlgren, R.M.T., H.T. Clifford and P.F. Yeo (eds.) 1985. *The Families of the Monocotyledons: structure, evolution and taxonomy*. Berlin: Springer-Verlag.
8. Bogner, J., and D.H. Nicolson (in press). A revised classification of *Araceae* with dichtomous keys. In J. Arditti and L. Nyman (eds) *The Biology of the Araceae*.
9. Croat, T.B. 1982. 'A study of Old World aroids'. *Aroideana* 5: 1, 13–25.
10. Gentry, A.H. 1982. 'Neotropical floristic diversity: phytogeographical connections between Central and South America, Pleistocene climatic fluctuations, or an accident of the Andean orogeny?' *Ann. Missouri Bot. Gard.* 69: 557–93.
11. Schott, H.W. 1860. *Prodromus systematis aroidearum* 1–602. Vienna: Congregationis Mechitharisticae.
12. Engler, A. 1879. '*Araceae*'. In *Monographiae Phanerogamarum* 2: 1–681. De Candolle. Paris: G. Masson.
    Engler, A. 1889. '*Araceae*'. in *Die natürlichen Pflanzenfamilien* 2(3): 102–53. Engler and Prantl. Leipzig: Engelmann.
    Engler, A. 1920. '*Araceae*'. In *Das Pflanzenreich* IV. 23F (Heft 73): 1–274. Leipzig: Engelmann.
13. Bogner, J. 1979. 'A critical list of aroid genera'. *Aroideana* 1: 3, 63–73.
14. Bogner, J. 1980. 'The genus *Scaphispatha* Brogn. ex Schott'. *Aroideana* 3: 1, 4–12.
15. Dilcher, D.L. and C.P. Daghlian 1977. 'Investigations of angiosperms from the Eocene of Southeastern North America: *Philodendron* leaf remains'. *Am. Journ. Bot.* 64, 526–34.
16. Ibid.
17. Mayo, S.J. 1986. *Systematics of Philodendron (Schott) Araceae with special reference to inflorescence characters*. University of Reading: Ph.D. thesis, 972pp.
18. Grayum, M.H. 1984. '*Palynology and Phylogeny of the Araceae*'. University of Massachusetts: Ph.D. thesis, 852pp. Ann Arbor, Michigan: University Microfilms International, xxiii.
19. Melchior, H. (ed.) 1964. *A. Engler's Syllabus der Pflanzenfamilien*. Berlin-Nikolassee: Gebrüder Bortraeger.
20. Hutchinson, J. 1934. *The Families of Flowering Plants*. Vol. 2 Monocotyledons. London: Macmillan.
21. Bogner, J. (pers. comm.)
22. Parmelee, J.A. and B.D.O. Savile 1954. 'Life history and relationships of rusts of *Sparganium* and *Acorus*'. *Mycologia* 46: 823–36.
23. Burger, W.C. 1977. 'The *Piperales* and the monocots. Alternate hypotheses for the origin of monocotyledonous flowers'. *Bot. Rev. (Lancaster)* 43, 345–93.
24. Hay, A. 1986. Cyrtosperma *Griffith and the origin of the aroids*. University of Oxford: Ph.D. thesis.
25. Dahlgren, R.M.T. and H.T. Clifford 1982. *The Monocotyledons – A Comparative Study*. London: Academic Press.

CHAPTER 2

1. Phillips, F.L. 1980. 'On coping with monsters'. *Aroideana* 3:2, 56–7.
2. White, R.R. 1985. 'Panama West'. *Aroideana* 5: 4, 116–21.
3. Mayo, S.J. 1980. 'Biarums for pleasure'. *Aroideana* 3: 1, 32–5.
4. Willis, J.H. 1952. 'Aroids foul and fragrant'. *Victorian Naturalist* 69: 4, 47–50.
5. Hooker, J.D. 1891. '*Amorphophallus titanum*', *Curtis's Bot. Mag.* 117, 7153–5.
6. Ryan, R. 1945. 'Punga Pung'. *Victorian Naturalist* 61, 219.
7. Meeuse, B.J.D. 1973. 'Films of liquid crystals as an aid to pollination studies'. In N.B.M. Brantjes and H.F. Linskens (eds.), *Pollination and Dispersal* 19. The Netherlands: University of Nijmegen.
8. Knoll, F. 1926. 'Die Arum-Blütenstände und ihre Besucher (Insekten und Blumen IV)'. *Abh. zool.-bot. Ges. Wien.* 12, 379–481.
9. Chen, J. and B.J.D. Meeuse 1971. 'Production of free indole by some aroids'. *Acta Bot. Néerl.* 20, 627–35.
10. Meeuse, B.J.D. and S. Morris 1984. *The Sex Life of Flowers*. London: Faber and Faber.
11. van der Pijl, L. 1937. 'Biological and

REFERENCES

physiological observations on the inflorescence of *Amorphophallus*'. *Recl. Trav. bot. néerl.* 34, 157–67.
12 Trelease, W. 1879. 'On the fertilization of *Symplocarpus foetidus*'. *Amer. Nat.* 13: 580–1.
13 Knutson, R. 1972. 'Temperature measurements of the spadix of *Symplocarpus foetidus*. (L.) Nutt.' *Am. Midl. Nat.* 88, 251–4.
14 Knutson, R. 1974. 'Heat production and temperature regulation in Eastern Skunk Cabbage'. *Science* 186, 746–7.
15 Ibid.
16 Walker, D.B., L. Sternberg and M.J. de Niro 1983. 'Direct respiration of lipids during heat production in the inflorescence of *Philodendron selloum*'. *Science* 220, 419–21.
17 Nagy, K.A., D.K. Odell and R.S. Seymour 1972. 'Temperature regulation by the inflorescence of *Philodendron*'. *Science* 178, 1195.
18 Gottsberger, G. and A. Amaral Jr. 1984. 'Pollination strategies in Brazilian *Philodendron* species'. *Ber. Deutsch. Bot. Ges. Bd.* 97, 391–410.
19 Nagy *et al.* 1972.
20 Chen and Meeuse, 1971.
21 Procter, M. and P. Yeo 1973. *The Pollination of Flowers*. London: Collins.
22 Ryan, 1945.
23 Dakwale, S. and S. Bhatnagar 1985. 'Insect-trapping behaviour and diel periodicity in *Sauromatum guttatum* Schott *(Araceae)*'. *Science* 54: 14, 699–702.
24 van der Pijl, 1937.
25 Madison, M. 1979a. 'Protection of developing seeds in neotropical *Araceae*'. *Aroideana* 2: 2, 51–67.
26 Dormer, K.J. 1960. 'The truth about pollination in *Arum*'. *New Phytol.* 59, 298–301.
27 van der Pijl, 1937.
28 Cleghorn, M.L. 1913. 'Notes on the pollination of *Colocasia antiquorum*'. *J. Proc. Asiat. Soc. Beng.* 9, 313–15.
29 Coleman, E. 1948. 'Interesting movement in scented Alocasia, *A. odora* (Roxb.) (C. Koch 1854) [Syns. *Arum odorum* Roxb., *Colocasia odora* Hort.].' *Victorian Naturalist* 65, 140.
30 van der Pijl, L. 1953. 'On the flower biology of some plants from Java, with general remarks on fly-traps'. *Ann. Bogor* 1, 77–99.
31 Meeuse and Morris, 1984.
32 Barnes, E. 1934. 'Some observations on the genus *Arisaema* on the Nilgiri Hills, South India'. *J. Bombay Nat. Hist. Soc.* 37: 630–9.
33 Meeuse and Morris, 1984.
34 Ramírez, B.W. and L.D. Gómez 1978. 'Production of nectar and gums by flowers of *Monstera deliciosa (Araceae)* and of some species of *Clusia (Guttiferae)* collected by New World *Trigona* bees'. *Brenesia* 14/15, 407–12.
35 Madison, M. 1977. 'A revision of *Monstera (Araceae)*'. *Contr. Gray Herb. Harvard Univ.* 207: 1–300.
36 Ramírez and Gómez, 1978.
37 Myers, N. 1985. 'Key links in the life chain'. *The Guardian*, Feb. 21.
38 Grayum, M.H. 1984. 'Palynology and phylogeny of the *Araceae*'. Ann Arbor, Michigan: University Microfilms International. xxiii, 852pp.
39 Ramírez and Gómez, 1978.
40 Grayum, 1984.
41 Madison, 1979a.
42 Grayum, 1984.
43 Ibid.
44 Ramírez and Gómez, 1978.
45 Madison, 1979a.
46 Madison, M. 1979b. 'Aroid profile No. 5: *Anthurium punctatum*'. *Aroideana* 2: 4, 126–7.
47 Beattie, A.J. 1982. 'Effects of ants on pollen activity'. In E.G. Williams (ed) *Pollination '82: Proceedings of a Symposium*, University of Melbourne: Plant Cell Biology Research Centre.
48 Sheffer, R.C., W.L. Theobald and H. Kamemoto 1980. 'Taxonomy of *Anthurium scandens (Araceae)*'. *Aroideana* 3: 3, 86–93.
49 Madison, 1979a.
50 Shaw, D.E., A. Hiller and K.A. Hiller 1985. '*Alocasia macrorrhiza* and birds in Australia'. *Aroideana* 8: 3, 89–93.
51 Madison, 1979a.
52 Hambali, G.G. 1979. 'The dispersal of taro by common palm civets'. *Proc. 5th Symp. Inter. Soc. Trop. Root Crops*, Manila and Baybay, Philippines.
53 Gottsberger and Amaral, 1984.
54 Koach, J. 1984. 'Methods of protection in the reproductive cycle in Israeli *Araceae*'. *Bull. Israel Plant Info. Center* 12, 13–19.
55 Killian, C. 1933. 'Développement, biologie et répartition de *l'Ambrosinia bassii* L. Deuxième partie'. *Bull. Soc. Hist. Nat. Afrique N.* 24, 259–94.
56 Madison, 1979a.

CHAPTER 3

1 Mayo, S.J., and M.G. Gilbert 1986. 'A preliminary revision of *Arisaema (Araceae)* in tropical Africa and Arabia'. *Kew Bull.* 41: 2.
2 Ohashi, H. and J. Murata 1980. 'Taxonomy of the Japanese *Arisaema (Araceae)*'. *J. Fac. Sci. Univ. Tokyo.* Sect. 3, 12: 281–336.
3 Murata, J. 1985. '*Arisaema taiwanense* J. Murata. A new species from Taiwan'. *J. Jap. Bot.* 60, 12: 353–60.
4 Bull, W. 1884. A retail list of new, beautiful,

REFERENCES

and rare plants. Chelsea, London: W. Bull.
5. Kingdon-Ward, F. 1956. *Return to the Irawaddy*. London: A. Melrose.
6. Bierzychudek, P. 1982. 'The demography of Jack-in-the-pulpit, a forest perennial that changes sex'. *Ecological Monographs* 52, 4: 335–51.
7. Lovett Doust, J. and P.B. Cavers 1982. 'Sex and gender dynamics in Jack-in-the-pulpit, *Arisaema triphyllum (Araceae)*'. *Ecology* 63: 797–808.
8. Lovett Doust, J. and P.B. Cavers 1982. 'Resource allocation and gender in the green dragon, *Arisaema dracontium (Araceae)*'. *Am. Midland Nat.* 108: 144–8.
9. Bierzychudek, 1982.
10. Polunin, O. and A Stainton 1984. *Flowers of the Himalaya*. Oxford: Oxford University Press.
11. Li, H. 1979. '*Arisaema*'. In C.Y. Wu and H. Li (eds) *Flora Reipublicae Popularis Sinicae* Vol. 13, part 2: 116–94. China: Institutum Botanicum Kunmingense Academiae Sinicae.
12. Herbarium, Royal Botanic Gardens, Kew: specimen collected by R.W.G. Hingston (No. 43).
13. Herbarium, Royal Botanic Gardens, Kew: specimen collected by E.M. Saunders.
14. Zang, J. (ed.) 1982. *The Alpine Plants of China*. Beijing: Science Press. New York: Gordon and Breach.
15. Mayo, S.J. 1985. '*Araceae*'. In R.M. Polhill (ed.), *Flora of Tropical East Africa*. Rotterdam/Boston: A.A. Balkema.
16. Herbarium, Royal Botanic Gardens, Kew: specimen collected by G. Forrest (No. 20496).
17. Herbarium, Royal Botanic Gardens, Kew: specimen collected by J.F. Rock (No. 24013).
18. Grayum, M.H. 1984. *Palynology and Phylogeny of the Araceae*. University of Massachusetts: Ph.D. thesis, 852pp. (Ann Arbor, Michigan: University Microfilms International, xxiii.)
19. van der Pijl, L. 1972. *Principles of Dispersal in Higher Plants*. Berlin–Heidelberg–New York: Springer-Verlag.
20. van der Pijl, 1972.
21. Mayo, 1985.
22. van der Pijl, 1972.
23. Bogner, J., S.J. Mayo and M. Sivadasan 1985. 'New species and changing concepts in *Amorphophallus*'. *Aroideana* 8, 1: 14–25.
24. Leigh, J., R. Boden and J. Briggs 1984. *Extinct and Endangered Plants of Australia*. South Melborne: Macmillan.
25. Revolutionary Health Committee of Hunan Province (eds.) 1978. *A Barefoot Doctor's Manual*. London: Routledge and Kegan Paul.
26. Li, 1979.
27. Delendick, T.J. 1985. 'Aroids at Brooklyn Botanic Gardens'. *Aroideana* 8, 1: 4–12.
28. Herbarium, Royal Botanic Gardens, Kew: specimen collected by G. Forrest (No. 8152).
29. Herbarium, Royal Botanic Gardens, Kew: specimen collected by P. Cribb.
30. Li, 1979.
31. Vogel, S. 1973. 'Fungus gnat flowers and fungus mimesis'. In N.B.H. Brantjes and H.F. Linkens (eds.), *Pollination and Dispersal*. The Netherlands: Dept. of Botany, Univ. of Nijmegen.
32. Ibid.
33. Arcangeli, G. 1891. 'Sull' *Arisarum proboscideum* Savi'. *Nuovo Giorn. Bot. Ital.* 23: 545–9.
34. Prime, C. 1960. *Lords and Ladies*, London: Collins.
35. Ibid.
36. Ibid.
37. Ibid.
38. Ibid.
39. Ibid.

CHAPTER 4

1. Street, H.E. and H. Opik 1970. *The Physiology of Flowering Plants: their growth and development*. London: Edward Arnold.
2. Hope, C.W. 1902. 'The "Sudd" of the Upper Nile'. *Ann. Bot.* 16: 63, 495–516.
3. Kingsley, M. 1897. *Travels in West Africa*. London: Macmillan.
4. Hope, 1902.
5. Fisher, J., S. Noel and J. Vincent 1969. *The Red Book – Wildlife in Danger*. London: Collins.
6. Aston, H.I. 1973. *Aquatic Plants of Australia*. Melbourne: Melbourne University Press.
7. Pancho, J.V. and M. Soerjani 1978. *Aquatic Weeds of Southeast Asia*. Quezon City: National Publishing Cooperative Inc.
8. Kokwaro, J.O. 1976. *Medicinal Plants of East Africa*. East African Literature Bureau.
9. Maheshwari, S.C. 1958. '*Spirodela polyrrhiza*, the link between the aroids and the duckweeds'. *Nature* 181: 1745–6.
10. Heywood, V.H. (ed.) 1978. *Flowering Plants of the World*. Oxford: Oxford University Press.
11. Dahlgren, R.M.T., H.T. Clifford and P.F. Yeo (eds.) 1985. *The Families of the Monocotyledons: structure, evolution and taxonomy*. Berlin: Springer-Verlag.
12. Plowman, T. 1969. 'Folk Uses of New World Aroids'. *Econ. Bot.* 23: 2, 97–122.
13. Ridley, H.N. 1930. *Dispersal of Plants throughout the World*. England: Reeve.
14. Croat, T.B. and N. Lambert 1986. 'The

REFERENCES

    Araceae of Venezuela'. *Aroideana* 9: 1–4, 3–213.
15 Hope, 1902.
16 Ibid.
17 Richards, P.W. 1979. *The Tropical Rain Forest*. Cambridge: Cambridge University Press.
18 Madison, M. 1979. 'Notes on some aroids along the Río Negro, Brazil'. *Aroideana* 2: 3, 67–77.
19 Hay, A. 1986. Cyrtosperma *Griffith and the origin of the aroids*. Ph.D. thesis: University of Oxford.
20 Ibid.
21 Ibid.
22 Ibid.
23 Ibid.
24 Ibid.
25 Ibid.
26 Ibid.
27 Lawrence, T., J. Toll and D.H. van Sloten 1986. *Directory of germplasm collections*. Vol. 2. *Root and tuber crops (revised)*. Rome: International Board for Plant Genetic Resources.
28 Hay, 1986.
29 Bogner, J. 1985. '*Jasarum steyermarkii* (Bunting), an aquatic aroid from Guyana Highland'. *Aroideana* 8: 2, 55–63.
30 van Steenis, C.G.G.J. 1981. *Rheophytes of the World*. Alphen aan den Rijn, The Netherlands/ Rockville, Maryland: Sijthoff and Noordhoff.
31 Ibid.
32 Wit, H.C.D. de 1982. *Aquariumplanten*. Baarn: Uitgeverij Hollandia.
33 Wit, H.C.D. de 1978. 'Revisie van het genus *Lagenandra* Dalzell *(Araceae)*'. *Meded. Landbouwhogeschool Wageningen* 78–13: 1–48.
34 Jacobsen, N. 1976. 'Notes on *Cryptocoryne* of Sri Lanka (Ceylon)'. *Bot. Notiser* 128: 179–90.
35 Jacobsen, N. 1980. 'Does *Cryptocoryne ferruginea* flower at full moon?' *Aroideana* 3: 4, 111–16.
36 Jacobsen, 1976.
37 Rataj, K. 1975. *Revision of the genus Cryptocoryne*. Praha: Fischer.
38 Jacobsen, N. 1979. *Aquarium Plants*. Poole, Dorset: Blandford Press.
39 Jacobsen, 1980.
40 Nicolson, D.H. 1969. *A Revision of the genus Aglaonema (Araceae)*. Washington: Smithsonian Institution Press.
41 Welsh, S.L. 1974. *Anderson's Flora of Alaska and adjacent parts of Canada*. Provo, Utah: Brigham Young University Press.
42 Chen, J. and B.J.D. Meeuse 1971. 'Production of free indole by some aroids'. *Acta Bot. Neerl.* 20, 627–35.

43 Hultén, E. and H. St. John 1931. 'The American species of *Lysichitum*'. *Svensk. Bot. Tidskr.* 25: 453–64.
44 Knutson, R. 1974. 'Heat production and temperature regulation in Eastern Skunk Cabbage'. *Science* 186: 746–7.
45 Camazine, S. and K.J. Niklas 1984. '*Symplocarpus foetidus*: interactions between spathe and spadix'. *Am. J. Bot.* 71, 843–50.
46 Grayum, M.H. 1984. *Palynology and Phylogeny of the* Araceae, Ph.D. thesis, 852pp. Ann Arbor, Michigan: University Microfilms International xxiii.
47 Bogner, J. and D.H. Nicolson (in press). 'A revised classification of *Araceae* with dichotomous keys'. In J. Arditti and J. Nyman (eds.), *The Biology of* Araceae.
48 Harrison, R.E. 1978. '*Zantedeschia* hybrids'. *J. Royal Hort. Soc.* 103, 78.
49 Letty, C. 1973. 'The genus *Zantedeschia*'. *Bothalia* 11, 5–26.
50 Whitfield, P. (ed.) 1984. *Longman Illustrated Animal Encyclopaedia*. Harlow, Essex: Longman.
51 Crusio, W. 1979. 'A revision of *Anubias* Schott *Araceae*'. *Meded. Landbouwhogeschool Wageningen* 79–14. 1–48.
52 Grayum, 1984.
53 Ibid.
54 Wei, F., and Y. Li 1985. 'A new spice, *Acorus macrospadiceus* from South China'. *Guihaia* 5: 3, 179–82.
55 Grieve, M. 1931. *A Modern Herbal*. London: Jonathan Cape.

CHAPTER 5

1 Mayo, S. 1980. 'Biarums for pleasure'. *Aroideana* 3:1, 32–5.
2 Meeuse, B. and S. Morris 1984. *The Sex Life of Flowers*. London: Faber and Faber.
3 Webb, D.A. and P.F. Yeo 1984. '*Dracunculus* Schott'. In S.M. Walters *et al.* (eds.), *The European Garden Flora*. Cambridge: Cambridge University Press.
4 Bailey, L.H. 1939. *The Standard Cyclopaedia of Horticulture* Vol. 1: 1071.
5 Meeuse and Morris, 1984.
6 Huxley, A. and W. Taylor 1977. *Flowers of Greece and the Aegean*. London: Chatto and Windus.
7 Davis, P.H. 1984. *Flora of Turkey*. Edinburgh: Edinburgh University Press.
8 Huxley and Taylor, 1977.
9 Riedl, H. 1980. 'The importance of ecology for generic and specific differentiation in the Araceae-Aroideae'. *Aroideana* 3: 2, 49–54.
10 Procter, M. and P. Yeo 1973. *The Pollination of Flowers*. Glasgow: Collins.
11 Feinbrun-Dothan, N. 1986. *Flora Palaestina*

Part 4. Jerusalem: The Israel Academy of Sciences and Humanities.
12 Alpinar, K. (pers. comm.)
13 Mayo, 1980.
14 van der Pijl, L. 1972. *Principles of Dispersal in Higher Plants*. Berlin-Heidelberg-New York: Springer-Verlag.
15 Herbarium, Royal Botanic Gardens, Kew: specimen collected by Davis (No. 26152).
16 Feinbrun-Dothan, 1986.
17 Boyce, P. (pers. comm.)
18 Herbarium, Royal Botanic Gardens, Kew: specimen collected by Guest (No. 13362).
19 Feinbrun-Dothan, 1986.
20 Killian, C. 1933. 'Développement, biologie et répartition de l'*Ambrosinia bassii*. L.' Deuxième partie. *Bull. Soc. Hist. Nat. Afrique N.* 24: 259–94.
21 Ibid.
22 Schauenberg, P. 1965. *The Bulb Book*. London: Frederick Warne.
23 Meeuse, B.J.D. 1966. 'The voodoo lily'. *Scient. Am.* 215: 80–8.
24 Ibid.
25 Herbarium, Royal Botanic Gardens, Kew: specimen collected by E.A. Robinson.
26 Herbarium, Royal Botanic Gardens, Kew: specimen collected by B. Goldsmith.
27 Herbarium, Royal Botanic Gardens, Kew: specimen collected by R.D. Meikle (No. 1294).
28 Mayo, S.J. 1985. '*Araceae*'. In R.M. Polhill (ed.), *Flora of Tropical East Africa*. Rotterdam/Boston: A.A. Balkema.
29 Herbarium, Royal Botanic Gardens, Kew: specimen collected by B. Goldsmith.
30 Herbarium, Royal Botanic Gardens, Kew: specimen collected by F.R. Irvine (No. 4601).
31 Herbarium, Royal Botanic Gardens, Kew: specimens collected by H.G. Faulkner (Nos 362 and 104).
32 Herbarium, Royal Botanic Gardens, Kew: specimen collected by J. Webb and S. Bullock (No. 495).
33 Grayum, M.H. 1984. *Palynology and Phylogeny of the Araceae*, (852pp.). Ph.D. thesis: University of Massachusetts. Ann Arbor, Michigan: University Microfilms International, xxiii.
34 Mayo, 1985.
35 Sivadasan, M. and D.H. Nicolson 1981. 'A new species of *Theriophonum* Bl. (*Araceae*) from India'. *Aroideana* 4: 2, 64–7.
36 Mayo, S. 1978. 'Aroid-hunting in Bahia, Brazil'. *Aroideana* 1: 1, 4–9.
37 Mayo, S. and G.M. Barroso 1979. 'A new pedate-leaved species of *Philodendron* from Bahia, Brazil'. *Aroideana* 2: 3, 82–94.
38 Bogner, J. 1979. 'A critical list of the aroid genera'. *Aroideana* 1: 3, 63–73.
39 Croat, T.B. 1985. 'Aroid profile No. 10: *Taccarum weddellianum*', *Aroideana* 8: 3, 94–7.
40 Madison, M. 1981. 'Notes on *Caladium* and its allies'. *Selbyana* 5: 342–77.

CHAPTER 6

1 Ayensu, E.S., V.H. Heywood, G.L. Lucas and R.A. Defilips 1984. *Our Green and Living World*. Washington: Smithsonian Institution Press. Cambridge: Cambridge University Press.
2 Dressler, R. 1981. 'A new name for the dwarf purple anthurium'. *Aroideana* 3: 2, 55.
3 White, R.R. 1983. 'Panama West'. *Aroideana* 5: 4, 116–21.
4 Madison, M. 1978a. '*Anthurium lilacinum*'. *Aroideana* 1: 3, 86–7.
5 Croat, T.B. 1983a. 'A revision of the genus *Anthurium* (*Araceae*) of Mexico and Central America'. Part 1: Mexico and Middle America. *Ann. Missouri Bot. Gard.* 70, 211–420.
6 Ibid.
7 Croat, T.B. 1986. 'A revision of the genus *Anthurium* (*Araceae*) of Mexico and Central America'. Part 2: Panama. *Monographs in Systematic Botany*. Vol. 14. St Louis: Missouri Botanic Gardens.
8 Madison, M. 1978b. 'The *Anthurium leuconeurum* confusion'. *Aroideana* 1: 1, 17–19.
9 Croat, T.B. 1983b. 'The origin of *Anthurium leuconeurum*'. *Aroideana* 6: 4, 132–4.
10 Croat, 1986.
11 Ibid.
12 Croat, T.B. 1980. 'Flowering behavior of the neotropical genus *Anthurium* (*Araceae*)'. *Amer. J. Bot.* 67: 6, 888–904.
13 Dortort, F. 1980. 'In the forests of Costa Rica'. *Aroideana* 3: 2, 39–48.
14 Croat, T.B. and R.D. Sheffer 1983. 'The sectional groupings of *Anthurium* (*Araceae*)'. *Aroideana* 5: 2, 85–123.
15 Croat, 1983a.
16 Grayum, M.H. 1984. *Palynology and Phylogeny of the Araceae* (852pp.). Ph.D. thesis Ann Arbor, Michigan: University Microfilms International, xxiii.
17 Ibid.
18 Brown, N.E. 1885. '*Aglaonema acutispathum*'. *Gard. Chron.* 24, 39.
19 Jervis, R. 1980. *Aglaonema Grower's Notebook*. Clearwater, Florida: Roy N. Jervis.
20 Ibid.
21 Brown, R.F. 1984. 'The new aglaonemas of Thailand'. *Aroideana* 7: 2, 43–52.
22 Jervis, 1980.
23 Grayum, M.H. 1982. 'The aroid flora of

REFERENCES

Finca La Selva, Costa Rica: a lowland wet forest locality'. *Aroideana* 5: 2, 47–57.
24 Sweet, R. 1839. *Hortus Britannicus*. London: Ridgeway.
25 Birdsey, M. 1951. *The Cultivated Aroids*. Berkeley, California: Gillick Press.
26 Ibid.
27 Nicholson, G. 1885. *The Illustrated Dictionary of Gardening*. London: Upcott Gill.
28 Rochford, T. and R. Gorer 1961. *The Rochford Book of Houseplants*. London: Faber and Faber.
29 Nicholson, 1885.
30 Bogner, J. 1985. 'A new *Chlorospatha* species from Colombia'. *Aroideana* 8: 2, 48–54.
31 Bogner, J. (pers. comm.)
32 Bogner, J. and D.H. Nicolson (in press). A revised classification of *Araceae* with dichotomous keys. In J. Arditti and L.J Nyman (eds) *The Biology of Araceae*.
33 Bogner, J. 1980. 'On two *Nephthytis* species from Gabon and Ghana'. *Aroideana* 3: 3, 75–85.
34 Bogner, J. 1972. 'Revision der *Arophyteae (Araceae)*'. *Bot. Jahrb. Syst.* 92: 1, 1–63.
35 Bogner, J. and D.H. Nicolson (in press).
36 Madison, M. 1981. 'Notes on *Caladium (Araceae)* and its allies'. *Selbyana* 5 (3 and 4), 342–77.
37 Croat, T.B. (pers. comm.)
38 Madison, 1981.
39 Bogner, J. 1985 (1986). 'A new *Xanthosoma* species from Pará, Brazil'. *Aroideana* 8: 4, 112–17.
40 Lightbody, J.P. 1985. 'Distribution of leaf shapes of *Piper* species in a tropical cloud forest: evidence for the role of drip tips'. *Biotropica* 17: 4, 339–42.
41 van Steenis, C.G.G.J. 1969. 'Plant speciation in Malesia, with special reference to the theory of non-adaptive saltatory evolution'. *Biol. J. Linn. Soc.* 97–133.
42 Fairchild Tropical Garden, Miami, Florida: specimen collected by Stanley Kiem 76–664 55.
43 Hotta, M. 1985. 'New species of the genus *Homalomena (Araceae)* from Sumatra with a short note on the genus *Furtadoa*'. *Garden's Bull.* 38: 1, 43–54.
44 Nicholson, 1885.
45 Birdsey, 1951.
46 Burnett, D. 1984. 'The cultivated *Alocasia*'. *Aroideana* 7: 3 and 4.
47 Findon, B. (pers. comm.)
48 Croat, T.B. and D. Burnett (in press).
49 Burnett, 1984.
50 Ibid.
51 Threatened Plants Committee of the International Union for Conservation of Nature and Natural Resources (IUCN), 1980: report to the Convention on International Trade in Endangered Species of Wild Fauna and Flora (CITES).
52 Hotta, M. 1967. 'Notes on Bornean plants'. *Act. Phyt. & Geobot.* 22, 159.
53 Burnett, 1984.
54 Threatened Plants Committee (IUCN), 1980.
55 Nicolson, D.H. 1968. 'The genus *Xenophya* Schott *(Araceae)*'. *Blumea* 16: 1.
56 Johns, R.J. and A. Hay (eds.). 1981. *A Student's Guide to the Monocotyledons of Papua New Guinea*. Papua New Guinea Forestry College Training Manual, Vol. 13.
57 Burnett, 1984.

CHAPTER 7

1 Fittkau, E.J. and H. Klinge 1973. 'On biomass and trophic structures of the central Amazon rainforest ecosystem'. *Biotropica* 5: 1, 2–14.
2 Mitchell, A.W. 1986. *The Enchanted Canopy*, London: Guild Publishing.
3 Birdsey, M. 1951. *The Cultivated Aroids*. Berkeley, California: Gillick Press.
4 Madison, M. 1977. 'A revision of *Monstera (Araceae)*'. *Contr. Gray Herb. Harvard Univ.* 207: 3–100.
5 Strong, D.R. and T.S. Ray 1975. 'Host tree location behaviour of a tropical vine (*Monstera gigantea*) by skototropism'. *Science* 190, 804–6.
6 Bogner, J. 1985. 'One new name and five new combinations in *Araceae*'. *Aroideana* 8: 3, 73–9.
7 Birdsey, M. 1962. '*Pothos aureus* transferred to *Rhaphidophora*'. *Baileya* 10: 159.
8 Birdsey, 1951.
9 Ibid.
10 Nicolson, D.H. 1984. 'A second collection of *Pedicellarum (Araceae)*'. *Aroideana* 7: 2, 56–7.
11 Madison, 1977.
12 Ibid.
13 Brunton, J. 1777. *Catalogue of plants*. Birmingham: J. Brunton.
14 Ramírez, B.W. and L.D. Gómez 1978. 'Production of nectar and gums by flowers of *Monstera deliciosa (Araceae)* and of some species of *Clusia (Guttiferae)* collected by New World *Trigona* bees'. *Brenesia* 14/15, 407–12.
15 Madison, 1977.
16 Ibid.
17 Ibid.
18 Madison, M. 1979a. 'Notes on some aroids along the Río Negro, Brazil'. *Aroideana* 2: 3, 67–77.
19 Spruce, R. 1853. 'Botanical objects communicated to the Kew Museum from

REFERENCES

the Amazon River in 1851 and 1852'. *Hooker's J. Bot.* 5, 247.
20 Madison, M. 1979b. 'Protection of developing seeds in neotropical *Araceae*'. *Aroideana* 2: 2, 52–61.
21 Madison, M. 1978. 'The genera of *Araceae* in the northern Andes'. *Aroideana* 1: 2, 31–57.
22 Madison, M. 1976. '*Alloschemone* and *Scindapsus (Araceae)*'. *Selbyana* 1, 325–7.
23 Madison, 1979a.
24 Croat, T.B. 1985. 'A new collection of the rare *Alleschemone occidentalis* (Poepp.) Engl. and Krause'. *Aroideana* 8: 3, 80–2.
25 Croat, T.B. 1981. 'A revision of *Syngonium (Araceae)*'. *Ann. Missouri Bot. Gard.* 68: 565–651.
26 Croat, T.B. 1983. 'A revision of the genus *Anthurium (Araceae)* of Mexico and Central America'. Part 1: Mexico and Middle America. *Ann. Missouri Bot. Gard.* 70, 211–420.
27 Croat, T.B. 1986. 'A revision of the genus *Anthurium (Araceae)* of Mexico and Central America'. Part 2: Panama. *Monographs in Systematic Botany* 14 (Missouri Botanical Garden).
28 André, E. 1877. '*Anthurium andraeanum*, J. Linden'. *Ill. Hort.* 24: 43.
29 Madison, M. 1980. '*Anthurium andreanum*'. *Aroideana* 3: 2, 58–60.
30 Kamemoto, H. and R. Sheffer, 1982. '*Anthurium wendlingeri* x *Anthurium scherzerianum*'. *Aroideana* 5: 4, 114–15.
31 Veitch, J. 1906. *Hortus Veitchii*. Chelsea, London: Veitch and Son.
32 Engler, A. 1905. '*Araceae-Pothoideae*'. In *Das Pflanzenreich* IV. 23B (Heft 21).
33 Banta, J. (pers. comm.).
34 Croat, T.B. and R.D. Sheffer 1983. 'The sectional groupings of *Anthurium (Araceae)*'. *Aroideana* 6: 3, 85–123.
35 Sheffer, R., W. Theobald and H. Kamemoto 1980. 'Taxonomy of *Anthurium scandens (Araceae)*'. *Aroideana* 3: 3, 86–93.
36 Altschul von Reis, S. 1973. *Drugs and foods from little-known plants*. Massachusetts: Harvard Univ. Press.
37 Mayo, S.J. 1982. '*Anthurium acaule* (Jacq.) Schott (*Araceae*) and West Indian "Bird's Nest" anthuriums'. *Kew Bull.* 36: 4, 691–719.
38 Croat, T.B. (description in press).
39 Mayo, S.J. 1986. *Systematics of Philodendron Schott (Araceae) with special reference to inflorescence characters*. Ph.D. thesis, Univ. of Reading. 972pp.
40 Ibid.
41 Ibid.
42 Madison, M. 1979a.
43 Aiton, W.T. 1813. *Hortus Kewensis*. 5: 310.
44 Barrière, J. de 1896. 'Les *Philodendron*'. *Ill. Hort.* 43: 278.
45 Bogner, J. and G.S. Bunting, 1983. 'A new *Philodendron* species (*Araceae*) from Ecuador'. *Willdenowia* 13: 183–5.
46 Croat, T.B. 1984. '*Philodendron rugosum*'. *Aroideana* 7: 1, 18–20.
47 Rochford, T. and R. Gorer 1961. *The Rochford Book of Houseplants*. London: Faber and Faber.
48 Bunting, G.S. '*Philodendron* Schott'. In S.M. Walters *et al.*, (eds.), *The European Garden Flora*, 80–5. Cambridge: Cambridge University Press.
49 Mayo, 1986.
50 Gottsberger, G. and A. Amaral Jr. 1984. 'Pollination strategies in Brazilian *Philodendron* species'. *Ber. Deutsch. Bot. Ges. Bd.* 97, S. 391–410.
51 Ibid.
52 Ibid.
53 Mayo, 1986.

CHAPTER 8

1 Letter translated in: Hooker, J.D. 1891. '*Amorphophallus titanum*'. *Curtis's Bot. Mag.* 117, t.7153–5.
2 Ibid.
3 Ibid.
4 Anon. 1889. 'A vegetable titan'. *Gard. Chron.* 3: 5, 748, 750, 755.
5 Blunt, W. 1978. *In for a Penny*. London: Hamish Hamilton.
6 Bogner, J. 1981. '*Amorphophallus titanum* (Becc.) Becc. ex Arcangeli'. *Aroideana* 4: 2, 43–53.
7 Gandawijaja, D., S. Idris, R. Nasution, L.P. Nyman and J. Arditti 1983. '*Amorphophallus titanum* Becc.: a historical review and some recent observations'. *Ann. Bot.* 51, 269–78.
8 Everett, T.H. 1955. 'Interesting observations on *Amorphophallus titanum* contained in certain garden correspondence'. *Journ. N. Y. Bot. Gard.* 5, 122, 125–6.
9 Camp, W.H. 1937. 'Notes on the physiology and morphology of *Amorphophallus titanum*'. *Journ. N. Y. Bot. Gard.* 38, 190–7.
10 Gandawijaja *et al.* 1983.
11 Fenzi, E.O. 1878. 'Another gigantic aroid'. *Gard. Chron.* NS 9, 596, 781, 788.
12 Everett, T.H. 1937. 'The cultivation and development of *Amorphophallus titanum*'. *Journ. N. Y. Bot. Gard.* 38, 181–5.
13 Anon. 1939. 'Another *krubi* from Sumatra blooms at the Garden'. *Journ. N. Y. Bot. Gard.* 40, 179–81.
14 Widjaja, E. (pers. comm.).
15 Camp, 1937.
16 Gandawijaja *et al.* 1983.
17 Woodward, C.H. 1937. 'From tropical

REFERENCES

18 mountain slope to northern conservatory'. *Journ. N. Y. Bot. Gard.* 38, 185-7.
18 Gandawijaja *et al.* 1983.
19 Camp, 1937.
20 Hooker, 1891.
21 Report in the *Christian Science Monitor* 1926.
22 Veendorp, H. 1956. 'A bizarre giant in flower'. *Gard. Chron.* 139: 9, 199.
23 Gandawijaja *et al.* 1983.
24 Woodward, C.H. 1937. 'The largest flower in the world'. *Journ. N. Y. Bot. Gard.* 38, 177-81.
25 Anon., 1939.
26 Bogner, 1981.
27 Dakkus, P. 1924. '*Amorphophallus titanum* Beccari'. *Gard. Chron.* 3: 76, 301-4.
28 Bogner, 1981.
29 Everett, 1955.
30 Dalton, B. 1985. *Indonesia Handbook*. Chico, California: Moon Publications.
31 Meijer, W. 1973. 'Endangered Plant Life'. *Biol. Conserv.* 5: 3, 163-7.
32 Alderwerelt van Rosenburgh, C.R.W.K. van 1920. 'New or noteworthy Malayan Araceae'. *Bull. Jard. Bot. Buitenzorg* ser. 3, 1: 369.
33 Bogner, J. (pers. comm.)
34 Nicolson, D.H. (pers. comm.)
35 Bogner, J., S.J. Mayo and M. Sivadasan 1985. 'New species and changing concepts in *Amorphophallus*'. *Aroideana* 8: 1, 14-25.
36 Stapf, O. 1923. '*Amorphophallus cirrifer*'. *Curtis's Bot. Mag.* 149, t. 9000.
37 Nicholson, G. 1885. *The Illustrated Dictionary of Gardening*. London: Upcott Gill.
38 Nicolson, D.H. 1984. '*Amorphophallus konjac* vs. *A. rivieri* (*Araceae*)'. *Aroideana* 7: 1, 6-8.
39 Phillips, F.L. 1980. 'On coping with monsters'. *Aroideana* 3: 2, 56-7.
40 Ibid.
41 Theus, L. 1983. 'A tour of Selby Botanic Gardens'. *Aroideana* 6: 2, 56.
42 McAlpin, B. (pers. comm.)
43 Nicolson, 1984.
44 Gimlette, J.D. 1929, *Malay Poisons and Charm Cures*. London: J. and A. Churchill.
45 Madison, M. 1980. Cover photo and caption. *Aroideana* 3: 1, 2.
46 Grayum, M.H. 1982. 'The aroid flora of Finca La Selva, Costa Rica: a lowland wet forest locality'. *Aroideana* 5: 2, 47-57.
47 Engler, A. 1911. '*Lasioideae*'. In A. Engler (ed.), *Das Pflanzenreich* IV. 23C (Heft 48).
48 Mayo, S. J. 1985. '*Araceae*'. In R.M. Polhill (ed.), *Flora of Tropical East Africa*. Rotterdam/Boston: A.A. Balkema.
49 Knecht, M. 1980. 'The uses of *Araceae* in African folklore and traditional medicine'. *Aroideana* 3: 2, 62-4.

CHAPTER 9

1 Cable, W.J. 1984, 'The spread of taro (*Colocasia* sp.) in the Pacific'. In S. Chandra (ed.), *Edible Aroids*, 28-33. Oxford: Clarendon Press.
2 Plucknett, D.L. 1976. 'Edible aroids'. In N.W. Simmonds (ed.), *Evolution of Crop Plants*, 10-12. London: Longman.
3 Cable, 1984.
4 Plucknett, 1976.
5 Onwueme, I.C. 1978. *The Tropical Tuber Crops*. New York: Wiley.
6 Records of taro cultivars in the National Biological Institute, Bogor, Indonesia.
7 Plucknett, 1976.
8 Information sheet: Waimea Arboretum, Oahu, Hawaiian Islands.
9 Maiden, J.H. 1889. *Useful Native Plants of Australia (including Tasmania)*. London: Trubner.
10 Purseglove, J.W. 1972. *Tropical Crops*. Vol. 1: *Monocots*. London: Longman.
11 Ghani, F.D. 1984. 'Preliminary studies on flowering in taro cultivars in Malaysia'. In S. Chandra (ed.), *Edible Aroids*, 169-72. Oxford: Clarendon Press.
12 Jackson, G.V.H. and I.D. Firman 1984. 'Guide-lines for the movement of germplasm of taro and other aroids within the Pacific'. In S. Chandra (ed.), *Edible Aroids*, 194-211. Oxford: Clarendon Press.
13 de la Peña, R.S. and D.L. Plucknett 1967. 'The response of taro to N, P, and K fertilisation under upland and lowland conditions in Hawaii'. 1st International Symposium on Tropical Root Crops, Trinidad.
14 Parkinson, S. 1984. 'The contribution of aroids in the nutrition of people in the south Pacific'. In S. Chandra (ed.), *Edible Aroids*, 215-24. Oxford: Clarendon Press.
15 Plucknett, D.L. and R.S. de la Peña 1971. 'Taro production in Hawaii'. *World Crops* 23: 244-9.
16 Coursey, D.G. 1968. 'The edible aroids'. *World Crops* 20: 3.
17 Whitney, L.D. 1937. 'Some facts about taro – staff of life in Hawaii'. *Parad. Pacif.* 49 (3): 15, 30.
18 Degener, O. 1930. *Plants of Hawaii National Parks illustrative of Plants and Customs of the South Seas*. Michigan: Braun-Brumfeld.
19 *Hawaii Taro Annual Summary*. 1986. Hawaii Department of Agriculture.
20 Haudricourt, A. 1941. 'Les colocasiées alimentaires (taros et yautias)'. *Rev. Internat. de Bot. Appliqué et d'Agric. Trop.* 21 (233-4), 40-65.
21 Morton, J.F. 1972. 'Cocoyams (*Xanthosoma caracu*, *X. atrovirens* and *X. nigrum*), ancient

247

22. Plucknett, 1976.
23. Ghani, F.D. 1984. 'The potential of aroids in Malaysia'. In S. Chandra (ed.), *Edible Aroids*, 10–16. Oxford: Clarendon Press.
24. Hay, A. 1986. Cyrtosperma Griffith and the origin of the aroids. Ph.D. thesis, Univ. of Oxford.
25. Thaman, R.S. 1984. 'Intensification of edible aroid cultivation in the Pacific islands'. In S. Chandra (ed.), *Edible Aroids*, 102–22. Oxford: Clarendon Press.
26. Grimble, A. 1952. *A Pattern of Islands*. London: John Murray.
27. Cable, 1984.
28. Grimble, 1952.
29. Nicolson, D.H. 1984. '*Amorphophallus konjac* vs. *A. rivieri (Araceae)*'. *Aroideana* 7: 1, 7–8.
30. Sastrapradja, S., G.G. Hambali and T.K. Prana 1984. 'Edible *Amorphophallus* and related species in Indonesia'. In S. Chandra (ed.), *Edible Aroids*, 17–23. Oxford: Clarendon Press.
31. Ibid.
32. Ibid.
33. Ibid.
34. Maiden, 1889.
35. Plowman, T. 1969. 'Folk uses of New World aroids'. *Econ. Bot.* 23:2, 97–122.
36. Ibid.
37. Prime, C. 1960. *Lords and Ladies*. London: Collins.
38. Plowman, 1969.
39. Labroy, M.O. 1908. 'Le ceriman du Mexique (*Monstera deliciosa* Liebm.), espèce fruitière?' *J. Agric. Trad. Bota. Appl.*, 8: 169–71.
40. Chandra, S. and P. Sivan 1984. 'Taro production systems studies in Fiji'. In S. Chandra (ed.), *Edible Aroids*, 93–101. Oxford: Clarendon Press.
41. Sivan, P. and M.B. Tavaiqia 1984. 'First taro variety developed for breeding programme released for commercial production in Fiji'. *Fiji Agr. Journ.* 46: 1, 1–4.
42. Carpenter, J.R. and W.E. Steinke 1983. 'Animal Feed'. In J.-K. Wang (ed.), *Taro: a review of* Colocasia esculenta *and its potentials*. Honolulu: University of Hawaii Press.
43. Parkinson, 1984.
44. Reppun, C. and P. Reppun (taro growers, Oahu, Hawaii): (pers. comm.)
45. Bradley, R. 1716. *Historia Plantarum Succulentarum*. London.

CHAPTER 10

1. Williams, C.A., J.B. Harborne and S.J. Mayo 1981. 'Anthocyanin pigments and leaf flavonoids in the family *Araceae*'. *Phytochem.* 20, 217–34.
2. Duke, J.A. and E.S. Ayensu 1985. *Medicinal Plants of China,* Vol. 1. Michigan: Reference Publications.
3. Tang, C. and W.S. Sakai 1983. 'Acridity of taro and related plants'. In J.-K.Wang (ed.), *Taro: a review of* Colocasia esculenta *and its potentials*. Honolulu: University of Hawaii Press.
4. Sakai, W.S., M. Hanson and R.C. Jones 1972. 'Raphides with barbs and grooves in *Xanthosoma sagittifolium (Araceae)*'. *Science* 178, 314–15.
5. Middendorf, E. 1982. 'The remarkable shooting idioblasts'. *Aroideana* 6: 1, 9–11.
6. Wright, W. 1787. 'An account of the medical plants growing in Jamaica'. *London Med. Journ.* 3: 3, 217–95.
7. Plowman, T. 1969. *Folk uses of New World aroids*. *Econ. Bot.* 23: 2, 97–122.
8. Arditti, J. and E. Rodriguez 1982. '*Dieffenbachia*: uses, abuses, and toxic constituents: a review'. *Journ. Ethnopharmac.* 5, 293–302.
9. Barnes, R.A. and L.E. Fox 1955. 'Poisoning with *Dieffenbachia*'. *Journ. of the Hist. of Medicine and Allied Sciences* 10, 175–81.
10. Ibid.
11. Walter, W.G. and P.N. Khanna 1972. 'Chemistry of the aroids 1. *Dieffenbachia seguine, amoena* and *picta*'. *Econ. Bot.* 26, 364–72.
12. Barnes and Fox, 1955.
13. Arditti and Rodriguez, 1982.
14. Barnes and Fox, 1955.
15. Morton, J.F. 1977. 'Poisonous and injurious higher plants and fungi'. In C.G. Tedeschi et al. (eds.). *Forensic Medicine*. Vol. 3, Philadelphia: Saunders.
16. Plowman, 1969
17. Carmichael, T.H. 1939. *The first supplement to the Homeopathic Pharmacopoeia of the United States,* 5th edn. Boston: Otis Clapp.
18. Frohne, D. and H.J. Pfänder 1984. *A Colour Atlas of Poisonous Plants*. London: Wolfe Scientific.
19. Hegnauer, R. 1963. *Chemotaxonomie der Pflanzen 2. Monocotyledoneae*. Basel and Stuttgart: Birkhäuser Verlag.
20. Willaman, J.J. and B.G. Schubert 1961. 'Alkaloid-bearing plants and their contained alkaloids.' *Technical Bulletin* 1234, Agr. Res. Service, USDA, Washington.
21. Arditti and Rodriguez, 1982.
22. Walter and Khanna, 1972.
23. Walter, W.G. 1967. '*Dieffenbachia* toxicity'.

REFERENCES

Journ. Am. Med. Ass. 201: 2, 140–1.
24 Ladeira, A.M., S. Andrade and P. Sawaya 1975. 'Studies on *Dieffenbachia picta* Schott, toxic effects in guinea pigs'. *Toxicol. Appl. Pharmacol.* 34, 363–73.
25 Kuballa, B., A.A.J. Lugnier and R. Anton 1981. 'Study of *Dieffenbachia*-induced edema in mouse and rat hindpaw: respective role of oxalate needles and trypsin-like protease'. *Toxicol. Appl. Pharmacol.* 58, 444–51.
26 Tang and Sakai, 1983.
27 Milne, L. and M. 1967. *Living Plants of the World.* London: Nelson.
28 Grieve, M. 1931. *A Modern Herbal.* London: Jonathan Cape.
29 Röst, L.C.M. 1979a. 'Biosystematic investigations with *Acorus* L. (*Araceae*) I. Cytotaxonomy'. *Proc. of the Koningklijke Nederlandse Akademie van Wetenschappen, Amsterdam,* series 33, Vol. 82: 1.
30 Röst, L.C.M. 1979b. 'Biosystematic investigations with *Acorus* L. A synthetic approach to the classification of the genus'. *Planta Medica (Journ. of Medicinal Plant Research)* 37: 4.
31 Duke and Ayensu, 1985.
32 Röst, 1979b.
33 Duke and Ayensu, 1985.
34 Moskalenko, S.A. 1986. 'Preliminary screening of far-eastern ethnomedicinal plants for antibacterial activity'. *Journ. Ethnopharmac.* 15, 231–59.
35 Plowman, 1969.
36 Salmon, W. 1710. *Botanologia. The English Herbal; or History of Plants etc.* London.
37 Arctander, S. 1960. *Perfume and Flavor Materials of Natural Origin.* Elizabeth, New Jersey: S. Arctander.
38 Grieve, 1931.
39 Schauenberg, P. and F. Paris 1977. *Guide to Medicinal Plants.* Guildford and London: Lutterworth Press.
40 Gibbs, R.D. 1974. *Chemotaxonomy of Flowering Plants.* Montreal: McGill-Queen's University Press.
41 Knecht, M. 1980. 'The uses of *Araceae* in African folklore and traditional medicine'. *Aroideana* 3: 2, 62–4.
42 Fielder, M. 1975. *Plant Medicine and Folklore.* New York: Winchester Press.
43 Quisumbing, E. 1951. *Medicinal Plants of the Philippines.* Manila: Bureau of Printing.
44 Plowman, 1969.
45 Ibid.
46 Elliot, S. and J. Brimacombe 1985. *The Medicinal Plants of Gunung Leuser National Park.* University of Edinburgh: project report available from the authors.
47 Duke and Ayensu, 1985.
48 Quisumbing, 1951.
49 Hartwell, J. 1982. *Plants used against Cancer.* Massachusetts: Quarterman Publications.
50 Fielder, 1975.
51 Plowman, 1969.
52 Gunther, E. 1945 (reprinted 1977). *Ethnobotany of Western Washington.* Seattle and London: University of Washington Press.
53 Knecht, 1980.
54 Duke and Ayensu, 1985.
55 Ibid.
56 Plowman, 1969.
57 Reffstrup, T. and P.M. Boll 1985. 'Allergenic 5-alkyl- and 5-alkenylresorcinols from *Philodendron* species'. *Phytochemistry* 24: 11, 2563–5.
58 Frohne and Pfänder, 1984.
59 Kokwaro, J.O. 1976. *Medicinal Plants of East Africa.* East African Literature Bureau.
60 Plowman, 1969.
61 Schultes, R.E. 'Plants as oral contraceptives in the Northern Amazon'. *Lloydia* 26, 67–74.
62 Weiner, M.A. *Earth Medicine, Earth Food: Plant Remedies, Drugs, and Natural Foods of the North American Indians.* London: Collier Macmillan.
63 Gunther, 1945.
64 Duke and Ayensu, 1985.
65 Plowman, 1969.
66 Duke and Ayensu, 1985.
67 Knecht, 1980.
68 Hartwell, 1982.
69 Duke and Ayensu, 1985.
70 Revolutionary Health Committee of Hunan Province 1977. *A Barefoot Doctor's Manual.* London: Routledge and Kegan Paul.
71 Duke and Ayensu, 1985.
72 Knecht, 1980.
73 Plowman, 1969.
74 Schultes, R.E. 1976. *Hallucinogenic Plants.* New York: Golden Press.
75 Sastri, B.N. (ed.) 1959. *The Wealth of India.* Vol. 5: *Raw Materials.* New Delhi: Council of Scientific and Industrial Research.
76 Duke and Ayensu, 1985.
77 Emboden, W. *Narcotic Plants.* London: Studio Vista.
78 Krukoff, B.A. and A.C. Smith 1937. 'Notes on the botanical components of curare'. *Bull. Torrey Bot. Cl.* 64, 401–9.
79 Plowman, 1969.
80 Gunther, 1945.
81 Alkire, W.H. 1968. 'Porpoises and taro'. *Ethnology* 7, 280–9.
82 Plowman, 1969.
83 Altschul, S. von Reis 1973. *Drugs and foods from little-known plants.* Massachusetts: Harvard Univ. Press.
84 Knecht, 1980.
85 Plowman, 1969.

REFERENCES

86 Williams *et al.*, 1981.
87 Burkill, H.M. 1985. *The Useful Plants of Tropical West Africa*. Vol. 1. Richmond, Surrey: Royal Botanic Gardens, Kew.
88 Gerard, J. 1597. *Herball or Generall Historie of Plantes*.
89 Grieve, 1931.
90 Plowman, 1969.
91 Maiden, J.H. 1889. *Useful Native Plants of Australia (including Tasmania)*. London: Trubner.
92 Knecht, 1980.
93 Griffin, G.J.L. and J.-K. Wang 1983. 'Industrial Uses'. In J.-K. Wang (ed.), *Taro: a review of* Colocasia esculenta *and its potentials*. Honolulu: University of Hawaii Press.

Index

Numbers in *italic* refer to pages on which illustration captions appear

Acoreae (tribe) 208
Acoroideae (subfamily) 31
Acorus 23, 25, 29, 31, 35, 98–9, 208
  calamus 54, 98–9, *193*, 213–15, *214*, 223
    var. *americanus* 214
    var. *angustatus* 214
    var. *calamus* 214
    var. *vulgaris* 214
  gramineus 23, 98
    var. *macrospadiceus* 98
  heeri (fossil) 34
  'Pusillus' 98
  tatarinowii 98
  medicinal uses 98–9, 213–15
Adder's tongue *See Arum maculatum*
Aglaodorum griffithii 30, 53, 91
Aglaonema 14, 23, 91, 127–30
  commutatum 128
    forma *elegans* 129
    'Pseudobracteatum' 129
  costatum 129–30
  crispum (*A. roebelinii*) 128
  elegans 128
  modestum (*A. acutispathum*) 127–8, 216
    var. *medio-pictum* 128
  'Green Goddess' 128
  'Shingii' 128
  'Variegatum' 127–8

  nitidum (*A. oblongifolium curtisii*) 27, 129
    pictum 129
    rotundum 130
Alismatiflorae 36
Alloschemone 53
  occidentalis 24, 157–8
Alocasia 25, 47, 137–43, 217
  x *amazonica* 32, 140
  alba 139
  bullata 139
  cucullata 142, 200
  cuprea 24, 138–9
    'Green Cuprea' 139
    'Green Shield' 139
    'Green Velvet' 139
  fornicata 200
  guttata var. *imperialis* 139
  'Hawaii' 139
  korthalsii 140
  longiloba 139
    'Magnifica' 139
  lowii 19, 139–40
    'Veitchii' 140
    'Grandis' 140
  macrorrhiza (*A. indica*) 53, 141, 199–200, *201*, 217
    'Variegata' 141
  magnifica 138
  maquilingensis 41
  micholitziana
    'African Mask' 141, *161*
    'Green Goddess' 141, *161*

  'Green Velvet' 141, *161*
  'Maxkowskii' 141
  odora 39, 47, *48*, 141
  portei 142
  puber (*A. pubera*) 41
  putzeysii 140
  'Quilted Dreams' 139
  'Reticulata' *129*, 143
  robusta 142
  sanderiana 26, 140, *141*
    'Gandavensis' 140
  scabriuscula 142
  thibautiana 140
  'Tigrina Superba' 143
  watsoniana 140
  wenzelii 142
  zebrina 142, 143
  food crop 199–200, *201*
Ambrosina 21
  bassii 54, 109—110, *110*
Amorphophallus 20, 21, 26, 36, 39, 127, 173–84
  blumei 204
  bulbifer 30
  cirrifer 22, 183
  decus-silvae (*A. brooksii*) *182*, *182*
  galbra *38*, 39
  konjac (*A. rivieri*) 23, 38, 184, 202–4, *203*, 216, 227
  napalensis 24, 67
  oncophyllus (*A. blumei*) 204

INDEX

paeoniifolius (*A.*
  campanulatus) 22, 39–40,
  44, 184, 202–4
 var. *hortensis* 202
 var. *sylvestris* 202
pendulus 22, 183
prainii 184
rivieri, 38, 184, 202–4, *203*,
  216, 227
titanum 21, 29, 39, 47, 81,
  173–81, *175*, 183, *193*
variabilis 45
food crops 202–4, *203*
Amydrium 153
 medium 153
Anadendrum 153
Anaphyllum 26
Anchomanes 26, 153
 difformis 26, *128*, 186, *186*
 welwitschii 217
André, Edouard 159, 169
Anthurium 14, 20, 21, 23, 25,
  27, *28*, 31, 37, 40, 52,
  53
 affine 115, 165
 amnicola (*A. lilacinum*) 38,
  120–2
 andraeanum 13, 27, 159–60,
  161
 armeniense 122
 bakeri 163
 berriozabalense 123
 clarinervium *24*, 122–3
 clavigerum 25
 clidemioides 163
 concinnatum 124
 corrugatum 123
 crassinervium 165
 croatii 25
 crystallinum 123
 dolichostachyum 53
 dressleri 27, 163
 flexile 163
 flexuosum 225
 forgetii 123
 formosum 125
 fragrantissimum 124
 globosum 159
 gracile *19*, *160*, 163
 hacumense 164
 halmoorei 165
 hookeri 164
 laciniosum 27, *33*
 leuconeurum (*A. cordatum*)
  122–3
 lilacinum, 38, 120–2
 lindmanianum 122
 magnificum 123
 nizandense 165
 nymphaeifolium 125, 225
 oblanceolatum *24*

oxycarpum 224
pallidiflorum *24*
palmatum 225
papillilaminum 123
pedatoradiatum 25, *124*
pentaphyllum 25
podophyllum 124–5, *125*
polydactylum 25
polyschistum 25
punctatum 51, 161–2
radicans 163
reflexinervium 165
rupicola 121
salviniae (*A. enormispadix*)
  164, 164
scandens 52, 163, 225–6
scherzerianum 22, 160–1
schlechtendalii 164–5
 ssp. *jimenezii* 165
section
 Calomystrium 27, 159–60
 Cardiolonchium 163
 Chamaerepium 163
 Pachyneurium 162, 164
 Polyphyllium 163
 Porphyrochitonium 161, 164
 signatum *24*
 spectabile 161, 162
 superbum 165
 sytsmae 121
 tetragonum 165
 tilaranense *24*
 trinerve 163
 veitchii *24*, 25, 162
 warocqueanum 25, 162, 167
 watermaliense 123
 wendlingeri 23, *24*, 161
 bird's nest type 164–5
 epiphytic and climbing 123,
  158–65
 lithophyte 120–2
 pendent 161–2
 pollination 49, 50
 scents 124
 terrestrial 122–5
Ants, epiphytes living in
  association with 51, *160*,
  167–8
Anubias 96
 barteri 96
  var. *nana* 96
 giganteus *64*, 96
 hastifolia 96
Aphyllarum 116
 tuberosum 116
Aponogeton madagascariensis
  (Aponogetonaceae)
  (Madagascan lace plant)
  26, 153
Aquatic plants 21, 30, 65,
  73–99

 flowers 73–4
 foliage 74
 free floating 76–81, *77*
 fruit dispersal 53, 54
 rheophytes 30, 85–7
 roots 28, 75, 85
 seeds 88
 submerged 84–5
 tidal zones 89–91
Arales (duckweeds and
  aroids) 35
Arcangeli, G. 69
Areae (tribe) 31
Arecales (palms) 35
Aridarum 23, 86
Arinae (subtribe) 31, 100,
  111
Arisaema 31, 45, 56, 57–65
 addis-ababense 56
 auriculatum 59
 biauriculatum 59
 candidissimum 65
 consanguineum 22, 59
 costatum 60
 dracontium 55, 63, 68
 exappendiculatum 25, 59
 filiforme 57
 fimbriatum 58, 59
 flavum 63–4
 griffithii 60–1, *60*
 hunanense 43
 limbatum 59
 mooneyanum 56
 propinquum (*A.
  wallichianum*) 59
 rhizomatum 57, *58*
 ruwenzoricum 64
 sikokianum 65
  var. *serratum* (*A. sazensoo*)
  var. *magnidens*) 26
 speciosum 59
 taiwanense 58
 thunbergii (*A. urashima*) 59
 tortuosum 22, 58
 triphyllum 13, 25, 55, 61–3,
  62, 74, 205, 217
 undulatifolium 58
 wallichianum 59
 yamatense 58
 food crop 205
 gender variation 48, 61–3, *62*
 whiplash 59–61, *60*
Arisarum
 proboscideum 21, *22*, 43, 55,
  69, 96
 vulgare 54, 110–11
Aroideae (subfamily) 31, 40,
  100
Arophyteae (tribe) 133
Arophyton 133
 buchetii 133

INDEX

Arrow arums 96–7, 205
Arum 31, 32, 45, 104–6, 213
  conophalloides 104–5
  creticum 33, 105–6
  cyrenaicum 105
  dioscoridis 22, 105
  hygrophilum 105
  italicum 13, 72, 106
    var. *neglectum* 205
  maculatum 13, 37, 44, 53, 55, 70–2, 70, 106, 205, 225
  nigrum 40, 46–7, 105
  palaestinum 105
  pictum 105
  food crop 106, 205
  medicinal uses 105, 106
Arum lily *See Zantedeschia aethiopica*
Arvi *See Colocasia esculenta*

Banta, John 162
Baraquin, M. 131
Bartlett, H.H. 130
Beccari, Dr Odoardo 86, 173–4, 175, 179
Biarum 23, 25, 38, 106–7
  davisii 24, 107
  eximium 107
  gramineum 106
  olivieri 109
  pyrami 107
  syriacum 106
  tenuifolium 32, 107
    var. *abbreviatum* 107
Black arum *See Arum palaestinum*
Blunt, William 175
Bog arum *See Calla palustris*
Bogner, Josef 32, 116, 182
Bowles, E.A. 100
Brown, N.E. 130, 139, 146
Bucephalandra motleyana 86–7
Bulbils 30, 33, 54, 68, 135
  hooked 66–7, 66
Bull, William 59, 140, 156, 159
Bunga bangkai See Amorphophallus titanum
Burbidge, F.W. 146

Caladieae (tribe) 134
Caladiinae (subtribe) 116
Caladiopsis *See Chlorospatha*
Caladium 27, 134
  bicolor 117–18, 217
  humboldtii 118
  lindenii (*Xanthosoma lindenii*) 134
  schomburgkii 118
  seguinum See Dieffenbachia seguinum
  ternatum 118
Calamus *See Acorus calamus*
Calla 31, 94, 99, 217
  palustris 65, 94, 205
Calla lily *See Zantedeschia aethiopica*
Calloideae (subfamily) 31, 94
Callopsis volkensii 129, 132
Carlephyton 133
  diegoense 133
Cataphyll 27, 28, 33
Cercestis 26, 148–9, 153
  afzelli 222, 226
  mirabilis 148–9
Chemical constituents 40, 42, 208–27
  calcium oxalate crystals 209
  indole 40, 42, 92
  oxalic acid 209
  raphides 209–11, 210
Chlorospatha 132, 134
  atropurpurea 129, 132
  corrugata 132
  kolbii 132
  mirabilis 132
Classification 32–4, 80–1
Climbers 29, 30, 33, 144–72, 160, 161
Cobra lilies *See Arisaema*
Cocoyam (new) *See Xanthosoma sagittifolium*
Cocoyam (old) *See Colocasia esculenta*
Colletogyne perrieri 133, 133
Colocasia 47, 217
  esculenta (*Colocasia antiquorum*) 22, 32, 47, 53, 135, 188–97, 190, 192, 207, 209, 213, 216, 224, 227
    var. *antiquorum* 190
    var. *esculenta* 190
    var. *globulifera* 190
    dasheen type 190
    eddo type 190
    medicinal use 196
    pests and diseases 192–3
    propagation 189, 191
    range of uses 196
    varieties and local names 188, 189–90, 194, 195
Colocasioideae (subfamily) 31, 40, 134, 158
Cordage (weaving, etc.) 225–7
Corms *See* Tubers
Corpse flower *See Amorphophallus titanum*

Cresta de gallo *See Anthurium andraeanum*
Croat, Dr Thomas 122, 158–9
Cryptocoryne 21, 23, 30, 54, 73, 75, 87–91
  beckettii 89
  ciliata 30, 53, 89, 90–1, 90
  ferruginea 91
  lingua 88
  nevillii 88, 88–9
  pontederiifolia 91
  retrospiralis 22, 89
  spiralis 64
  thwaitesii 88
  walkeri
    var. *legroi* 89
    var. *lutea* 89
    var. *walkeri* 89
  wendtii 89
Cuckoo pint *See Arum maculatum* 55
Culcasia 149
  parviflora 220
  rotundifolia 149
  striolata 220, 224
Curare 223
Cyclanthes (Panama hat plants) 35
Cyrtosperma 83–4, 187
  americanum 84
  chamissonis 84
  johnstonii 84
  lasioides 84
  merkusii (*C. chamissonis*; *C. edule*) 30, 84, 187, 200–2, 201
  senegalense 83–4
  spruceanum 84
  wurdackii 84
  food crop 200–2, 201

Damon, Samuel Mills 159–60
Darwin, Charles 120
Dead horse arum *See Helicodiceros muscivorus*
Devil's ivy *See Epipremnum pinnatum* 'Aureum'
Devil's tongue *See Amorphophallus konjac*
Dieffenbachia 14, 23, 28, 40–1, 52, 127, 130–2, 210–13, 210, 223, 225
  x *bausei* 131
  'Camille' 131
  'Exotica' 131
  leopoldii 131
  'Marianne' 131
  picta 27, 130–1, 210–11

INDEX

seguine 23, 27, 130–1, *161*, 210–11
'Noblis'
  'Amoena' 131
  'Baraquiniana' (*D. verschaffeltii*) 131
  'Jenmannii'131
  'Memoria Corsii' 131
  'Tropic Snow' 131
  'Wilson's Delight' 131
Dioscoreaceae (yams) 35
Distribution 31
Dormancy 30, 116, 118, 133, 176, 187
Dracontioides 26, 83, 153, 187
  desciscens 83
Dracontium 26, 84, 127, 153, 185, 216
  asperum 205
  gigas 185, *185*
  pertusum See Monstera deliciosa
  pittieri 185
  polyphyllum 24
  food crop 205
Dracunculus 45, 102–3
  canariensis 102
  vulgaris 21, *22*, 42, 102
    var. *creticus* 103
Dragon arums *See Dracunculus*
Dressler, Dr Robert 121
Duckweed *See Lemnaceae*
Dumb cane *See Dieffenbachia*
Dyes 224–5

Edible aroids 14, 188–207, 205
  Acorus calamus 215
  Alocasia 141, 199–200, *201*
  Amorphophallus 184, 202–4, 203
  Arum 106
  Colocasia esculenta 14, 188–97, *192*, 209
  Cyrtosperma merkusii 200–2, *201*
  Eminium spiculatum 109
  Monstera deliciosa 206
  Philodendron bipinnatifidum 172
  Spathiphyllum 206
  Xanthosoma 197–9, *198*
Elephant yam *See Amorphophallus paeoniifolius*
Eminium 43, 107–8
  intortum 108
  lehmannii 108
  spiculatum

  ssp. *negevense* 109
  ssp. *spiculatum* 108, 109
Engler, Adolf 32, 33, 35, 98, 128, 131, 133, 134, 158, 162, 170, 172, 185, 208
Epiphytes 29, 30, 51, 66, 120, 123, 144–72, *157*, *160*
  litter-basket 165
Epipremnum 149–50, 153
  aureum 149–50
  pinnatum 149–50
    'Aureum' 149–50
    'Marble Queen' 149–50
Ereriba See Homalomena

Flamingo flower *See Anthurium andraeanum*
Flowers 15, 18–23
  bisexual 18, 19–20, *19*, 21, 23, 127
  underground 113, *113*
  unisexual 18, *19*, 20, 21, 114
Flytrap *See Arum maculatum*
Foliage *24*, 23–7, *28*, 32, 58, 114, 176
  aquatic plants 74
  fenestration 26, 152–3
  pouched 135–6
  shapes *24*, 25–6
  variegation 26–7
  venation patterns 23, 87
Forbes, H. 174
Forrest, George 65
Friar's cowl *See Arisarum vulgare*
Fruit 52–4
  camouflage 52, 107
  capsular 90
  dispersal 52–4, 181
  edible 205–6
  underground 113, *113*

Giant krubi *See Amorphophallus titanum*
Golden club *See Orontium aquaticum*
Golden pothos *See Epipremnum pinnatum* 'Aureum'
Gonatanthus 65–7
  pumilus 66, 217
Gonatopus 26, 114
  boivinii 97, 114
  clavatus 114
  marattioides 114
Goosefoot plant *See Syngonium podophyllum*
Green dragon *See Arisaema dracontium*

Gymnostachys 31, 98, 208
  anceps 226, *226*

Habitat 30
  arid and seasonally dry 100–18
  cloud forest 124
  desert 108–9
  mountainous 63–4, 68
  tidal zones 89–91
  tropical rain forest 30, 55, 56, 119–43
  wetlands and water 30, 73–9
  woodlands 55–72
Hairy arum *See Helicodiceros muscivorus*
Hallucinogenic aroids 223
Helicodiceros 21, 45
  muscivorus 96, 100–1
Heteroaridarum 86
Heteropsis 28, 156
  integerrima 53, 156
  jenmani 226
  spruceanum 156
Holochlamys guineensis 24, 127
Homalomena 27, 31, 136–7, 223
  aromatica 223
  cordata 136
  lindenii 128, 137, 223
  megalophylla 136
  minutissima 136
  occulta 223
  rubescens 223
  wallisii 136–7
Hooker, Sir Joseph 39, 159, 174
Hotta, Mitsuro 182
Hottarum 86
Humbertina See Arophyton
Hunter's robe *See Epipremnum pinnatum* 'Aureum'
Hutchinson, J. 35

Imo *See Colocasia esculenta*
Indian turnip *See Arisaema triphyllum*
Inflorescence 15, 18–23, *19*, *22*, 81
  Amorphophallus titanum 175, 176–7, 179–80
  as food 206
  odour 37–40

Jack-in-the-pulpit *See Arisaema triphyllum*
Jasarum steyermarkii 73, 75, 84–5, *85*
Jenman, G.S. 131

253

INDEX

Kalo *See Colocasia esculenta*
Keladi *See Colocasia esculenta*
Kingdon-Ward, Frank 60
Konjac *See Amorphophallus konjac*
Kris plant *See Alocasia sanderiana*

Lagenandra 21, 87
  thwaitesii 87
Lamarck J.-B. 40
Lasia
  concinna 82–3
  spinosa 82–3
Lasioideae (subfamily) 26, 31, 40, 187
Lasiomorpha senegalensis 84
Leeuwenhoek, Anton van 209
Lemnaceae (duckweed) 35, 80–1, 209
Leopard lily *See Dieffenbachia*
Leopard palm *See Amorphophallus konjac*
Liliaceae 35
Linden, Jean 159
Linnaeus, C. 95, 154
Lithophytes 120, 120–2, 139
Lobb, Thomas 138, 150
Lords-and-Ladies *See Arum maculatum*
Low, Hugh 139
Lysichiton 31, 43, 64, 99
  americanus 42, 91–2, 93, 217, 224
  camtschatcensis 64, 91–2, 93

Macintyre, Daniel 160
Madison, Michael 134
Magic and ritual uses 224
Malanga *See Xanthosoma sagittifolia*
Marcgravia
  Marcgraviaceae 146
  paradoxa *See Rhaphidophora korthalsii*
Marsh calla *See Calla palustris*
Martius 57
Mayo, Dr Simon 100
Medicinal uses 14, 67, 68, 80, 98–9, 208
  Arum 105, 106
  Colocasia esculenta 196
  Stylochaeton 114
  anti-cancer aroids 220–2
  contraceptives 219
  expectorants 218
  healing aroids 216–17
  insecticides 219–20
  sedatives 222

sterilization 211
stimulating aroids 217–18
Microcasia pygmaea See Bucephalandra motleyana
Monarch of the east *See Sauromatum venosum*
Monstera 26, *32*, 36, 52, 146, 152–5, 217
  acuminata 154
  adansonii 154
  deliciosa 13, 14, 23, 27, 49, 50, 153–4, 206
  dubia 147, *147*, 155
  latevaginata *See Rhaphidophora korthalsii*
  obliqua 24, 152, 154
  tuberculata 155, *155*,
  pollinators 49
  section
    Echinospadix 146–7
    Marcgraviopsis 146
  skototropism 146–7, *147*
  stolons 146–7, *147*
Monsteroideae (subfamily) 31
Montrichardia
  arborescens 30, 50, 81–2, 204, 205, 226
  linifera 81
  paper making 226
Mouse plant *See Arisarum proboscideum*

Neoteny 81, 155
Nephthytis 30, 132–3
  afzelli 132–3
  hallaei 133
  picturata *See Cercestis mirabilis*
  poissonii 133
  swainii 133
Nymphaeales 35–6

Odour 15, 37–43, 71, 104, 107, 112, 121–2, 124, 176, 177
  dispersal 37–43, *58*, 61, 93, 104, 112, 172
Orontium aquaticum 21, 50, 65, 93–4, 205

Pandanales (screwpines) 35
Pauella sivagangana *See Theriophonum sivaganganum*
Peace lily *See Spathiphyllum wallisii*
paedomorphosis (see Neoteny)
Pedicellarum paiei 152
Peltandra 34, 96, 97
  virginica 96, 205

Pentagonia (Rubiaceae) 26
Philodendroideae (subfamily) 31
Philodendron 14, 40, 46, 52, 54, 165–72
  acuminatissimum 46
  andreanum 169
  angustisectum (*P. elegans*) 170
  asperatum 169
  bipinnatifidum (*P. selloum*) 26, 41–2, *41*, 53, 115, *161*, 172, 205, 226
  cannifolium 167–8, 171
  cordatum 171
  craspedodromum 224
  domesticum (*P. hastatum*) 171
  dubium 170
  elegans (*P. radiatum*) 24, 26
  erubescens 27, 170
    'Black Cardinal' 170
    'Brandy Wine' 170
    'Golden Erubescens' 170
    'King of Spades' 170
    'Prince of Orange' 170
    'Red Duchess' 170
    'Red Emerald' 170
  fragrantissimum 167
  goeldii 156, 172
  haematinum 225
  ilsemannii (*P. sagittifolium*) 170–1
  imbe (*P. sanguineum*) 170, 171, 226
  insigne 171
  lacerum 170
  leal-costae 24, 115–16, 167
  limnestis (fossil) 34
  linnaei 167
  mamei 169
  martianum (*P. cannifolium*) 167–8, 171
  melanochrysum 167, 169
  myrmecophilum (*P. deflexum*; *P. megalophyllum*) 168
  ornatum 169
  pedatum (*P. laciniatum*) *128*, 170
    'Florida' 170
  pertusum *See Monstera deliciosa*
  pterotum 169
  radiatum 166, *166*, 170
  rugosum 160, 169
  sagittifolium 170–1
  saxicolum 115
  scandens (*P. cordatum*; *P. oxycardium*) 24, 27, 168, 226
  ssp. *scandens* forma *micans* 168
  ssp. *oxycardium* 168

254

INDEX

selloum 172
sodiroi 169
speciosum 25, 34
squamiferum 160, 169, 170
tripartitum 25
verrucosum 169
'Wend-imbe' 171
wendlandii 171
climbers and epiphytes 165–72
foliage 166–7, 169–70
leafless shoots (flagella) 167
myrmecophiles (ant plants) 167–8
pollination 167
rosette-forming 171
self-heading 171–2
stalks 160, 169–71
subgenus *Meconostigma* 30
Phymatarum 86
Pig lily *See Zantedeschia*
Piko lehua apii 135
Pinellia 68, 213, 222
  cordata 68
  pedatisecta 222
  ternata 43, 68, 222
  tripartita 222
  tuberifera 222
Piperales 35
Piptospatha 86
  ridleyi 65, 86
Pistia 20, 21, 28, 30, 31, 36, 54, 216, 217, 227
  stratiotes 23, 65, 73, 76–81, 77
Pistioideae (subfamily) 31
Plowman, Timothy 135, 227
Plumier, Charles 154
Pollen
  starchless 127
  structure 50
Pollinators 49–51, 61, 69, 71, 81, 88, 92, 94, 104–5, 122, 127, 167
  attraction of 40–4, 45, 47, 61, 69, 71, 88, 101, 104–5, 107, 124, 154, 172, *175*, 177
  detention of 21, 43–8, *48*, 71, 88, 101–2, 177
  'trap-flower syndrome' 44
Pothoideae (subfamily) 31, 114, 150
Pothoidium 20, 150–2
  beccarianus 152
  lobbianum 151, 152
  remotiflorus 152
Pothos 23, 28, 30
  celatocaulis *See Rhaphidophora korthalsii*
  repens 151

scandens 22, *24*, 150, 152, 161, 222
seemannii 151–2
Priest's hood *See Arum maculatum*
Projeto Flora Amazonas 157–8
Protarum sechellarum 134
Pseudobulbs 171
Pterostylis coccinea (Orchidaceae) 43
Punga pung *See Amorphophallus paeoniifolius*

Rafflesia arnoldii (R. titan) 179, 181
Refractive tissue 44, 47
Remusatia 65–7
  hookerana 66
  vivipara 54, 66, 66
Reproduction 37–54, 30, 54, 66–7, 66
  apomicts (self-fertilizing) 37, 51–2, 160
  attraction of pollinators 40–3, 45, *58*, 61, 69, 71, 88, 93, 101, 104–5, 107, 124, 154, 172, *175*, 177
  bisexual flowers 18, 19–20, 19, 21, 23, 94
  bulbils 30, 54, 66–7, *66*, 68, 135
  cross-pollination 45–8, *48*, 167
  detention of pollinators 21, 43–8, *48*, 70, 71, 88, 101–2, 177
  fruiting 32, 52–4, 71, 81
  gender variation 48, 61–3, 62
  mimicry as attractant 43–4, 69, *96*, 99, 100, 101, *175*
  non-fertilized seeding 128
  non-sexual 37, 51, 54
  odour 37–40, 124
  odour dispersal 37–43, *58*, 61, 93, 112, 172
  pollen structure 50
  stolons 29, 54, 77
  'trap-flower syndrome' 44
  unisexual flowers 18, 19, 20, 21, 61–3
Rhaphidophora 33, 149, 153, 224
  celatocaulis 146, 149, 224
  decursiva 149
  korthalsii 146, 149, 224
  merillii 224
Rhektophyllum 148–9
  camerunense 148

mirabile *See Cercestis mirabilis*
Rheophytes 30, 85–7
Rhizomes 29, 54, 57, 186, edible 204–5
Rhodospatha 156
  blanda 156
  picta 156
Richardia See Zantedeschia
Roots 28–30
  aerial 29, 116, 155–6
  contractile 93
  edible 188–205
  tuberous 113–14
  uses 156
  water-borne 28
Runners (*See* Stolons)

Satin pothos *See Scindapsus pictus*
Sauromatum venosum (S. guttatum) 42, 44, 111–12, *111*
Scaphispatha 32, 116
  gracilis 116
Schismatoglottis 31, 137
  homalomenoidea 86
  neoguineensis 137
  parviflora 86
  picta (S. calyptrata) 137
  'Silver Heart' 137
Schizocasia See Xenophya
Schizocasia portei See Alocasia portei
Schott, Heinrich 32, 33, 116, 128, 139
Scindapsus 146, 150
  borneensis 32
  pictus 150
Seed
  dispersal 32, 52–4, 71, 81, 94, 106, 107, 185
  vivaparous 90
Settlers' twine *See Gymnostachys anceps*
Sheath 27, *28*, 58
Shingle plant 146–7, 149
Skototropism 146–7
Skunk cabbage *See Lysichiton; Symplocarpus foetidus*
Solomon's lily *See Arum palaestinum*
Spadix 13, 15, 18–23, 58–61, *58*
  heat producing 40–2, 93, 104, 112, 172
  tails 43, *58*, 104
  whiplash *Arisaemas* 59–61, 60
Sparganiaceae (bur-reeds) 98

255

INDEX

Spathe 13, 15, 18–24, 92
  constriction to imprison pollinators 47, *48*
  extendable 90–1
  refractive tissue 44, 47
  tails 43, *58*, 104
Spathicarpa sagittifolia 22, *32*
Spathiphyllum 23, 25, 31, 50, 125–7, 206
  cannifolium 39, *97*, 125–6
  cochlearispathum 217
  commutatum 126
  floribundum 54, 126
  wallisii 22, 126
Spruce, Richard 156
Starch 225
Starchwort *See Arum maculatum*
Stenospermation 32, 157
  multiovulatum 157
Stolons 29, 54, 77
Stylochaeton 29, 43, 113–14
  bogneri 114
  crassispathus 113
  euryphyllus 114
  hypogaeus 114
  lancifolius 113, *113*, 225
  natalensis 114
  puberulus 113, 114
  zenkeri 114
  intoxication produced by 114
Sweet flag *See Acorus calamus*
Sweetheart vine *See Philodendron scandens*
Swiss cheese plant *See Monstera deliciosa*
Symplocarpus foetidus 23, 41, 93, *192*, 205, 216, 223, 224
Synandrogyne See Arophyton
Synandrospadix 29
  vermitoxicus 97, 116–17
Syngonium 50, 52, 158
  podophyllum 158
  'White Butterfly' 158

Taccarum, 117
  weddellianum 117, *117*, *128*
Talas *See Colocasia esculenta*
Tannia (Tanier) *See Xanthosoma sagittifolium*

Taro *See Colocasia esculenta*
  giant *See Alocasia macrorrhiza*
  swamp *See Cyrtosperma merkusii*
Thaumatophyllum spruceanum See Philodendron goeldii
Theriophonum 115
  fischeri 115
  minutum 25, 115
  sivaganganum 29, 115
Thomsonia napalensis See Amorphophallus napalensis
Titan arum *See Amorphophallus titanum*
Toxicity 14, 116–17, 130, 205, 208–13, 223
  as defence 52, 57, 104
  hunting potions 223–4
Tropical rain forest, 30, 55, 56, 119–43, 120
  climbers 144–72
  epiphytes 144–72, 157
  lithophytes 120–2, 139
  terrestrials 120–43
Tubers 28, 29–30, *32*, 54, 57, 113, 116–18, 186
  dormancy in 30, 116, 118, 133, 187
  edible 188–207
  propagation by 118
Typhales (reedmaces) 35
Typhonium 32, 67–8
  acetosella 67
  alpinum 68
  brownii 67, 204
  diversifolium 68
  eliosorum 67–8
  giraldii (*T. giganteum*) *33*, 68, 222
  roxburghii *38*, 39
  trilobatum 25, 47, 67
  food crop 204
Typhonodorum 34
  lindleyanum 53, 97

Ulearum 132
Uromyces sparganii (rust infection) 35
Urospatha 53, 83, 84, 187
  caudata 217
  sagittifolia 21, 83

Veitch, John Gould 130, 131, 138, 142, 146, 162
Voodo lily *See Sauromatum venosum*

Wallis, Gustav 162
Water lettuce *See Pistia stratiotes*
Watson, W. 174
White, Robert 120–1
White sails *See Spathiphyllum wallisii*
Wolffia (duckweed) 80–1

Xanthosoma 46, 50, 134–5, 217
  atrovirens 135–6, 199
    'Appendiculatum' 135
    'Variegata Monstrosa' 135
  auriculatum 216
  belophyllum 199
  brasiliense 199
  caracu 199
  jacquinii 199
  lindenii See Caladium lindenii
  mafaffa 199
  plowmanii 135
  pubescens 134–5
  robustum 227
  roseum 227
  sagittifolium 40, 197–9, *198*, 209
  striatipes 134
  violaceum 199
  viviparum 54, *33*, 135
  bulbils 135
  food crops 197–9, *198*
Xenophya 143
  brancaefolia 143
  lauterbachiana 143

Yautia *See Xanthosoma sagittifolium*

Zamioculcas 26
  zamiifolia 97, 114
Zantedeschia 29, 95
  aethiopica 13, 21, *24*, 94, 95, 217, 225
  albomaculata 225
  elliottiana 95
  pentlandii 95
  rehmannii 95